The Twelve Elements

The Twelve Elements

Cosmographical Roots of Traditional Chinese Medicine

William Wadsworth

To order additional copies of this book, contact:
Xlibris Corporation
1-888-795-4274
www.Xlibris.com
Orders@Xlibris.com
36413

CONTENTS

Dedication

With profound gratitude and respect to the saints of Beas, Maharaj Charan Singh Ji and Baba Ji Gurinder Singh who live what they teach and who act as all others might act in a sacred world. Their word is the unwritten word, the intelligence that is awake in the void behind all appearances, the source of number, and the root of human language and meaning. Everything of merit in this book derives from them.

Preface

This book is about recovering a traditional worldview for purposes of effective clinical treatment using acupuncture. The graphic on the frontispiece of Sebastian Munster's seventeenth-century book about cosmography shows all the elements operating in an idealized harmony and balance of nature between heaven and hell with fire above, water below and with earth, air, and all living things between. To study a traditional system is to search for this kind of philosophical coherence and natural comprehension of order so that the aims of treatment can be just as clear. To learn any traditional science, particularly Chinese acupuncture, as an English-speaking American is to struggle through a sea of strange terminology.

Every specialization requires such language, but the cultural context in which terms like *shao yang*, terrestrial branches, ghost points, *shen* and *hun* developed, mostly is a mystery. When my training cracked the mystery, the situation was worse because I had definitions that didn't add up to a coherent system. You could learn it, but you couldn't fully believe it or do it. Many of the terms had a useful clinical reference, but the philosophy behind the system remained obscure. Perhaps the reason is explained by Levi-Strauss's insight, "The meaning of a word depends on the way in which each language breaks up the realm of meaning to which the word belongs; and it is a function of the presence or absence of other words denoting related meanings."[1] The common assumptions and concepts that support systems of language or that make vocabulary a decent transmitter of thought were missing.

For me, it required ten years to feel that I understood from a wide perspective what the descriptions implied in the clinic and what the basic terminology intended to say about human beings. It was then in the middle of a living practice of acupuncture that the real search for a satisfactory empirical medical philosophy behind clinical acupuncture began. I found that what students are generally taught is a system that is mostly fragments, anecdotes, and hand-me-downs. In particular, for instance, in acupuncture there are eight extra meridians, which roughly refer to the major anatomical planes of the body that divide it into eight quadrants. Do the meridians and their eight confluence points refer to the divisions or the spaces they delineate? Then, the midline plane of the body is divided into two separate meridians, but the plane of the horizontal or waist vessel is a single undivided meridian while the vertical plane that divides back from front has four meridians associated with it. So the eight confluence points for these meridians could potentially be sixteen or thirty if the implications of these meridian divisions are completed. Clearly, these points and meridians represented a shorthand or code for a larger system, and the shorthand was compressed to fit into a number-driven conception, but what was that system? What is unmanifested or implied in the system? The system of acupuncture clearly was a part of a larger pattern of language and culture different from my training.

I completed a doctorate degree as part of my search. My PhD thesis was about *The Anatomy of Melancholy*, which is a famous and entertaining seventeenth-century English scholastic work about traditional Western medicine. It was written from a cosmographer's perspective. Research showed me that academic writers in the history of science and literature had each mined the *Anatomy* for confirmation of their own historical assumptions, neither recognizing the controlling Aristotelian philosophy and world outlook that dominated the book. From observing this tendency to project one's own worldview, I developed a desire to learn what people historically really were thinking and to get past the chaos of textual variation to the basic structures. I developed a particular interest in the relationship between cosmography and literature, which, in Europe, comprised a vibrant tension between traditional Aristotelianism and the emerging three-dimensional worldview brought on by voyages of discovery and the recovery of Ptolemy's *Geographia*. The translations of Plato and fanfare around the *Pimander* of Hermes Trismagistus also stimulated change in the philosophy of science. These changes affected how whole peoples organized their thinking. The result of this study was the rediscovery and reconstruction of the background naturalistic worldview that Robert Burton was promoting and defending. By similarly focusing on the bits and pieces of cosmography in traditional Chinese medicine, I have been able to reconstruct in a tattered whole cloth a satisfactory cosmographical basis for most clinical practices, a basis that was largely dismantled with the emergence of modern medicine in China in the eighteenth and nineteenth centuries.

Writing as a generalist, I may tread roughly on tradition and specialization, but sometimes, a fresh look will yield unforeseen benefits. I'm not asserting a grand common system below all cultures, a moiety, but I'm taking some intellectual risks by thinking through the clinical implications of the syncretic Chinese conception of number. Paul Unschuld, Elizabeth Rochat, Dan Bensky,Stephen Birch, and other great scholars are doing the other essential job of textual and cultural conservancy.

The twelve elements simply are the first twelve whole numbers that provide, in traditional thought, the basis for twelve prismatic cosmographical glasses through which an acupuncturist can view his or her patient. If the whole is symbolized by a circle or sphere, each number divides the whole in a different way. These produce geometric forms that, when reapplied to the cycle of the day, month, or year, results in conceptually important spatial divisions of heaven. These, in turn, result in definite phases of time that were first perceived in the apparent regular motions of the stars, planets, sun, and moon.

The basic premise of *Twelve Elements* is that every window on the world is a window on the same world. This means that each of the geometrically driven models—whether five elements, eight principles, or twelve meridians—must display congruent associations and meanings for same periods of time or in relation to symbolic structures such as the eastern horizon or midheaven. This obvious point only needs to be emphasized in a book like this because it has been neglected historically in clinically defined acupuncture practice.

It is out of this pristine and perhaps simplistic first glimpse of the order of the world that the system of acupuncture emerged. Nearly the whole of acupuncture practice depends on this simple ancient numerical model. Moreover, it is likely that the sum of the benefits of clinical acupuncture derives its result from the natural and parallel structural symmetries of the body and their phase relations to human physiology in time.

Numbers naturally arise as first concepts when we look at the body or the world with innocent eyes. We have ten fingers and toes, two eyes, one head, three divisions of the fingers, two legs, not four. When we gaze outwardly, we observe that we are part of a larger whole that surrounds us in all directions with infinite particular forms packed into three dimensions. During daylight, we see one world; but a few hours later, in darkness, we see another. The one world has become two, each distinctly different. These basic observations mean something in a world where Timaeus can affirm, "The visible world has been declared to be a living creature made after likeness of an eternal original"[2] or Huang fu Mi in the persona of Su Wen can say, "There are three hundred and sixty-five qi points (on the human body) that correspond to a single year. Please tell me whether the minute connecting vessels and ravines and valleys have similar correspondences as well."[3] Attempts to understand the relations between day and night lead inevitably to the horizon and distinctions between heaven and earth. Day and night yield two very different kinds of experience, as do the derivative transitions and peaks of the month and year. Numbers help to understand these basic categories of experience and to differentiate among entire domains of experience, and the same spatial standards apply to the body.

In this book, I am sharing the result of a personal, empirical journey on the fringes of medicine. You cannot glimpse the beauty of a traditional cosmology without having some of it rub off. Slowly, the acupuncturists realize that it is they who are one with the universe, they who experience yin and yang or day and night, they who stand in the matrix of heaven and earth and who carry the responsibility of understanding life and guiding patients. The animistic unity of the simple whole numbers is very appealing and instructive in a fragmented world of calculus and fractals. The system shows to an ordinary person that there is an order and intelligence built into the shape of the human body and the world. What's missing in the teaching of clinical Chinese medicine and medicine generally is the fundamental fact that there is a one-to-one relationship between the human being and the world.

In my view, the human body is part of the world. For a world that craves simplicity, this is a simple statement that, if taken at face value, has important ramifications. If we are spatially part of the world and we belong in it and have sprung from it, then our body is an outcome of laws that govern the whole. The symmetry in our body must also be the symmetry of the world; the functions of our body must also be functions of nature. If there are implications for form and space, there also are implications for time. Our behaviors can and must conform to the laws of nature, or we die. We jump off a cliff or walk unclothed into a blizzard, there will be predictable outcomes, similarly with food, exercise, and behavior. Culturally, if we abuse nature by overplanting, digging excessively, rerouting water, poisoning the soil, or burning in excess, we destroy an essential, fragile equilibrium, and in so doing, we endanger ourselves. If the human body developed in this world and is of this world, then the greater equilibrium is in us as well. We futurists, too, can affirm, as the ancient seers did, that there is a one-to-one relationship between our bodies and the world. Contemporary overanalysis, materialism, nihilism, and loss of faith are barriers to this perception.

As many philosophers have noticed, the body is a microcosm containing all the substance, functions, and qualities visible in nature. Consciousness or intelligence permeates the whole, and the shape of the whole implies both order and consciousness. More than any other single factor, whole numbers helped to organize a rational approach to the body and the world. When numbers are applied to the body and the world, they define bodies in distinctive ways. The number two, for instance, makes gross distinctions like left and right or up and down, and making these distinctions is useful to anatomy and topography. The lens of number may focus certain perceptions, and even when the object under observation is the same, the vista created by each number is different.

The literalness of the traditional mind is hard to replicate after graduating from an American high school. With typical schooling, you are likely to lose the capacity to understand the godlike moral responsibility and capability possessed by every human being. In the traditional model, sages recognized that our sensitivities are such that we can track the ways that our physiology adapts continuously to the three great light cycles: the annual seasonal progression, the monthly lunation, and the cycle of day and night. These three cycles, their phases, and numbers, simply put, are the foundation of Chinese medicine and most other naturalistic systems. Our schooling should include sensitizing the subtle perceptions to experience these cycles.

When the annual cycle is divided in four by a solstice or equinox and the ancients identified the cardinal points of the year, they assigned to each a prime quality—hot, cold, moisture, or dryness. Winter is cold, summer is hot, and their interaction defines the seasons. When this pair interacts, moisture or dryness result. When the equilibrium of moisture and dryness changes inevitably, temperature changes. When four qualities are paired as hot-moist, hot-dry, cold-moist, and cold-dry a division of eight results. The humors were the basis of Western medicine for 1,500 years and the spatial division; eight, is the basis of most point selection strategies of traditional Chinese medicine since it defines the quadrants of the body conceived as back to front or above to below. These whole numbered divisions of the body and the world create geometrical patterns on the window to the body and world, and these patterns have meaning (fig. 2).

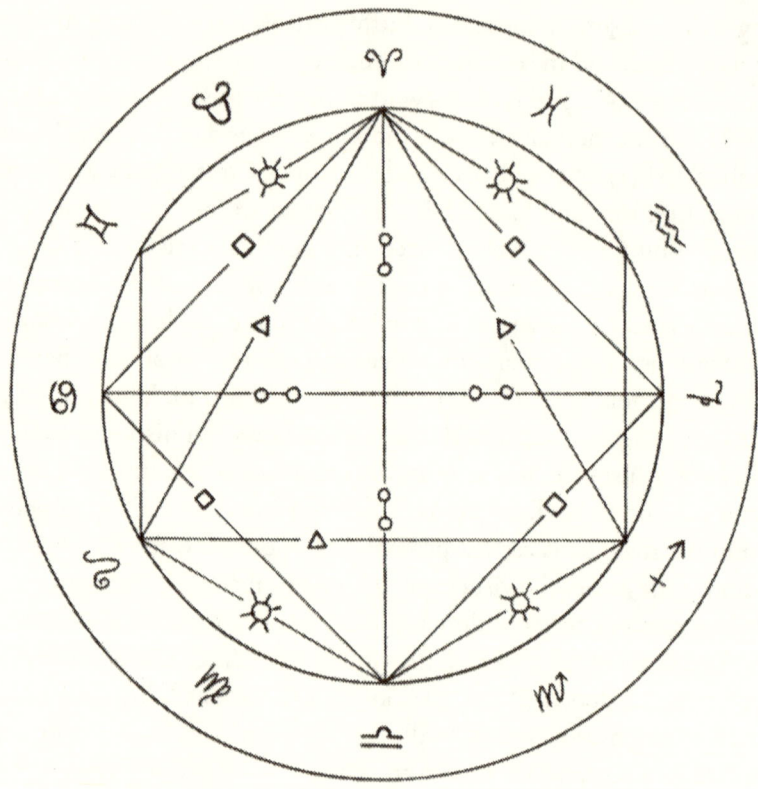

Figure 2. The Geometrical Aspects from Firmicus Maternus, a Roman Astrologer

When the moon circulates through its twenty-eight-day circuit through the heavens, seven divisions can be applied to each quadrant, and together, they form the twenty-eight Chinese constellations. What has been forgotten in most clinical practice is that the new, waxing, full, and waning phases of the moon, which determine whether it is visible or invisible, determine as well what zones of the body below or above the waist should be treated during those phases.

Much of tradition is lost oral tradition, legend, and maxim. Lao Tzu considered the Tao to be dim to human understanding. In the oldest tradition therefore, I bow to the creator begging that he may transmit to the reader his own primordial wisdom, which is the basis of all medicine for the flaws of language, culture, and intelligence in the author make it nearly impossible for the pure wisdom to be communicated. However windy the following text, the purpose is sincere.

It is my experience that each number and its divisions have a natural, living cosmographical connection to each of us, and that connection is hinted at throughout the Chinese number system and medical tradition. It is numbers that explain the five-element system, the three yins and the three yangs, the six external and internal causes of disease, the "eight element or principle acupuncture," the "nine divisions of China." Numbers are an important code for philosophic constructs, and they form the basis for the mnemonic systems in traditional medicine and astronomy, but they are more than that. When applied to the three light cycles, they unlock the riddle of existence and bring the practitioner and the patient closer to their birthright as happy and legitimate coparticipants with the supreme intelligence that designed the world.

Acknowledgments

This project has taken many years to reach publication and only through the support of those close to me has it reached the light of day. With love and appreciation to Jane Clarke Wadsworth for unfailing support in this endeavor.

Danni and Anna Gold contributed directly and gave brilliant insights along with unfailing supporting for the endeavor financial and other.

Lonny Jarrett has helped immeasurably by reading early drafts. He has inspired this effort with his published genius and clinical mastery. He made it clear that we need both scholarship and holistic thinking to treat effectively with acupuncture. He included myth, psychology, geography, and cosmology in *Nourishing Destiny*. My work is merely a footnote to these important contributions.

Duncan Laurie has shown the limits even of cosmology and the need for broad energetic understanding of medicine. I appreciate his patient commiseration, reading of dim copy, and support.

With gratitude to Brad who is living proof that some of the wisest men embrace solitude, study nature, and live outside the system.

With appreciation to my brilliant students at Pacific College of Oriental Medicine, NY who inspired this version of the book.

To Sr. Anne O'Donnell of Catholic University for patience and training.

With love and appreciation to the others who gave love and sustained me in dark years

Introduction

I have devised this mirrour, or cosmograhical glass, in which men will behold not one or two persons, but the heavens with her planets and starres, the earth with her beautiful Regions and the seas with her merveilous increases.

—William Cunningham, *The Cosmographical Glasse (1598[4])*

Thinkers from every country and generation have considered the relations between humans and their environment. The terms of their explanations have changed, but the degree of intelligence has not. While the advance of mathematics and science has increased the sum of specialized knowledge, important vistas have been termed "esoteric" and neglected. A historical perspective shows a bias in early modern science created by a general distrust of the senses and a philosophical rejection of qualitative standards from traditional philosophy and medicine. The qualities could not be materially or mechanically isolated for quantitative study, and statistical mathematics were as yet undeveloped. Scale and structure continue to dominate modern science despite the blending of disciplines and the recognition of certain limitations in the scientific model. Expanding complexity is bounded by what is and what is not nonpolynomial—complete or NP-complete for short or what allows analysis: "There is no known algorithm that totally solves certain problems, that yields a mathematically provable minimum solution, other than just running through all these cases."[5] Despite the advantages of computers, problems involving demographics, higher levels of complexity, intelligence, and consciousness itself are difficult to solve definitively. That is where rational qualitative standards remain useful as means for consolidating complex trends using standard terms for observable conditions that bridge all disciplines. Studies of cosmic cycles as "esoteric causes," as defined below, are viewed respectfully as the origins of the history of rationality. The models of human physiology and cyclic phases were intertwined; and as such, neither the senses nor empirical conditions were excluded from consideration. Moreover, they continue to generate valid models for a cosmos that generates regular meteorological conditions with obvious physiological, cultural, and subjective parallels.

Anthropologists and historians postulate that in ancient societies, humans did not differentiate themselves from the cosmos. This unity is tallied today in studies of primitive societies in anthropology—the "Large Hadron Collider" of the origins of culture as analyzed by many and reflected upon by figures like Levi-Strauss and Julian Jaynes.[6] Eastern traditions of contemplation always have reduced human consciousness to a point of awareness in a vast continuum, and concepts of emptiness or void have allowed contemplation in every period to induce a perspective in which objective conditions supersede personal bias, projection, or delusion.[7] From that perspective, particularly in the case where knowledge of prebirth and postdeath contexts inform knowledge of terrestrial conditions, the question of being advanced or primitive is moot.

Thus, the argument in all early cosmographical texts that explained that the body was a microcosm or miniature of the larger macrocosm: the epitome of everything visible in the outer creation now is achieving renewed vitality through modern advances in scholarly and contemplative study. Part of the argument of this book is that there is a one-to-one mapping between individual and cosmos. This is both a matter of consciousness and fact. Studies of cause as defined below are viewed as the origins of the history of rationality not because in ancient societies people did not differentiate themselves from the cosmos, but because they retained knowledge of their connection at various stages of personal or social complexity.[8] The language of numbers and qualities in ancient geometry and cosmographical systems allowed the person to remain in the formulation in a way similar to the semiology of Charles Pierce.[9] The advantage of the qualities (hot, cold, moisture, dryness, etc.), when treated as signs, was that they were not merely theoretical stances or images, but participants in the continuum of human experience.

However, the early cosmologies also suggest that individual differentiation was equally essential, that a person was the measure of the universe, owing to the divisions of human bodily topography that in turn generated number theory. Moreover, the dignity of humans lie in their grasp of the cosmic order; without which, the world would be a mere chaos or labyrinth filled with vicious, incongruous events. Shamans, saints, priests, healers, and scientists always shared knowledge to help and to give meaning and order to their peers. Human consciousness was and is the mediator between the world and the implicate realm, the mystery from which all life emerges and to which it returns. From these positions, primitives may be considered different from people with secular education simply because they didn't forget their connection with the cosmos nor consider knowledge merely as a means to power or as an antidote to mystery.

Humans certainly have electrochemical parallels and connections with other earth dwellers; while in the more uncharted world of the mind and spirit, I propose that there are similar rational connections between man's physical and mental traits and the cosmos. Traits have directionality and attributes in common with the forward flow of nature. The solar wave operating on the rotating and orbiting earth have much to do with this. The three great cycles of light have operated all through the course of human evolution; and they define the typical periods of activity and rest, the agricultural cycles, and even periods of physiological processes that are tied to quantity and quality of light and associated subtle energy. They certainly are the roots of philosophy, and philosophers continue to affirm the likelihood of subtle intelligent connections between humans and the cosmos like those of Teilhard de Chardin, Chilton Pierce, Jeremy Narlay, Mihaly Csikszentmihaly, and others.[10] That this relationship between humans and the cosmos may operate in every cell and in our DNA is not far-fetched. Not only the dual-stranded structure of DNA that reflects the perpetual dualism of human thought and perception of material conditions and the evolutionary content in chain sequences, but also the living shape, the movements, and the electrical frequency responses suggest communication and adaptation with other cells and perhaps with environmental conditions. Today, the study of DNA entails not only what it is, but what it does—which, of course, implies a mechanism by which behavioral evolution is recorded on the molecular level. "Does" in the sense of being part of a feedback loop in which natural cycles and social influences affect our genetic make up and genetic changes affect our behavior. The position advanced in *Twelve Elements* is that the organization of the cosmic cycles determines the frequencies and electrical periods that organize the background possibility and limits for the assimilation of information. The waxing and waning moon, the increase and decrease of seasonal light, and the daily increase and decrease of light provide a typical rhythm of activity and rest—of ingestion and assimilation, of catabolism and anabolism—that optimizes conditions for evolutionary learning. From one perspective, these patterns symbolize how people assimilate and transmit information.

Leon Hammer and Jarrett have emphasized after Drs. Shen and Manaka that acupuncture is effective because it uses the body's information system.[11] Extended departures from these rhythms create a crisis in the biological information system that leads eventually to negative thinking, destructive emotions, disease, physical excesses or deficiencies, and, eventually, to death. When people stop developing, transforming, and learning, poor health follows soon after. Those frequencies that are beneficial to life have direct relation to the frequencies and conditions generated by the bodies and cyclic periods of the solar system and perhaps the galaxy.

The aim in *Twelve Elements* is to draw from certain recurring ideas in traditional cosmologies, those that remain useful to mastering the real conditions of life. Each of the twelve chapters treats one whole number and contains at its end clinical applications associated with the outlook of the world provided by the chapter number. Integrated ideas are invariably ecologically sound and socially moral because their logic derives from the human perception of symmetry and regularity in nature. Science has developed technology that probes ever deeper into that symmetry, and the aim of this book is not to dispute or diminish the value of resulting discoveries. The links to advances in modern science are implicit, not explicit. My focus is not new science, but recovering a coherent wide view of a healing system from the ancient experts based on the cycles of light rather than on their typically rigid inherited systems of qualities and standard element assignations. The philosophical rationale for the study of cycles must be resurrected from dead associations and placed on a secure footing of human experience before more advanced analytical, statistical, and experimental methods are applied to the model. In this book, I recover and celebrate the orderly cyclic processes of nature and the information available directly to the senses and mind that yield in humans a concordant, cohesive sense of belonging of rational purpose and social value.

The last writings where these values could be found, to my satisfaction, were writings on cosmography, science, medicine, and philosophy in the Renaissance through Newton. I am resurrecting the term "cosmography" because of the whole view of the world that it contained historically and because it expresses a spatial sense of the human life placed in the widest context of meaning. Robert Burton relied on it to express the ethos of the Renaissance.[12] This wide view, the view of a generalist, always is important to retain if someone wishes to realize healthily his or her full potential. The modern notion of cosmography is well summarized by E. Patricia Vicari: "Geography, known in the sixteenth century as 'cosmography,' was treated as a subdivision of mathematics or an aid in understanding history."[13] Thomas Blundeville, author of *Cosmographia* (1608), summarizes authoritatively the view held in this book that cosmography is a subdivision of mathematics and geometry with the following parts:

- Astronomy is a science, which considereth and describeth the magnitudes and motions of the celestiall or superior bodies.
- Astrologie is a Science, which by considering the motions, aspects, and influences of the stares, doth foresee and prognosticate things to come.

- Geographie is a knowledge teaching to describe the whole earth, and all the places contained therein, whereby universall Maps and Cardes of the earth and sea are made.
- Chorographie is the description of some particular place, as Region, Ile. Citie, or such like portion of the earth severed by it selfe from the rest.[14]

Chorography describes local area according to the balance of elements and harmony between elements and culture and was an early form of epidemiology. Astrology was used in medicine to show the balance of elements and humors in a patient and, at the time of treatment, in order to aid with diagnosis and prognosis. This book simply submits that the light cycles and the conclusions of historical physicians should be studied before being rejected. The aim here is to fascinate and to invite further study of these direct relations between people and the world ultimately so these understandings may become a basis for physiological diagnosis and treatment.

A Glimpse of Traditional Assumptions

Behind the geometrical worldview of traditional cultures is human experience. One of the larger aims of this book is to invite people to learn from the past and to try to use their own instincts and subtle senses to live vividly with a direct sense of the motions of the earth, moon, and planets. To live in connection to the living world is a worthy endeavor and a rewarding one. I turn to common sense and the symmetries of the human body that suggest a number of straightforward connections between humans and their world. Body topography is a good place to start. The body can be divided roughly in half at the level of the waist. If we visualize the waist as a mathematical plane, that plane compares closely to the plane of the horizon. A line passing through the center of the body from top to bottom will point toward a unique point in heaven and the center of the earth below, thus, creating a zenith and nadir, which in turn can become the basis for defining two more planar great circles, one that divides front from back similar to the prime vertical and another one side from the other similar to the prime meridian. The four topographic quadrants of the body, thus created with the basic division between up and down, help to orient the individual in the world. The back is a source of strength and protection, the motor functions dominate; the front is softer and more vulnerable and contains major cavities that hold internal organs. These divisions have empirical value because such divisions identify important zones of bodily function and influence. For instance, both urine and feces (liquid and solid) evacuate below the waist. The centers of breath and bodily warmth are generated above the waist in a way analogous to those of wind and sun above the plane of the horizon. Lungs, heart, and kidney functions are bilateral—as are eyes, ears, and limbs. Three grand vertical divisions of the legs, three on the arms with three matching ones on the abdomen, and three more on the face reflect the three grand divisions of sky, earth, and the terrestrial zone of the living in between. If we face south, the back is north and the reverse, while east and west correspond to left and right accordingly.

By arbitrarily selecting a standard position of reference, as in the Orient, standing facing south with hands in the air or as in the west standing with the hands forward and at the sides, the result is a basic way of identifying everything in the bodily landscape that otherwise would be difficult to distinguish one from the other, much less to value in a context. We also have, implicitly, a way of identifying and relating to conditions in the outer world that have reference to bodily experience: cold winds blow from the north, tightening shoulder muscles; warmth comes from heaven in the south, striking the chest and warming the heart; or sunset is like autumn, decline, and death. Even without a deep knowledge of the heavens, early human societies would select an important feature of the landscape, a mountain or mound, prominent in their area to define an objective special center with which they could align their personal directional sense. Alternatively, they would create a mound or pyramid at some sacred site to act simultaneously as tomb and residence for ancestors as well as a symbolic repository for cultural memory and ceremonial connection to heaven and its life-giving cycles and the underworld with its mysteries.

The mountain would refer to that part of the body above the horizon and waist; the heaven and below the mountain would be caves of initiation; the lower parts of the body, the roots and legs, the underworld; while the four directions imposed on the mountain (six when up and down is included, eight when transitions between seasons are involved, ten when the solar year and concept of infinite cyclic repetitions in time are explained, and twelve will be the transformations of the moon and nature and the variations of space and form) define the other quadrants of the cosmos. The quadrants in turn house all the meteorological, vegetative, mythical, and celestial content of tribal activity. Thus, the mountain or tree at the center of any cultural world was simply a way of linking the individual to the landscape. Cultural memory then becomes personal memory and experience. These practical measures quickly developed metaphorical and moral implications. If the enemy always came from the north and if the south was always arid, these symbols and maxims will define the sensible way to live and become the basis for transmitting useful knowledge from one generation to the next.

The core of all observable regularity from such a center of the earth lies in the three great cycles of light—the annual, the monthly, and the daily. This book identifies how humans transform these cycles into the standards of human perception that lead to legitimate cosmographical, medical, and social organization. The astronomical structures behind these three cycles are described in appendix 1. The appendix explains that the three planes—the ecliptic, the equator, and the horizon—each form the base for a three-dimensional (triangular) system and, in a way, are a shorthand way of referring to seasonal, monthly, and daily cyclic meanings.

These cycles also are the basis for the conception of heaven (sun), earth (moon), and horizon (human). This means that when sages refer to heaven, earth, horizon, or the number three, it is code for understanding these three whole cycles that portray the whole celestial sphere from a particular vantage point that either expands or compresses time. It is within this framework that each generation, by adapting to new conditions, invariably loses some cultural memory, but necessity in turn drives each to recover those truths that remain undiminished by trends and fads. The loss of cultural memory now is accelerating, owing to contemporary and unprecedented induction into a psychological, confusing commercial world of television, which replaces the power of hard-won cultural experience, imagination, and myth with marketing fantasies. Oral traditions and early texts remain, in contrast to specialized and fractured postmodern visions, wonderfully direct about the relations between humans and their world. Much of the aim in traditional astrology was to act as a repository for cultural meaning. This is evident in references to gods with their physical descriptions and stories rendered in great detail.[15] Realism and the validity of feelings and sense experience are not questioned; they are investigated.

The imagination is harnessed to real conditions, conditions that persist through eons of conquest and ages of ice. Few writers before the seventeenth century disputed the idea that humans are an integral part of a living world, whose meteorological conditions and seasonal transformations provide mirrors for the ebb and flow of the human spirit. They all assumed that human affairs, like those evident in nature, were cyclic rather than progressive. However, it is naïve to think that perennial philosophy denied that each year was unique and that the people and the creation did not have a developmental purpose. What we are asserting is that traditional cultures were not linear thinking and tied to a notion of history bound hand and foot to notions of progress, manifest destiny, and Judeo-Christian historicism. If a man was active and aggressive, his counterpart in nature could be seen in storms and thunder. The shared conditions helped to project the images of gods and goddesses that expressed qualities shared with heaven and earth. For a person to become great, he had to live and act according to the mandates of the character he was born with, whether as a warrior or healer. Every hero or character would have his time, only to be defeated in the cycle of life by his counterpart.

In Norse mythology, the god of thunder, Thor, could be outwitted by the trickster figure, Loki, just as periods of storm would be preceded by deceptive calm while, in their turn, both gods, as symbols for rational intelligence, had to contend with the giants—the brute forces of time, space, ocean, air, and land. Historically, owing to this foundation of associative thinking, scientists and philosophers before the seventeenth century assumed that our behavior made a difference and that morality and material conditions were interdependent. Some today continue to assert the importance of personal and environmental right action, value, and meaning, insisting that personal responsibility is linked to survival of our living, dynamic world. Contexts and language change, but the offspring cosmologies of earlier systems resonate with people today because they conform to obvious facts of human experience.

The links are particularly specific and apparent in the syncretist writings of early Chinese medicine, notably the *Ling Shu* or "Divine Pivot":

> The Yellow Emperor inquired of Bogao, "I would like to hear how the limbs and joints of the body correspond to sky and earth."
>
> Bogao replied, "The sky is round, the earth rectangular, the heads of human beings are round, and their feet rectangular to correspond. In the sky, there are the sun and moon; human beings have two eyes. On earth, there are the nine provinces; human being shave nine orifices. In the sky, there are wind and rain; human beings shave their joy and anger. In the sky, there are thunder and lightning; human beings have their sounds and speech. In the sky, there are Five Sounds; human beings have their five-yin visceral systems. In the sky, there are the Six Pitches; human beings have their six-yang visceral systems . . . In the sky, there are the ten-day weeks; human beings have ten fingers on their hands."[16]

The body and the world specifically are linked through number and geometry, and the enumeration of parts begs for systematic analysis as to how geometry helped give treatment insights to early practitioners of stone acupuncture.

To understand the real observable conditions behind ancient myths, it is useful to understand traditional numbers and why they carried certain kinds of interpretive meaning. However, what often is lost in the clutter of numbers, instances, and historicism is the importance of the light cycles. All the ancients refer to light as the source of influence in astrology; but ultimately, it is the seasonal, monthly, and daily increase and decrease in light to which they refer. All the more, esoteric theories about influence from the light of stars and planets at their root derived from a reapplication of the phase concepts of the cycles to other bodies.

Ultimately, the influence of a body is simply its placement in the context of the light cycles and its geometrical relations or connections with the primary context—namely, the position of the sun, the moon, and the horizon and what that implies about the phase relation of the seasons, the lunation, or the day. I will introduce here the basic architecture behind the light cycles and the specialized numbers that the Chinese used to differentiate between conditions having to do with celestial bodies and conditions having to do with earthly transformations (ten celestial stems and twelve terrestrial branches). These two systems lay a foundation for the understanding of why traditional societies found it necessary to apply qualities to numbers and objects. These numerological and geometrical facts of life had consequences in human experience; they explained satisfactorily the subtle differentiations between time and space, and they offered a rational basis for understanding the personal impact of medical conditions.

Chapter 1 ―

The One-to-One Relation between Humans and the World

From the widest possible vantage point, perhaps today, the view from space, people can see the earth and all the people, creatures, and forms in it as a single whole. This one, this whole, collectively has a sum of intelligence; it operates according to mathematical and physical laws, and it has, in its cumulative motions, life. An individual is a small part, a derivative part of that whole; and as such, he or she participates in all the same processes and constraints of the whole. A person like the world can be designated by the number one; and to the innocent observer, that one, taken out of context and placed against the vacuum of space, may be seen as a miraculous summation of the larger whole, containing all its molecules, functions, and even intelligence. Humans are composed of fluids (water), of substances and forms (earth), of growth and decline (wood), of breath and gases (air), of independent equilibrium between heat and cold (fire). However, many elements, functions, or component are identified; the whole is in a way an epitome, a unique something that is partially a process in time and partially a form located in space. The human being is the ultimate site for the drama to unfold between the universal and the particular or the world and its parts.

From the clinical perspective, to see the individual with fresh, innocent eyes as a unique homeostasis frozen for contemplation in a timeless moment of encounter, to see in that individual at the point of encounter the result of a continuous history of personal and planetary evolution, and to see as well in the explicit condition of the patient the prognosis of likely outcomes with the best possible outcome of healing, that innocent whole view may produce the best insights into treatment intervention. At least from this perspective, the compass is set toward the goal of spiritual and comprehensive understanding. We may intuitively grasp the nature of a person, as we often do on our first encounter, and this can organize a mountain of disconnected analytical details, symptoms, or signs.

The reason for this is that an open heart and mind in the clinician will allow more useful information to be considered. In a way, cultivating openness is, on a primitive level, cultivating our peripheral or hunter's vision, the kind of vision that can see at night or that can distinguish danger in a complex or darkened background or sense movement and proportion where there is no substantial recognition. In a more advanced framework, cultivating openness of this kind can be to use the third eye—the eye of the mind, the eyes of a lover who must understand everything about the beloved, the eyes of a mother concerned about her child, the mind of a mathematician visualizing a pattern in chaos or emptiness. This kind of cultivation in a trained mind can lead to the immediate, direct perception and consolidation of complex layers of information, layers otherwise buried under piles of complicated chains of cause and effect that bury most simple facts. The clinician will see in a glance the balance of all oppositions, the degree of vitality, the state of the spirit, the flash in the eye, the proportionality of the whole, the effect of shock or grief or babies or toil.

Every human being, every creature and part, and the world as a whole share unambiguously a discernible line of energetic congruence at any given moment. Being open with our broadest and widest capacities of perception to this riverlike flow in time and space can help the practitioner to meet the needs evident in the patient's unique expression of the totality. Thoughts appear and disappear, emotions appear and disappear, bodies appear and disappear, the sun appears and disappears; thus, the *Book of Changes* or Edmund Spenser's *Mutability Cantos*.[17] Were we to understand the phases and flows of these movements, we would understand much that are essential to life and health.

Early agricultural societies understood the rhythms of the day, month, and year because the crop cycle was dependent on them. Even an unusually warm autumn would eventually succumb to the will of heaven and the inevitable effects of the equinox and the shortening of days. Consequently, the heavenly bodies were thought to have a bearing on the structure, proportionality, and behavioral traits of the human form. As the following quotes show, key traditional systems found meaningful ways to link the periods and structures of human experience to the heavens, and these links apply equally today to the conceptual structures of traditional acupuncture.

Megalithic

> The builders of castle Rigg found a site with a convenient horizon
> for measuring the exact sun-setting axis on Midsummer Day. This is
> the key axis of the year for it marks the turning point of the light of
> the sun. The winter solstice is of equal cosmic importance, but at the
> summer solstice the Sun is at the height of its power, clearing the skies
> for uninterrupted measurement.[18]

In chapters 4, 5, and 6, this central imperative of early astronomy will
be admitted as a possible cause for the application of four branches and their
associated acupuncture meridians to the fire element whose location is in the
upper central part of the body that naturally would be associated with midsummer
and midday.

Early Greek, Hesiod

> When the Pleiades, daughters of Atlas, are rising, begin your harvest,
> and your ploughing when they are going to set. Forty nights and days
> they are hidden and appear again as the year moves round, when first
> you sharpen your sickle. This is the law of the plains . . . strip to sow
> and strip to plough and strip to reap, if you wish to get in all Demeter's
> fruits in due season.[19]
>
> When the piercing power and sultry heat of the sun abate, and
> almighty Zeus sends the autumn rains, and man's flesh comes to feel
> far easier—for then the star Sirius passes over the heads of men, who
> are born to misery, only a little while by day and takes greater share
> of night—then, when it showers its leaves to the ground and stops
> sprouting the wood you cut with your axe is least liable to worm.[20]

Hesiod's beautiful writing places the actions of heaven and the play of the
god's side by side with the phases of agricultural necessity and detail. The rising
and setting of stars clearly are associated with agricultural periods, and Zeus
(heaven, masculine) and Demeter (earth, feminine) are interacting dynamically
throughout the growth cycle. Human life is distinguished by tools, by work, and
by comprehension of the law of the plains or the horizon and the patterns of the
growth cycle on it. These are similar perennial perceptions driving the Chinese
conceptions of number and cosmic relationship.

Plato

> The sight of day and night, the months and the returning years, the equinoxes and solstices, has caused the invention of number, given us the notion of time, and made us enquire into the nature of the universe; thence we have derived philosophy . . . We should see the revolutions of intelligence in the heavens and use their untroubled courses to guide the troubled revolutions in our own understanding, which is akin to them . . . all audible musical sound is given us for the sake of harmony, which has motion akin to the orbits in our soul and which, as any one who makes intelligent use of the arts knows, is not to be used, as is commonly thought, to give irrational pleasure, but as a heaven sent ally, in reducing to order and harmony and disharmony in the revolutions within us.[21]

Plato internalizes the experience of number and cosmology to an unusual degree, turning the contemplation of heaven into a means to magic and self-mastery; but Plato works from the foundation of element and cyclic thinking that pervades Greek thought. To some degree, he reveals the esoteric side of numbers suggested in the Elysian mysteries; but the introspective experience of nature is part of all traditional cultures, where man was perceived as an integral part of the cosmos and, naturally, his philosophical perceptions and understanding would, if aligned with the cosmos, produce internally the harmony and balance so evident in nature.

Roman Astrology

> The sun and moon, when configured with any one of the planets, also co-operate: the sun adds a greater nobleness to the figure, and increases the healthiness of the constitution; and the moon, especially when holding or delaying her separation, generally contributes better proportion and greater delicacy of figure, and greater moisture of temperament; but, at the same time, her influence in this latter particular is adapted to the proper ratio of her illumination; as referred to in the modes of temperament mentioned in the beginning of this treatise.[22]

Vestiges of this notion of the importance of lunar phases to human temperament may be found in the natural calendrical links between four seven-day weeks in a month that originally were tied directly to the complete twenty eight-day lunation. Before artificial lighting, these lunar phases played a distinctive role in differentiating one day from another and anticipating what seasonal hunting or agricultural actions should be taken. Much of Chinese astrology came from the Mediterranean; and in all medical systems, the lunar phases formed the basis for diagnosis and treatment since they established primitive links between the onset, crisis, and resolution of disease and a natural time frame. In chapter 7, the way that the lunations were used to support acupuncture treatment is examined.

Early Chinese, Su Wen

> Yellow Emperor: Some treatises explained to the Emperor by the scholar Kwee Chu relates that the movement and quietude are outlined by the divine brightness, by the Sun and the Moon and that the upward and downward movements of Yin and yang are expressed by cold and summer heat . . .
>
> Chi Po: It represents his clear awareness of Tao which is Yin and yang of Heaven and the Earth. The number, which can be counted, is the Yin and yang within the human body. The number of Yin and yang may be ten, or they may be extended to ten thousand. Yin and yang of heaven and the earth cannot be counted by number, they can only be expressed in symbols.[23]
>
> Therefore, it is maintained that there is Yin within Yin and there is yang within Yang. From dawn until noon is hot the period of yang of Heaven and it is yang within Yang; from noon until sunset is the period of yang of Heaven and it is Yin within Yang; from sunset until crowing of the cock is the period of Yin of Heaven and it is Yin within Yin; from crowing of the cock until dawn is the period of Yin of Heaven and it is yang within Yin.[24]

Chi Po produces a graphic with four quadrants that express ratios of light and darkness (fig. 1) that he clearly applied to any light cycle—whether daily, monthly, or seasonal. Yang and light is above, and yin is below the horizon. Rising light is relatively more yang than declining light. Thus, rising light in the morning with the sun above the horizon is yang within yang, and the waxing moon moving toward the full moon also is yang within yang in phase. He links these periods to activity and rest and expresses a similar notion to that of Roman astrology.

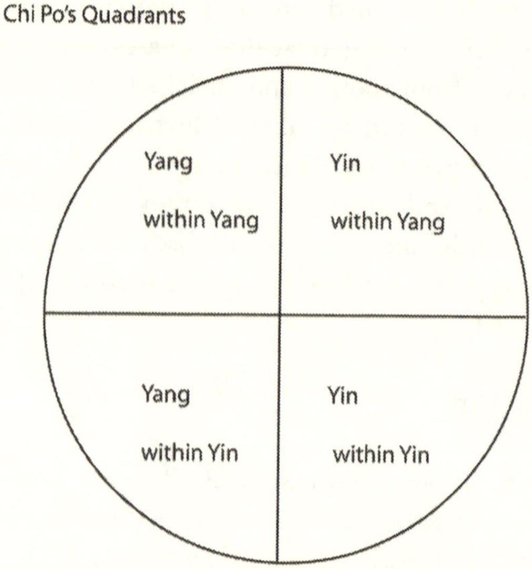

Chi Po's Quadrants

Figure 1. The Quadrant Relations between Yin and Yang

As Plato maintained, cosmology is the basis of all systems of learning, not only medicine, and numbers provide the foundation. The Chinese number characters suggest as well that that numbers always had a cosmographical component in which the bodies' anatomical divisions play a part. All numbers on the body that refer to parts of the cycles of heaven refer ultimately back to the one, the totality, which always is more than the sum of its parts. Numbers are, in the end, merely symbols, and this explains why the Chinese developed different symbols to express the concept of number better in the context of heaven or the context of earth and why they kept the symbolic links between their standard numbers and cosmic origins. This fact is useful to remember in the clinic because the patient always is more than the sum of his or her problems or conditions, and as Plato suggests, a diagnosis is a rough approximation of a problem whose roots may run deeply into the psyche.

The scholars Needham and Liu concluded from their study of early Chinese medical texts that acupuncture points then were selected on the basis of cosmographical speculation.

Anatomically oriented (*Ling Shu* and *Huang Di Nei Jing*), they differentiate the acupoints and name many of them, grouping them out of the totality of points all over the surface of the body. One might see a parallel here with the identification of constellations from among the assembly of stars.[25]

Stars are unlimited, and among all the numbers, one is the only symbol that attempts to encompass them.

I. The Number One

Numbers as used in Chinese cosmography had something to do with quantities and much to do with relations, processes, and qualities. Traditional cosmography was a geometrical and, therefore, graphic notion that included geography, chorography (early epidemiology), astronomy, and astrology.

The standard number representations are radicals with their own intrinsic meanings that are revealing about Chinese cosmology. Scholars, like Manfred Porkert, downplays the importance of numerology and cosmic clock systems, but some careful extended, valid generalizations about the relations between Chinese numbers and cosmology are possible.[26] The structures of characters for whole numbers from one to twelve and the additional characters for the ten-stem and twelve-branch phase representations make clear connections between number representations and observable conditions in the world. These representations are so old and the insights so primary to Chinese cultural thinking as to condition subsequent theories about the Chinese naturalistic cosmos. Each chapter will summarize the basic character meanings as communicated in Wieger's *Chinese Characters*, then offer some limited intuitive observations.

[一], Wieger, "One; alike; to unite; the whole of."[27]

The horizontal line probably derives from the line of the horizon, which is a basis for all rational differentiation. The horizontal line implies above and below but does not visually designate it, thus, retaining its core identity while generating from implication the concept of two parts. This idea that each number generates its next in sequence is an intrinsic Chinese philosophical perception about both number and nature itself. It implies, in every instance, a relationship between observer and object and a similar relationship between any object and its context. Even the number one is an arbitrary designator for the whole and is, therefore, a limiting delineation of what is, but it also is the beginning of differentiation that leads to understanding and reason. Beyond this number, like the mimetic world to which it refers, is fundamentally generative. One begets two, two begets three; there is movement, just as the present actions and circumstances inevitably produce an outcome. This insight is the basic premise behind the *I Ching* and attempts by Chinese astrology to anticipate outcomes on the basis of a clear perception of what is manifested now and, therefore, implied for the future.

The radical one appears in other ancient characters for ten, ten thousand, twenty, and thirty (which symbolizes a generation); and it is in *ping* (3), the third celestial stem, and the second terrestrial branch *ch'ou* (3). Because of the cosmographical reference in the shape of the figure itself, these irregular extensions to other parts of the numbering system suggest the presence of a conceptual web of meanings and connections between numbers more than a mechanical extension.

Timaeus contemplating the number one asks, "Are we then right to speak of one universe or will it be more correct to speak of a plurality or a few? One is right, if it was manufactured according to its pattern; for that which comprises all intelligible beings cannot have a double."[28] The number one creates a tension between one object and all that is experientially intelligible, and this enters the domain of paradox when the numbers of objects multiply toward infinity. The only resolution is the acknowledgement of an encompassing conception of one.

II. The Cosmographic Structure

The figure for one is the open circle of existence, a complete cycle of light and darkness or an unmanifested whole, potentiality, or unity as a child in the womb (fig. 2). This drawing represents the plane of the prime vertical with its basic east-west orientation and with its implicit above-to-below relationship. In its simplicity, it is the first image that extends beyond a point, and it represents a baseline of self-conscious existence in an expanded context. The cosmic diagram has no separate identifying feature. The Chinese character for the number one, a horizontal line, gives a specific, limited location that when placed in the circle implies heaven and earth, two, even though it is one line. The one, as the original, has fundamentally no dualism, no differentiation, and is whole and complete. Even the circle that shows a complete cycle is a symbol that limits the one and should be conceived as purely geometrical or invisible to be accurate. Nevertheless, figure 3 elegantly represents the concept of one with one stroke.

Cosmic Man
As One or Conception

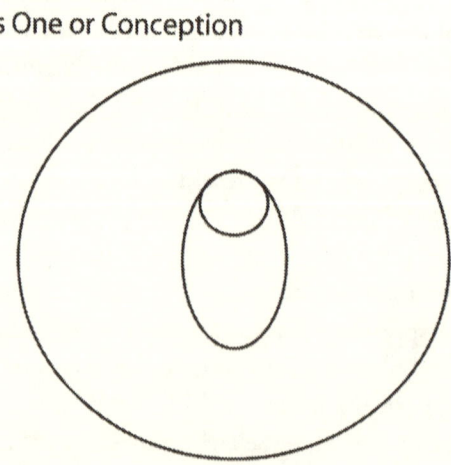

Figure 2. The Prebirth Cosmic Human or the Complete Cycle—Seasonal, Monthly, or Daily

One, identified by a line, in a way is a one that is recognizable and limited to normal perception and a particular point of view. It is this latter one, the one that is not the Tao but a particular whole vantage point or a particular whole world view, that is the drawing for two of the circle with a single line (fig. 2-2). This single line defines something that at least has the possibility of a postbirth existence, something individual that has come into being that is part of the one, indistinguishable in a way, but that nevertheless has reference to the plane of existence, the horizon. The Chinese number one implies the great circle but does not draw it. The prime vertical as drawn in figure 3 is a two-dimensional representation of a sphere more than a plane, and the sphere represents the whole. Therefore, the difference between the three-dimensional representations of the numbers and the Chinese character representations is the absence of this first stroke and the economical stripping down of the implied three-dimensional situation into symbolic form. From this point forward, it needs to be understood that the three-dimensional drawings show the astronomical and cosmographical preconditions that lead with the addition of strokes to the meanings of numbers. For simplicity, the reader can subtract the prime vertical and consider the horizon drawn in two as the root of the number one, and the three planes in three as the root of making a vertical differentiation in two (*erh*).

One

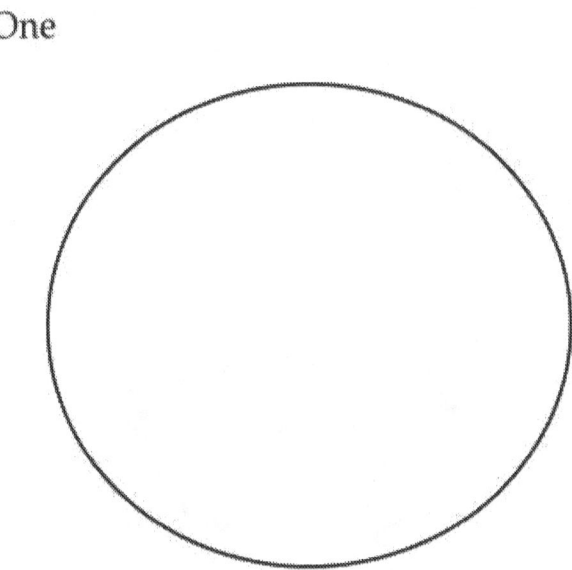

Figure 3. The Postbirth One of Existence

One

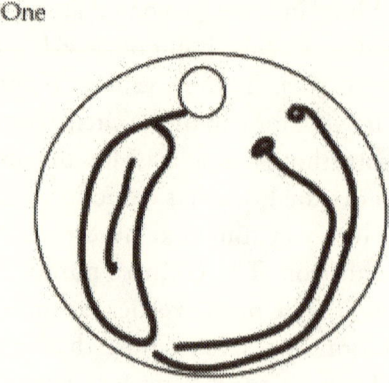

Figure 4. The Prebirth Representation of the Individual as the Cosmic Human

Figure 4 includes an inelegant stick figure to show the relation between the concept of one and the human figure. The figure is wrapped around the zodiac in a fetal position as a representation of the cosmic human, who is the human form projected onto heaven. This is prebirth imagery, as the ancients well knew. This is the mystery of the birth of the infant and the birth of the cosmos taken together as extensions of each other. Figure 5 shows a stick figure for the position of a human body viewed from the number one. The body is horizontal facedown flat on the horizon with the dimensionless horizon with the head to the east. The back, yang, is facing heaven; the front is facing yin, earth below. This is the position of animals as well as men, the creatures with four limbs below aligned with the natural order.

One

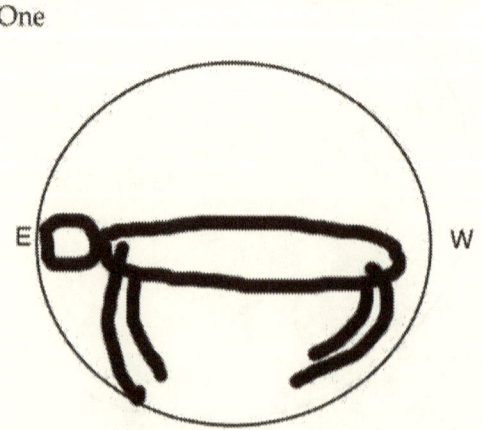

Figure 5. Stick Figure Showing the Anatomical Position for the Number One

III. Medical Theory

Patients need to be seen for what they would be if they returned to wholeness. To anticipate what they could be, we must see through the patient history and overt symptoms to the original bias that created a condition and the original nature that became vulnerable to the particular bias. The whole view gives us an impression of original nature, just as our social first impressions give us important clues about how to approach a person. We can work our way back to a whole view using analysis, but the original impression with all its intuitive information must be there to guide us. If first impressions are cultivated, the reconstruction is less necessary.

Treatment from a wide perspective always is treatment of a living person, not a condition. Though this statement may seem so obvious as to be silly, it's frequently missed in practice and important to underline. In a hospital setting, a patient who is hideously transformed—by cancer, burns, or injuries—always remains somewhat less shocking when seen in person than in a photograph because of this truth. Though we may do our anatomical training on a corpse, observing an identical pathology in a living person results in something different. The same medical remedies don't work identically on every patient, and they will have different side effects. These clinical observations apply to acupuncture where life and consciousness itself is the real object of treatment, not the pathology. As whole systems of acupuncture submit to the limitations of the scientific method, the resulting gains need to be weighed against a potential loss of perspective when the diseased part becomes the whole focus. Life and consciousness need to be harnessed for effective healing.

Treatment from the widest perspective always is a cosmographical treatment in which the phase of the day, the month, and the year combine with the phase of life and evolutionary trend of thought and action to yield a path for intervention. The intervention from this perspective is seldom intended simply to counter a disease process, like treating a fever with cold remedy; rather, it is opening a pathway or removing an obstacle to the patient's recovery. It involves a return to healthy origins or original nature, which invariably is a return to alignment with the ebb and flow of energy and circumstances in the cosmos. Reduced to basics, this means that the equilibrium in our bodies must conform easily with the equilibrium in the light cycles of nature. When it is night, we rest; when it is night and the dark of the moon, we rest more deeply; when it is night in the dark of the moon in the middle of winter, we rest for longer hours and recover more deeply, sleeping longer than in the summer. When we are truly healed, we live consciously in the flow of tonalities that result from the light cycles adjusting as we go.

Recovering this wide view means that neither the patient nor practitioner repeat what they did before. The patient never presents the same condition; the practitioner never applies the same treatment. They never stop, never get stuck in past accomplishments; and they always see new opportunities, new dimensions, new breath, and fresh understandings. If the psyche is blocked, there will be physical possibilities; if the physical is limited, there will be opportunities in the psyche. A wide view will quickly discern from conversation, even from a gesture, what level is working and what isn't.

When we assume that humans have a one-to-one relation with the world, every patient must be communicating their total condition all the time and in every situation. By training the mind to observe nuance of facial color and structure, subtleties of voice, odor, and hints of emotion, a practitioner can pick up on the signals being sent in one or all areas. The patient may express more clearly in one or other mode, or the practitioner may be more sensitive to one or another mode; yet training the senses and perceptions across the board is valuable so that the whole may be registered.

The most important single diagnostic tool is precisely the method that a practitioner has talent in or that he or she studies thoroughly—be it tongue, color, odor, palpation, or pulse. Every part is communicating the whole condition of the patient, but the part has to be understood in the widest perspective of experience and observation. If the imprint of a finger is left on the surface of the skin, you know there is some fluid accumulation and poor lymphatic circulation. The condition has to be evaluated as local or widespread, in a young person or old, and whether it is relatively thick or thin, or long lasting or temporary in relation to some kind of norm established by study and multiple applications. This single condition could be the aftermath of an infection or the onset of kidney failure, an indication either of temporary exhaustion or of chronic disease. In either case, it is a symptom of profound chi deficiency and poor circulation, but it is the context of multiple observations that brings perspective to the observation. This single symptom in context can show the relative balance of all other functions and the underlying strength or weakness of the patient if and only if it is linked to a comprehensive view of the patient in their world.

IV. Medical Practice—Diagnosis

The anatomical position of one is important to keep in mind. Animal origins and patterns continue to affect humans. In some ways, standing is an anomalous development. We can remember that the back has the same conceptually organizing and activating, protective, and strengthening roles as the sky. If a person is highly disorganized, chaotic, or debilitated, it is the back and the acupuncture points along the spine that will have the greatest possibility of restoring order. The nerve trunks that organize the body zones emerge from the spine and distribute downward to the abdomen. Treating the *shu* points on the back is essential for the treatment of *wei chi* for restoring order, for giving strength, and for reducing defensiveness and inducing adaptability. Similarly, abdomen points always have to do with particular local organs and zones with internal feelings and with private needs. You cannot treat psychological problems without considering points on the front or functional and social problems without considering points on the back. Crawling is a powerful tool to recovery from periods of debilitating illness, exhaustion, or injury. Crawling balances the hemispheres of the brain and integrates functions and organs. Organs hang internally in positions of least stress on the body in this position. So apart from seeing the patient as a whole, we can visualize them in the anatomical position of the number one and remember some essential root relations between back and front that otherwise might not occur to us when organizing treatment interventions.

The direct path to achieving a systemwide understanding of the patient's condition is to study all six-pulse positions at once. This forces synthesis of the primary traits applicable to all positions and, therefore, the whole system. When accompanied by openness and an intention to understand the whole person, the sensitive touch from the outermost contact with the skin to the deeper levels will show a great deal about a patient from the level of appearances to deeper relations including the general health of tissues. Every aspect of the contact is relevant, but ranking of qualities too can be important.

Pulse diagnosis on the whole system gives an overall impression, which can be broken down by divisions of two or three or twelve to give more precise meanings. Such divisions are important, but they will be important precisely because they apply to the whole system. Clearly, the qualities that have to do with space and structure are important such as whether the pulse is superficial or deep, thick or thin, hard or soft. The peculiarities of the pulse in qualities having to do with time—like rapid or slow, slippery, choppy, or interrupted—will also be important especially if they apply across the board.

In general, irregularities of heart rate and heartbeat show fundamental instability in the system, the effects of shock, organ weakness, or chronic emotional issues. The soft pulse will indicate a relative yang or chi deficiency when compared to blood or yin. A hard pulse will indicate a relative blood or yin deficiency when compared to chi. Both chi or blood, yang or yin, may be simultaneously weaker in an individual than what would be for them with a healthy normal pulse, even when one predominates, and this will only be perceivable when a whole view of the patient is achieved in some relation to the whole pulse and their mutual appearances. Other specific factors, such as thick and thin, will confirm or repudiate the condition indicated by one polarity. A young person with a soft, deep pulse will be more startling than an older person or someone suffering from chronic disease possessing the same pulse. If the soft pulse also is thin, we know there is not only yang and chi deficiency but also yin and blood deficiency. The two parts contribute to analysis that understands the whole; but the direct perception of the whole will instantly show an unusually soft, deep, and thin pulse along with other qualities that in the context of age and overall appearance will have great significance.

V. Medical Practice—Treatment

Treatment based on the whole view and the whole pulse should clear the effects of trauma that affect homeostasis. Clearing an interrupted heartbeat or one changing its rate will be a great improvement across the board as well as strengthening a soft pulse or softening a hard pulse. A few treatments rise to this general level.

Aggressive Energy

This treatment uses a shallow insertion in the back *shu* points of the five-*zhang* organs: lung, pericardium/heart, liver, spleen, and kidney. If other *shu* points are active or upper and lower *shu* points are active, other nearby points can also be activated. This treatment tests reactivity and indicates some degree of turbulence or suppression that restricts the normal function of a *zhang* organ. These problems generally have an emotional component because it is in the solid organs that deeper emotional problems accumulate to the level where they affect organ functions. Treating all three zones of the torso can be important, and the proximity of *shu* points to nerve endings along the spine makes this a more effective general intervention than treating the five organ-related *mu* points on the front of the body. The *mu* points are better served to affect local areas rather than whole zones of the body.

In a small number of patients, less than 20%, this can be the most important treatment you can ever do because in a patient without pathology, it allows the body to bounce back from shock or trauma and begin immediately to feel more open and relaxed.

Treating Tendencies

There is a tendency in traditional medicine, a tendency based on herbal and physical medicine to treat opposites. In acupuncture, this can be useful, but it is on a general level as, or more important than, treating wholes and whole systems that are directly involved and focusing treatments there.

Quadrants of the body and quadrants above and below using the eight extra meridians in balance can treat the whole person.

Three-yin and three-yang treatments can be effective on the whole. Any weakness of chi and yang will be greatly helped by treating the rising cycle *shao yang* and *jue yin* with its mother and son *shao yin/tai yang*. Any weakness of blood and yin will be greatly helped by treating the declining cycle *yang ming* and *tai yang*. Absolute deficiencies of yang will be treated by *tai yang* and *shao yin* hand; absolute deficiencies of yin will be treated by *shao yin* and *tai yang* foot, supported by either the rising or declining cycle as needed. A whole view will see instantly all the relations in space, and inquiry will yield knowledge of the trajectory in time and know where to intervene.

The uses of treatment systems based on whole number analysis will be developed in succeeding chapters; but in every case analysis, it must yield its useful information back to the number one with its implication of a one-to-one relation between humans and their world. This implies ultimately that there will be one point or intervention that, at any given moment, has the possibility of bringing a person into alignment with the whole or that has the possibility of restoring and individual to a primordial birth-level harmony that allows them to fulfill their potential as human beings.

A simple treatment using acupuncture that focuses on the whole and has a profound centering effect is to treat the centerline, front or back, using the major zones of consciousness in the head and torso. On the front, this is du 20 at the top of the head, du 21, ren 20, ren 17, ren 15, ren 12, ren 4, or ren 6. On the back, this is du 20, 16, 4.

More important than treatment interventions is the diagnostic core that organizes all analysis and allows for intuitive selection of the framework that will work best for an individual patient.

Chapter 2 二

The Number Two, Yin and Yang

Yin and yang are widely written about and understood even in popular culture, and these concepts are used in precise ways in acupuncture clinical practice. The clinical process narrows the idea of yin and yang to symptoms of heat or cold, to evidence of weak immune function or diminished energy. There remains a need to understand the original cosmographical conceptions of yin and yang because they strengthen powers of observation in the clinic.

The last chapter about the number one established the possibility of seeing a person as a whole being fully integrated into their context. This widest of possible perspective sustains a unique capacity for consolidation of complicated impressions into useful standards. It allows for the use of peripheral vision, imagination, instinct, and analysis in the same simultaneous framework. We see the whole person in an instant as we, the clinician, possessing pure intentionality and for a moment remove our filters and barriers to such direct perception. The number one allows trained people to retain their connection with humanity. The number one refers to reality, and treating on the basis of reality is a good idea.

The number two is a very daring number because it makes an analytical distinction between an object and its surrounding, intellectually separating one thing from another, in a world, which, in fact, possesses no such distinctions except through the mediation of human perception. Even human analysis cannot fully separate any two poles of opposition. Heat and cold can never be separated from their continuum; the object is inscrutable even when quantified, the standard being entirely contextual with absolute zero. Being neither relatively hot nor cold is merely an object distinguished from what is not absolute zero. Two is always difficult to speak of because language itself is structured around subjects and objects, and it must refer to two contextual interdependencies before uttering a sound. If we apply the number two to human anatomy, we simultaneously achieve side to side, back to front, up and down, inner and outer as basic spatial distinctions; and that is eight, not two. This is quite apart from hot and cold, two eyes, two hands, two feet, etc. Clearly, to study the number two is to open a dilemma that includes all even numbers and a variety of cosmographical complications. There is something to be said for keeping an eye on one and not forgetting it.

Nevertheless, the number two has to be understood as the basis of clinical anatomy, particularly with its geographical and topographical differentiations of up and down, side to side, back to front, and exterior to interior. It helps us to distinguish the locations of unique sensations and make general assessments about function and perception whether relatively hot or cold, moist or dry, higher or lower, hard or soft, etc.—all of which are vital to medical understanding. Be warned, however, that the physician will not be unchanged by abandoning synthesis when applying this analysis with the result that he or she succumbs to merely technical analysis.

Identifying any object distinguishes it from its surrounding. Singularity first identifies an object, then what it is not, and finally the perspective of an observer. One creates two creates three. We find what we look for. Therefore, we should think deeply about health rather than disease or work from outside the model.

Plato describes in *Timaeus* the emergence of yin and yang from a single point that forms the apex or inception of a triangle (fig. 1). On one strip are multiples of two and the other of three, the yin and yang implications of numbers. From this vantage point, two describes the combination of yin and yang numbers that form the basis of different types of triangles, which, in their turn, are the root structure of geometrical forms and, therefore, the ten thousand things of Chinese philosophy. Numbers emerging from two or yin and yang, then, are generative both in the sense that they explain the emergence of form from nothingness and in the sense that they generate each other, producing, or at least providing, the means for differentiating the vast array of objects in the creation.

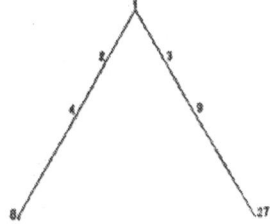

Figure 1. Timaeus's Image for the Yin and Yang Generation of Numbers and Concordant Triangles and Basic Geometrical Form behind the Creation[29]

Chi Po, the sage of the *Nei Jing*, explained above that all visible phenomena (singulars) are manifestations on earth of the dynamic interaction between heaven and earth:

> The manifestations of change consist in the perpendicular phenomena of the heavens and the physical shape of the earth (horizon), so that the seven planets (sun, moon and five visible planets) are moving according to the longitude and latitude of the cosmos and the five-elements are manifesting themselves on earth. The earth is a carrier of the categories of objects with physical shape; the universe is displaying the pure energy of heaven. The movements of physical objects and pure energy bear a resemblance to those of roots and branches.[30]

At the outset of Chinese medical philosophy, the first great treatise emphasizes the circulation of the planets and heaven and the abstraction of clear geometrical space as distinct from the physical forms on earth; and this distinction permeates all subsidiary divisions between medical categories such as locations of above and below or outer and inner, vital spirits such as *hun* or *po* or *shen* or *yi*, energetic density such as chi or blood/body organ character, and function such as *fu* and *zhang*, elements such as fire and water, distinctions between left and right kidney functions, and the two eyes. These dualistic distinctions based on physical symmetry and behavioral observation in a wide context suggests that Chinese medicine has its foundation in cosmology and numerology.[31]

Chi Po's description is very precise, yet there are some built-in paradoxes. The vertical is contrasted to forms, not to the horizontal. Thus, yang and yin are not oppositions only in a typical set of polarities, but very different across a spectrum of characteristics. Forms and the vertical are not a typical isolated continuum, but an observation about two fundamentally different aspects of experience. Though forms can be distinguished from the void of heaven and the air, the bodies in heaven too are objects while it is the intangible five elements, whose conception derives from the ten celestial stems, that in his words are the visible evidence of insubstantiality on earth.[32] In European astrological traditions as well, the heavens were thought to be pure and unchanging in character while nature was a muddy combination of functions and forms; but there is evidence that Chi Po is speaking past these conventions to communicate the degree of interdependency between yin and yang with forms in heaven characterized as pure chi and intangible elements on earth conceived as elements of change in forms.

These kinds of reversals are common in Chinese clinical approaches, where more commonly than not, the meridians and points are selected to treat problems of an opposite nature. This kind of thinking can celebrate ambiguity to the point of silliness, so it's worth returning to the light cycles that were the source of yin and yang and then on that basis to consider the basic functional relationships being invoked by Chi Po.

I. The Number Two

About number two (*erh*), [☰] Wieger says, "The number of the earth, because it makes the pair with heaven, the number of the two principles yin and yang (lesson 2, two strokes)."[33]

Those two strokes represent heaven and earth, yin and yang, which link the number system directly to relations in the observable world. Numbers are representations or signs with unmistakable reference, not abstractions or combinations of abstractions. This connection between world and numbers prevents a tendency toward rigidly dualistic thinking in the structure of the number philosophy. Dualistic thinking would propose a heaven abstracted from the observable sky and similarly independent of material forms and substance, which is the trend in Western thinking—to differentiate spirit from matter, mind from body, and so forth. The horizontal lines, one above the other, indicate the superior to inferior relation of heaven to earth in relation to the whole and its symbol, the horizon. The implication of horizon in *erh* mirrors in reverse the implication of heaven and earth from the single line for the horizon in (I). This clearly is the basis for generating the notion of substance in a strong line and emptiness for a broken line as a way of representing an unstated or implied condition in observed circumstances.

That Wieger claims two for earth is interesting since heaven and earth are represented; and perhaps, it simply derives from the notion that heaven is the origin of all things, including the earth, in an undifferentiated way. From this perspective, the single line above for heaven is an extension of the one, so a representation of two lines inherently emphasizes the earth as distinct from the heaven, yin as distinct from yang, etc. Opposition, then, implies differentiation and form, this within a totality, which suddenly is implied and subordinated rather than explicit. Nevertheless, there is a way back home to wholeness suggested by the emptiness at the level of the horizon between heaven and earth.

II. The Cosmographic Structure

Figure 2 shows two planes, the plane of the prime vertical representing the great sphere and the plane of the horizon/equator. The line of the horizon divides the sphere into heaven and earth defining up and down. If regarded as representing the greater sphere, the prime vertical disappears, and the line of the horizon divides heaven from earth. If regarded as a plane, the prime vertical divides back from front while also mapping an east and west division of vertical space. The pair of planes creates two basic zones (heaven and earth) and four quadrants. Two leads to four, which leads to eight.

The Chinese number *erh* with its two lines makes explicit what is implicit, that the addition of the horizon to the prime vertical (two planes) divides heaven from earth, creating an absolute distinction within a whole.

Two

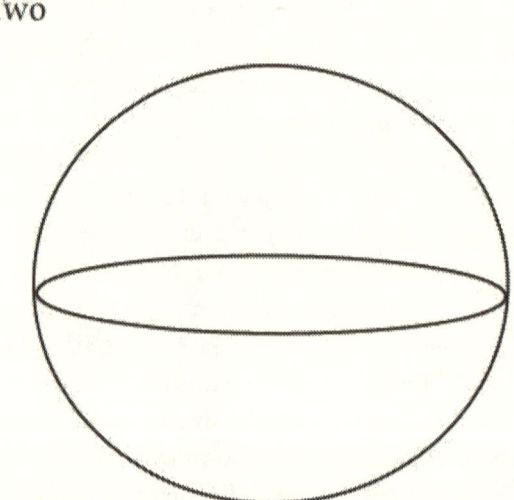

Figure 2. Two, the Prime Vertical, and the Horizon

The cycle of the day achieves a fundamental distinction at dawn and dusk between day and night. Figure 2 shows this absolute break from one state to the next. The sun in heaven generates daylight, and the moon too sheds some light episodically in the night. The sun and moon traditionally embodied yang and yin respectively. These bodies and the five planets occupy heaven, and therefore, heaven and light have forever been associated. Because the sun appears to descend below the earth at night and caves are dark, earth and darkness too traditionally have been associated with each other.

The number two then represents a dynamic and unstable equilibrium at any moment in time on the level of the horizon: equilibrium that has a bias moving toward one or another of its poles. This dynamic equilibrium and directionality are comprehensively driven by daily, monthly, and annual light cycles. Day is warm; night is cold. The terms "yin" and "yang," which distinguish the oppositions of all dual relations, derive ultimately from this distinction between above and below, day and night.

Figure 2 shows the circle of the day with light above and night below a horizon line in the middle. The horizon line defines an absolute break between day and night, light and darkness, and dawn and dusk. The circle in a way can be turned into a clock's face for the hourly periods or equally well for the lunations and seasons; in the former, the half-moon's phases of waxing and waning would align with the horizon with the dark of the moon below and the full moon above, similarly for the equinoxes and solstices for the year.

Figure 3 shows the image of an old graphic of the division of day and night, and figure 2 reduces this basic portrayal of a diurnal cycle to the fixed differentiation

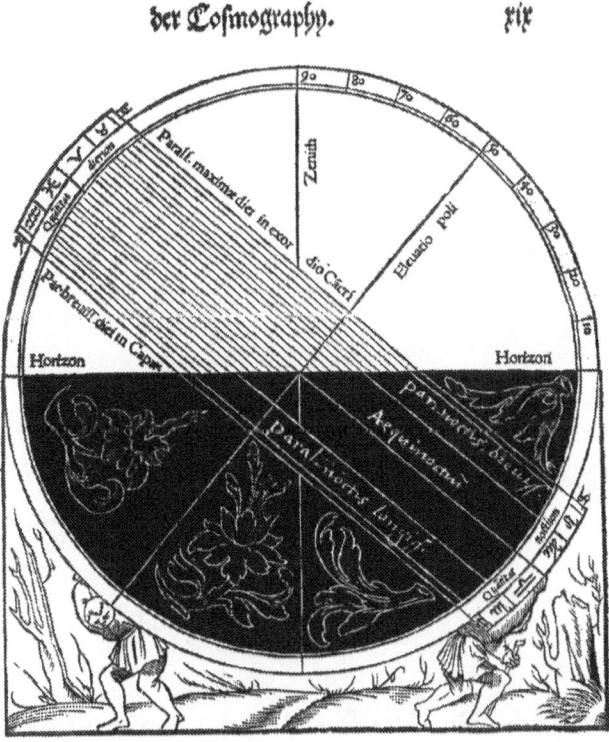

Figure 3. Sebastian Munster: A Cosmological Diagram for Day and Night

of day and night, the corollaries to the distinction between heaven and earth. The horizon is the plane where the rising and setting sun defines the distinction between day and night. It is a fairly clear distinction. Apart from a brief period of dawn or twilight and depending on cloud cover when the sun is up, its day and people can see to navigate; when it's dark, that's another matter. In terms of the light cycle, day and night is the effect of the sun being above or below the earth, in heaven or in the earth. Light from the sun and moon was seen as the pure energetic counterpart of the substance of the earth; and traditionally, the five planets were supplementary points of circling colored light—red, blue, cool white, bright white, and blue green—displayed natural affiliations with the seasons, changes in vegetation on earth, and correlated animal shapes and roles. In general, the Moon dominates the night sky, the sun dominates the day, and the planets function in both though more subordinate to the sun than the moon.

Figure 4 includes the human figure to show the one-to-one relationship when the first distinction is applied to the cosmic human. To the Chinese, the head—as the seat of consciousness and steady human purpose—was associated with heaven; the body with its mutating organs, topography, and functions was aligned with earth.

Two

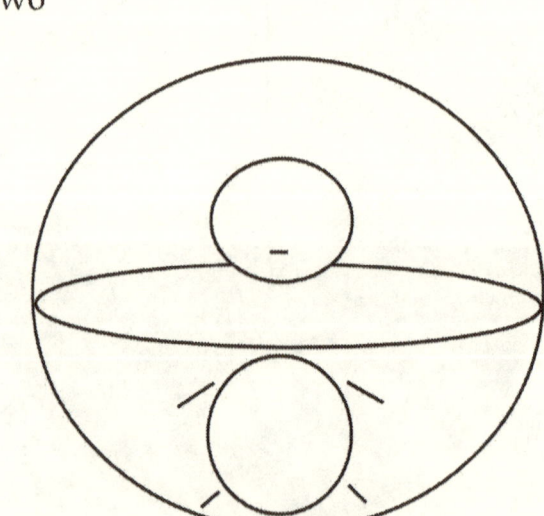

Figure 4. The Number Two: Heaven and Earth Showing the Head as Heaven and the Torso and Limbs as Earth

An alternative way to anatomically understand two was to divide the body midway at the waist or center of the body (fig. 5). This places the arms above and legs below with their different functions and associations, locating the head and heart above, and generative and eliminative processes below. This model divides the digestive processes in half and begs for the emergence of the number three with its three lines that would explicitly include the middle, horizon. As it is, the basic cosmographical diagram focuses on the horizon in the middle/waist as defining heaven and earth.

Two

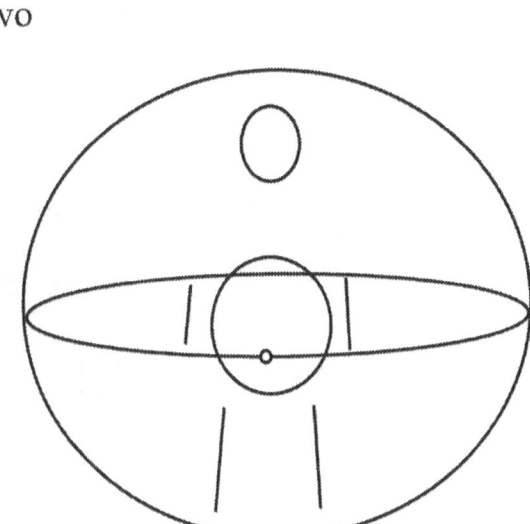

Figure 5. The Number Two: The Horizon/Middle Divides Heaven and Earth

With a person facing south in the Chinese anatomical position, the prime vertical cuts the great sphere and the body into two symmetrical halves back to front. This maintains in a standing figure, the distinction established in the cosmic position of one that showed the prone body with the back as yang facing heaven, the front as yin facing down. Now in the standing figure, the back is yang, the front is yin, and the three-dimensional sphere has been rotated ninety degrees well and good. In figures 3 and 4, the orientation of the prime vertical seems east to west, but it should be remembered that east and west have not been differentiated and that, as a plane standing in for the sphere, the orientation can be in any direction. This means that from any perspective on the level of the horizon or waist, a vertical plane can be drawn that will divide two opposing principles. The principle of yang and yin applied to heaven and earth is ambient and universal. The body itself suggests that inner and outer and the distinction of back to front to be primary because of the extreme differentiation between the them. Su Wen speaks, "Speaking of the yin and yang of man, then the outside is yang, the inside is yin. Speaking of the yin and yang of the human body, then the back is yang, the abdomen is yin."[34]

The symmetrical halves of the body also show another kind of opposing differentiation that begs elaboration. This latter symmetry is implied by the east-west orientation of the prime vertical, as related to the difference between morning and afternoon, rather than the absolute distinction of day versus night or summer versus winter, which is implied by the difference above and below or back to front respectively. The functional distinction between symmetrical halves of the body can only be fully elaborated with the number three and the introduction of the prime meridian that stands at ninety degrees to the prime vertical, but because of the omnidirectional nature of the prime vertical, both functional and structural, asymmetrical and symmetrical oppositions are included in the differentiations suggested by the prime vertical in figures 3 and 4. The absolute distinction in these figures is between heaven and earth.

A great contribution of the number two, when applied to cosmology and anatomy together, is that it differentiates above from below and identifies the two trends' vertical trends of the body between heat above and cold below, dryness above and moisture below, and the circulatory interactions between fluids and gasses that generates heat and cold. Two implies that structure and function are interdependent yet distinct categories of relationship.

Beyond that, two refers to important fundamentally different perspectives on the same greater cyclic whole. Yin and yang refer ultimately to space and time, respectively, and how they intermingle in the light cycles and processes of life.

One perspective results from the absolute distinction between heaven and earth that was originally implied by the horizon line. This is a thoroughly spatial distinction that has a matching visual result in time with the distinction of day and night. This distinction has to do with the actual structure of the light cycle and the timeless physical/geometrical order of the whole. This perspective is anatomical and allows cosmographical phases to be applied directly and literally to the cosmic man and mediated through that model the individual patient. If for instance the sun, the moon, or the season are approaching a peak of light, a practitioner would use upper-body points to augment functions in the upper body, chi, or yang—points to supplement or lower leg points to counterbalance an extravagance.

The second perspective makes a distinction between increasing and decreasing light, which are two distinct processes that unfold in time. The landmarks occur at noon and midnight, dawn and dusk, or more accurately near quadrant-based reference points; but these spatial reference points are organized to show phases in time more than absolute relations of one to another. This second perspective describes the trajectory of change and assists a cosmologist-physician in framing a medical history and achieving diagnosis, prognosis, and a treatment plan.

Using the same example as above, if the sun, the moon, or the season are approaching a peak of light, a practitioner would look first to the cluster of three-yin and three-yang meridians that organize this part of the cycle, having to do with rising light, namely *jue yin* and *shao yang*. Treatment would focus generally on the upper body and hand meridians to achieve equilibrium owing to the phase of the cycles but in particular would adapt treatments with greater leverage using the whole implied system in the extreme case to augment using in some balanced *shao yang* hand and to slow the excitation implied using *jue yin* foot. These meridian clusters have to do with process, the process of increasing light and how these processes achieve homeostasis in the body.

Using the absolute or spatial model that differentiates above from below in terms of the entire cycle, the principle of opposites either in time or space can be applied. The braking action would not use the internal or foot's rising cycle partner, *jue yin*, to regulate the rising system as just suggested, but would use the opposite time phase from the declining cycle supplemented by anatomical location. That is the opposite time phase to *shao yang* fire would be *tai yin* metal, but its structural counterpart that would give opposition by principle (*tai yin*) and opposition by structure (leg meridian opposite to *shao yang* hand) would be *tai yin* foot. This is opposite in every way—related to yin and yang, up to down, side to side, external to internal. Using *jue yin* foot to compensate for *shao yang* hand is external to internal and up to down and can be used appositionally across the body but is not symmetrically opposite or functionally opposite by principles of both space and time.

Chi Po communicates efficiently the combination of space to time, between geometrical absolutes and processes that are implied by the light cycles. It is from this distinction that all the lesser dualisms—having to do with effects such as light and dark hot and cold, hard and soft—come into being. When doing clinical assessment, a practitioner can ask of any dualism—such as hard and soft, "Where are they located?"(space) and "How did they come into being?" or "When was the symptom aggravated?"(time). Similarly, with a soft pulse, we can make an absolute determination based on visual appearance and depending on positional factors like depth, width, and degree to assign a label—chi deficient, severe chi deficient, or yang deficient. Alternatively, we can anticipate from the medical history and analytical perspectives applied to the same interactive factors, the likely crisis and the likely causes for the condition and the trajectory of the indicators.

The cosmographical model distinguishes effectively between space and time and allows for two kinds of analytical processes to be applied to what is essentially a single indissoluble combined matrix or whole cycle.

When combined space and time produce a composite drawing showing four quadrants, these quadrants show how yin and yang interact as shown in figure 6a where the horizon is the basis for a spatial system and in 6b where prime vertical is the basis for the temporal cycle. Figure 6c replicates chapter 1, figure 2 above that depicts Su Wen's combined a and b with gradations of yin and yang.

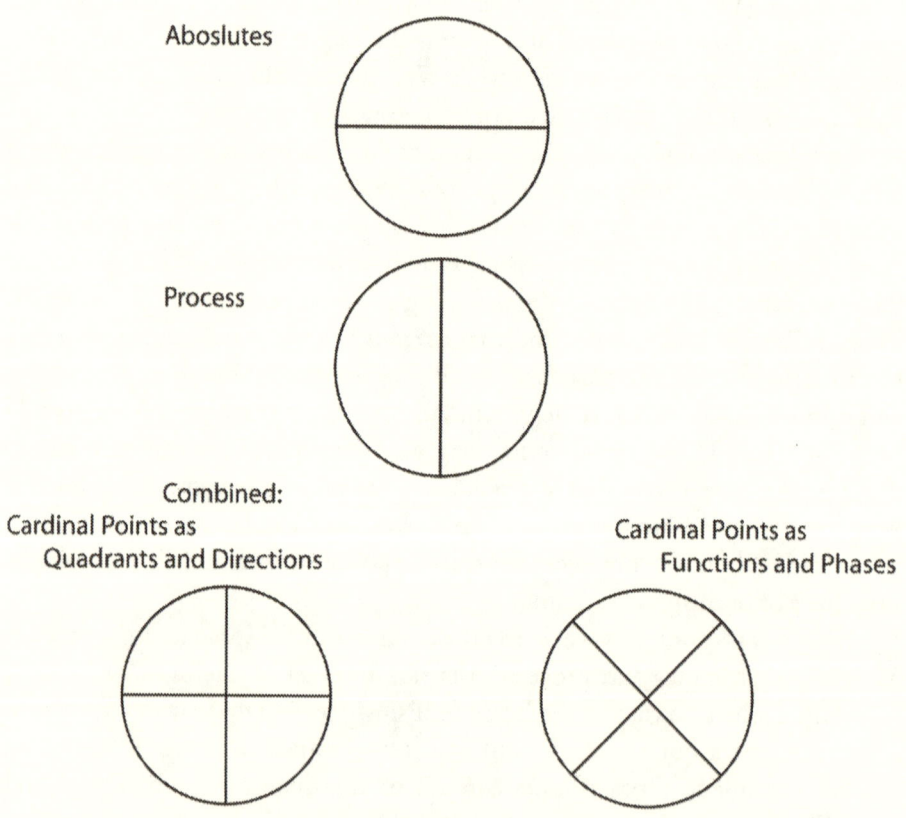

Figure 6. Yin and Yang as (*a*) Absolute, (*b*) Process, (*c*) Combined

Figure 6a is drawn to show that when the horizon divides heaven and earth into fixed divisions, the east and west points describe the processes of dawn and sunset, which are the equilibrium points between the poles of heaven and earth. The east and west poles are defined by the vector toward increased light or increased darkness more than by their absolute state of light or darkness. Similarly, figure 6b shows that when time is the primary perspective, the prime vertical becomes the basis for defining an absolute polarity between increase and decrease of light with dawn and dusk as the penultimate statement of the trend. In a reversal of roles, the ends of the plane of the prime vertical become points, where what were before absolute positions now become the points of transition or transformation from increase to decrease or the reverse.

III. Medical Practice—Theory

The drawing in figure 6 shows how two becomes four becomes eight automatically; but at the same time, each increment adds a better detailing of the dynamic of space and time, yin and yang. This will be elaborated in detail in chapters 4 and 8. However, at the outset, it is important to understand that what is absolute or prebirth for time is relative and postbirth for space and the reverse. The best use of these drawings for a clinician is to understand that there is a one-to-one relationship between the body and the world and that as a consequence, structure and function work together seamlessly but with different implications depending on the clinical situation.

One patient may show up with symptoms that always recur in one quadrant, let us say the shoulder area. The next patient may show up with the same shoulder symptoms that recurs once a month or whenever it rains. These may require different vectors of intervention. The first shoulder problem may respond well to geometrical, spatial, and positioning of the needles; the other to treatment of organ systems, process, and time. The *shao yang* gallbladder/three heater (increasing light cycle shoulder area) or *yang ming* large intestine (decreasing light cycle shoulder area) depending on which periods and phases activate the problem.

IV. Medical Practice—Treatment

Acupuncture tradition offers some ways to treat microcosmic problems involving dysfunctional spatial or temporal halves, whether in space—above or below, side to side, back to front, exterior to interior—or in time-rising cycle, declining cycle, collapse of yang and yin. These treatments are primary clearing treatments that involve whole systems and should precede interventions for local problems if the symptoms warrant it. Some of the theory behind these treatments will be explained in greater detail because they involve points from more complex systems based on more advanced whole numbers, but the aim here is to identify points fundamentally important to the balance between yin and yang at its most basic level of differentiation. Naturally, the interventions stimulate the nexus points between functions of time and space.

The basic use of yin and yang is instinctive and practical. If there is a pain on one side of the body, you can treat the area and strengthen your treatment by treating its opposite as well. If there is an excess or stagnation on one side, there will be a deficiency on its opposite. If one area is weak, its skeleto-muscular counterpart will overact upon it or suffer first. If above or below are empty, energy will have to be drawn from its opposite in order to create balance. Even without diagnosis or analysis, it always is clinically useful to palpate and sense the relations between planes and quadrants, side to side, up to down, back to front, outside to inside. This is all common sense and safe ground for treatment.

Based in figures 3 and 4 that show the one-to-one relation between the body and the world, this means in particular the points that directly control the planes of the prime vertical, (prime meridian is implicit) and horizon.

Table 1 (Appendix 1) derives from the planes shown in figure 4 that displays how the prime vertical defines back and front, and this division has integrated meanings with the horizon or waist vessels that it divides as the geographic division above and below and day and night. The human body is unique in its versatility and is reflected in the ability to place the plane of the prime vertical in any part of the great sphere that it symbolizes. Because the prime vertical can divide the body east and west (in four, the prime meridian) or north and south, oppositions are retained in whatever direction the body faces. If in anatomical position the left and right of the body exhibit symmetrical oppositions, the same symmetries will apply with the body facing north instead of south. Although the implications in these changes are philosophically significant, the standard differentiations between structure and process are retained in such 180-degree reversals. However, the plane itself that distinguishes left from right symmetries penetrates back to front and, thus, contains the meanings of back to front in its affiliation with the *chong mo*, the penetrating vessel. This shows the absolute interdependency of structure and function which allows for integrity even when the figure is rotated ninety degrees, which symbolically places absolute and structural distinctions between yin and yang on the body in relation to purely process or phase orientation in the cosmos symbolized by dawn or dusk.

The complex interdependency of primary planes that define the body and the world can also be seen in the *ren* and *du mai*, which are two halves of the only plane that has representative acupuncture points unique to it. These meridians symbolically manifest the prime meridian, which divides the body into symmetrical left and right halves. Nevertheless, the plane itself runs back to front and, therefore, connects with the interior to exterior relations of the *chong mo* while it defines the waist vessel/horizon plane north to south. This again shows the interdependency of planes, how a single plane to the early mathematical mind automatically has two sides (yin and yang); in this case, ren and du meridians create an implicit ninety-degree relationship to another plane by distinguishing in this case the left from right (4). Given the absolute contribution of the number two—which was to distinguish heaven from earth, up from down—these additional two directions yield six directions. Then the division of heaven and earth when applied to the four directions yields four directions above and below or eight. In basic geometry, any plane drawn in a three-dimensional space will imply these spatial relationships. Clearly, when the Chinese clinicians gave ren and du meridians their own points, they signaled that this vertical plane that extends from heaven to earth down the midline of the body is the principle distinction that is the root of the other eight extra meridian trajectories that in their turn remain implicit as they are borrowing points from other number-dependent meridian systems.

Each whole number shows the human figure in a particular ideal geometrical orientation, and the accumulated numbers will show how humans accumulate complex dimensionality that allows them to evolve and to adapt freely while learning from various layers of contemplative meaning that are built into the structures of the body and nature.

The ren and du meridian figure prominently in table 1 because it has the first and only eight extra meridian with externalized points. It expresses outwardly the basic vertical relationship in an upright human to the outer world, being of the compressed meanings of the prime vertical and prime meridian and each of their two halves yin and yang and these above and below the waist *dai* vessel above and motility vessel below.

Of similar interest in table 1 is the recognition that all three-yin and three-yang leg meridians and all the vertical planes and their extra meridian corollaries beyond ren and du connect with the waist vessel and symbolically with the plane of the horizon. It is this plane that differentiates above from below and heaven from earth.

All the three-yin and three-yang hand meridians have internal connections but no external points on the *dai* (waist) meridian. These internal connections that are attributed to meridians and are associated with heaven and yang contact and penetrate the horizon, but they do so invisibly. This important observation is about how yang penetrates yin invisibly and then manifests externally as forms. The bodily extension of this theory is that yang penetrates interiorly then manifests externally through the leg meridians. All three-yin and three-yang foot meridians have exterior points of contact with the horizon and waist vessels. It is noteworthy that the yin leg meridians that govern bodily form cease at the top of the chest and depend upon hidden internal pathways to connect with the sense organs on the head (heaven). Clearly, cosmographical theory has intruded to govern meridian, anatomical, and clinical theory. Cosmologists who noticed that forms appear on the earth and above the horizon but that yang or chi acts invisibly to generate this life and appearances observed the external form of this same process (See appendix 1 for Positions of points on the Planes or in a controlling relationship to planes defining spatial and temporal zones).

Yang and Yin Related to the *Dai Mo*

Vestiges of cosmology abound in the names of points associated with the waist and *dai* vessel. The waist and plane of the horizon are synonymous. Waist vessel points are essential for creating equilibrium between above and below between aspiration and instinct. The point names and indications support this notion.

Stomach 25 is called heavenly pivot, which is a clear expression of the role of the horizon as a pivot or mediator between heaven and earth. Located on the *yang ming* meridian that passes through the horizon from heaven downward this implies a relationship between heaven and the upper body and the lower front half of the body defined by the ninety-degree relation of the horizon and the prime vertical that divides front from back.

Spleen 15, the great horizontal, is clearly a reference to the horizon; this point links with the yin linking vessel, which deals with the upper-front half defined by the trajectory of the prime vertical symmetrically left and right. The yin meridian trajectory is upward on the front of the body, so this would relate to the upper-front half of the body.

Gallbladder 26 (27, 28), the girdling vessel, has the name of the waist meridian itself and, therefore, is a quintessential expression of the horizon and waist. The placement just below liver 13 and in a vertical line have no anatomical significance but have great cosmographical significance. It maps the prime vertical with its east-west orientation on the human body, showing the point where the yang appears from below the horizon. Because they are *shao yang* wood points, this clearly refers to dawn and the east. The unique feature of symmetry from left to right in the human body is that the function of dawn, the function of spring, and the function of the new moon works on the body when it faces either north or south. The meridians have symbolic value based on the baseline anatomical position facing south like the sage facing life or north like the mystic returning home.

The connection of gallbladder 25 to the prime vertical links it by definition to the linking and motility vessels, which in space (yin) defines the upper, above the waist and lower, below the waist zones respectively and, in function, the vertical relations. Being a foot meridian, the gallbladder links with the yang motility vessel, which in turn regulates the lateral balance in the body and has close ties to the *tai yang* of the prime meridian as indicated by the point UB 62. Once again, Chinese clinicians use one system to get leverage on another, yang on yin vertical on horizontal and here the center *tai yang/shao yin* to regulate the center or back to front to regulate side to side and back and front.

Liver 13 is the completion gate. In addition to being on the *dai mo* and symbolically on the horizon, this point also is a meeting point for all interior *zhang* or interior organs. This unification is accomplished because the waist vessel links all the yin and is fundamentally associated with yin, earth, and every form on it—where Chi Po claims the five elements circulate. The relative elevation of liver 13 above the horizon supports the role of the wood element as regulating growth and the rising cycle of light and life. Like the leaves of a tree, this point nudges against heaven, the diaphragm. The lateral position of this point and its role as the *mu* point for the spleen, which generally is associated with earth element and the declining cycle, suggests that the point acts as a counterpart of spleen 15, defining the rising of dawn just below the horizon versus dusk settling down below the horizon.

Urinary bladder 23 is the *shen shu* kidney/spirit *shu*. UB 23 is located at the level of du 4, *ming men*, that is the gateway to the center of the body. This point symbolizes the point closest to expressing the unique identity and individuality (*shen*) at a particular location on the horizon. Kidney is the leg meridian counterpart to the *shao yin* heart/fire meridian, and it is the kidney as a leg meridian that gives the heart and *shen* a body to be located in. *Ming men* is a profound and important concept with many meanings and associations; but from the geometrical and cosmographical perspective, *ming men* is the gate into the actual location where individual perspective resides, the point at which the prime vertical, prime meridian, and the horizon intersect symbolically in the human form. The prime meridian governs the internal to external relations, and therefore, it is on the GV that the outer gate that provides access to the internal center is located. *Shen shu* emphasizes the yang contribution of individualized consciousness available at the level of the horizon on the back yang side of the body.

KIdney 16, huang shu, the vitals *shu*, is given the same significance and identity as the primary organ *shu* points on the back even though it is situated on the front. Its location on the level of the horizon and in direct opposition to *shen shu* elevates its status as a major control point. Because it is the counterpart to UB 23, which focuses on yang and consciousness, *huang shu* focuses on the universal but unconscious vitality that sustains material activity and form. It is a point about drive and need that requires a person to seek completion and perpetuation by hunting and gathering, by ingesting food, and by seeking partnership. Sustaining form is its contribution to identity, not consciousness.

Treating the horizon points can unite splits that relate to heaven and earth, above and below. These symptoms may show up as fluid retention, patterns of heat and cold, or as behavioral splits such as hyperactivity versus lethargy.

Clearly, based on the last points mentioned, differentiations can be made useful to the strengthening of balance between back to front. People can lose their moorings by under—or overnourishing, by extremes of depravation or excess. People similarly can lose their moorings from shock or suffering that confuses them and makes them lose their basic sense of mission, or on the other hand, they may fall prey to neurotic preoccupations. The importance of horizon points in relation to back and front splits is that these points focus on reality, on real-world needs and real-world aspirations. There may be a basic differentiation between the *shen* applied to real conditions of life and the *shen* at the level of the heart, which may have to do more directly with emotional and relational factors than how to.

Yang and Yin Related to Divisions between Left and Right

Similarly, dualistic splits may occur in fundamental ways between left and right sides of the body. The horizon points have to do with balance in the world of affairs with feasible aspirations and with adaptation to change. Splits from side to side also can occur in the upper body and lower body. These can rapidly be discerned by using Akabani testing of response time on endpoints of hand and foot meridians. These can identify those meridians that have left to right splits. Balance can be achieved by needling the *luo* or source points on the side with the slower response time. If several meridians exhibit the pattern, the five elements diagram can be used to judiciously select the prior orb involved. If UB or kidney is involved, it generally is the most essential imbalance to correct.

Yang and Yin related to Chi Wild and the Husband-Wife Separation

Leon Hammer links a relatively weak left arm radial pulse when compared with the right as a form of chi wild. A chi wild condition disrupts homeostasis by reversing normal relations or exhibiting extreme separations of yin and yang. If the pulses of a chronically ill person float toward the surface, if the pulses collapse or show rapidly changing and unusual rate, rhythm, and combinations of qualities, and if a patient is cold above and hot below or has severe one-sided symptoms, all these represent fundamental disruptions. All these must be corrected before more standard kinds of equilibrium can be established. And I would maintain that there are a few key principles to correct such imbalances. All of the following use entry-exit points in combination with key transfer points between elements to reestablish the equilibrium of appropriate three-yin and three-yang blocks of related meridians.

1. Ensuring that the core circulation is working in the center can treat severe depletions of yin. Using spleen (SP) 20, liver (LV) 14, kidney (KI) 27 with heart 1 together will ensure that the heart is connected to all yin vessels and will maintain stability. Using urinary bladder (UB) 67, stomach (ST) 45, and gallbladder (GB) 41 with KI 1 will link all available yang energy to the yin root to ensure that the circuit that maintains basic equilibrium and structural integrity is working.

2. If all pulses are empty and depleted, treat ren 1 and du 1 together to bring up the pulse.

3. If extreme agitation and nervous depletion and a feeling of being overwhelmed is evident with a vibration on the pulse, treat the water points on the yang meridians particularly UB 67. This treatment can be accompanied by KI 10, UB 16 and KI 6 and UB 62 can substitute Du 20 or UB 67. With heart agitation, treat heart (HT) 7, pericardium (PC) 8 and LU 7 with small intestine (SI) 1 to ensure that *tai yang* reasserts control. With difficulty in accepting facts and a feeling of being dominated by circumstances, do SI 19 needle or ear pellets for twenty-four hours. This last opens the corpus colosum and allows for rapid assimilation and integration of information.

4. Extreme depletion and weakness from starvation or deprivation. Multiple moxa on the umbilicus using salt as a heat buffer.

5. With right pulses and declining cycle meridians and postnatal chi severely deficient relative to left pulses, treat to draw energy across the divide using entry-exit combinations: LV 14, KI 27 to Lung (LU) 1; heart 9 to large intestine (LI) 4; or transfer combinations that include spleen 2, stomach 41, spleen 1, lung 2. Another option is to use prime meridian connections on the proper level, like ren 9, 12, 14, or 17 in order to create connections side to side at the level of the heart/lung, liver/spleen,and kidney left and right.

6. Lonny Jarrett gives a detailed clinical perspective on the husband-wife treatment, which consists of trying to restore the fundamental balance of the yang left pulses and their associated element orbs. The key point for

this transition is KI 7 supported by UB 67, KI 3, LV 4, and HT 7, which draw from the right side pulses to the left side while stabilizing the heart and circulation throughout.

7. Divisions front to back are fundamental divisions that initially express themselves with weak yang, as chaos, and with weak yin, as rigidity and vulnerability. These traits may occur with any element or combination of elements in various states of disruption. Feelings of loss of control and behavioral chaos can be treated using points near the spine, *shu* points, and outer points on the back. Rigidity and suppression can be treated using kidney, lung, and spleen points on the relevant level physically and psychologically.

8. The heart/kidney axis. Lonny Jarrett speaks about the core relation between heart and kidney functions: "The heart/kidney axis is the foundational axis of alignment between the spirit and the potential virtue of each human being. It is the interpenetration of the shen and jing through this axis that allows us to know our selves through introspections and permits our original nature to flourish." From the cosmographical perspective, this is an internal equilibrium between heaven and earth that is associated with the prime vertical, prime meridian, and the vertical axis. Heaven and earth are in a dynamic equilibrium that allows the body to adapt to circumstances at noon or midnight, waking and sleeping, etc. Each draws on the other. Without sleep and recovery, strength disappears; without activity, the system becomes restless. All other cyclic functions borrow from this core equilibrium that helps the body to adjust to the impact of larger seasonal and social changes. A prolonged heat wave will eventually draw water from soil, lakes, and oceans, which will fall as rain to cool the land. Similarly, the body will draw on internal reservoirs of minerals, nutrients, and fluids to compensate for extreme exertion or externally imposed stresses of exposure or injury. If the core is weak or lacking in resources, the body has no means to adapt and cannot handle emotional and physical shocks. These in turn can unbalance the root equilibrium between mind and body, causing psychological or physical disruption. The acupoints HT 7 and KI 3 are key points to restoring order; using *shu* points for heart and kidney together can restore integrity. Using upper kidney meridian points as prescribed by Lonny Jarrett can reestablish connection with the *shao yin* core, particularly if the heart is absent or withdrawn and there is a loss of personal connection and capacity. Heart 1 and kidney 1 bring resources to bear on imbalances in their partner—that is, to treat vitality issues, fear, or physical shock, use heart 1; and to treat psychological stress, treat kidney 1. Each provides resources and reserves to serve its opposite.

9. For all issues of identity disruption and shock, the essential points to remember are source points, which always give the right amount of coherent chi that connects with a person's essence through the various element windows, which become more important when their phases are activated by season or the diurnal clock.

Chapter 3 三

The Number Three— Heaven, Earth, and Humankind

The number three offers a way to divide a whole cycle of light and darkness into its basic structural components. It defines the vertical axis by reintroducing the single horizontal line of the number one (一) that was conspicuously absent in the number two (二), which consisted of a line for heaven above a void and a line for earth below a void. The horizontal lines are stacked vertically, and they imply together the importance of the vertical; indeed, they depend on the vertical for their position. Yet the shape of three or *san* (三) originates in the horizon like all the first three numbers for the horizon acts as a mediator between heaven and earth. The line of the horizon is, in a way, a mere concept when compared to the vastness of heaven and the scale of the earth. Human beings who live on it are a small yet integral part of the mix, important in a way precisely because they can abstract their location on the earth as a geometrical plane and peer into the heights and depths. This identifies the human function with all living things but also with intelligence and consciousness.

Timaeus, the Greek geometer and cosmographer, extend two yin and yang numbered strips mentioned in the last chapter from point or apex downward (fig. 2-1). These become the sides of an infinite number of triangles, which are the first simplest geometrical forms. Simple whole number multiples create specific divisions and proportions for triangles of special importance to the creation of primary geometrical solids, some of these shapes are seen in nature. Desmond Lee states, "In Plato's description the numbers measure off corresponding lengths on a single strip of soul-stuff. Four and nine, eight and twenty-seven are square and cube numbers, which are thought of as two-dimensional and three-dimensional, planes and solids. The reason for stopping at the cube is that the cube symbolizes body in three dimensions" (fig. 1).[35] The progression of odd and even numbers generate when linked by the base of the pyramid at different levels of multiplication (dimension) eventually yield three-dimensional forms.

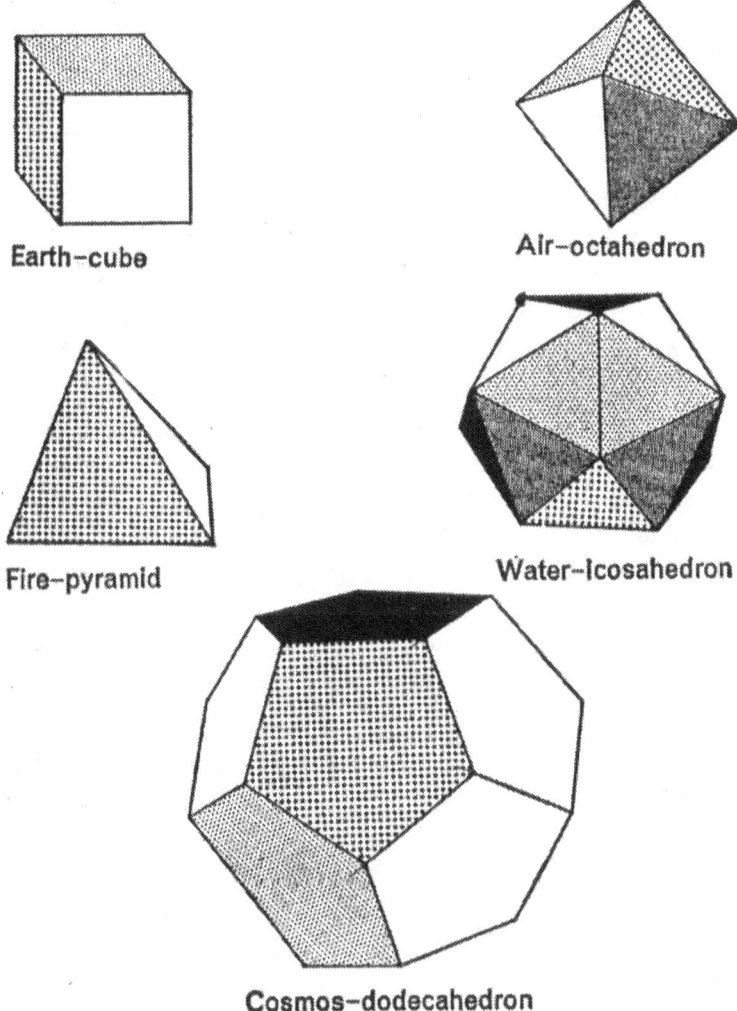

Figure 1. The Core Geometrical Solids That Result from Elementary Triangles[36]

F. M. Cornford concludes from studying Plato's conceptions behind the elemental forms in figure 1b, "We have now found a satisfactory meaning for Plato's statement that the existence of several varieties (grades) of each primary body is due to the way in which the elementary triangles of either pattern are put together to form larger or smaller triangles of the same pattern . . . Our hypothesis . . . satisfies Plato's declaration that symmetry and proportion were introduced down to the smallest detail."[37] For Timaeus, the triangle with its three sides is the basis of all form, though to him the triangles themselves are pure, unmanifest mathematical forms. In the cosmographical conception, the apex, a point, is in heaven; the sides yang and yin descend to a base, the earth and base of any triangle, thereby creating the heaven, earth, and the horizon with its yin and yang dynamic and expression in created beings. The figure of an equilateral triangle also works in the context of the circle of the day, month, or year as a structural foundation that divides 120-degree swaths of the year, and the apex of the triangle with its equal slices can be placed at any point or particularly at the four cardinal points (the square formed by the squares of whole numbers down the sides of Timaeus's soul cloth) and achieve the same harmonious, stable arrangement. This is why ultimately the division by twelve, which combines the cardinal points with the principle of the triangle and, therefore, fully links the vertical principle of heaven, earth, and horizon to the four seasonal directions on the horizon is so prominent as a template for viewing change and predicting fortunes.

The third-century text *Nei P'ien* of Ko Hung explains about the abstract and mythical Tai Yuan Mountain (triangle): "There is an administrative force of one hundred and twenty well coordinated. Situated on a line east and west, there are black growths by the thousands."[38] This is the same geometrical form, an equilateral triangle with each side of which occupies 120 degrees of the cosmic circle. The base of the triangle is defined by east and west, the vertical of the mountain defines heaven and earth, but also on the base on the horizon, north and south, is the foundation of the growth cycle, growths. The black growths combines the notions of night, winter, and water, which establish midwinter solstice, midnight period as the base or heel stone in the circle of time for the spatial earth base square of the pyramid/mountain. A heel stone in a stone calendar approximately indicates the direction of the midsummer sunrise, and if the midwinter full moon rises over it, the moon must be halfway between the major and minor standstills. If eclipses do occur that year, they will certainly be at midwinter and midsummer.[39] This pure mountain describes the basis of form and with great economy of the growth cycle and, therefore, of all action and ethics. As described further in the passage, actions, which do not recollect the heavenly cause and deviate from the ideal laws of nature, will be punished.

James Legge's representation of the *ba gua* or the endless interdependent circulation of yin and yang is encircled by trigrams that formally summarize the primary kinds of triangular relations possible between heaven and earth. These are in a way elaborations of different triangulations of the positions of the sun in relation to the horizon in either the diurnal or seasonal cycle. These relations establish the landmarks for the proportionality of light and darkness.

Figure 2. The *Ba gua* with the Trigrams of the *I Ching*

Yet in cosmographical terms, four must inevitably introduce description of the movement implied in three. Four reduces to a description of cardinal points that further differentiate the space of three but in so doing reduce the verticality to a single plane, whose cardinal points owe their existence to postnatal temporal relations. The four directions loses and now must imply the dimension of heaven and earth or through the notion of a cube encompass both with the focus on the plane of the horizon. Again, cosmic relations determine the shape, reference, and meaning of numbers.

By stars, the old form mentioned by Wieger would have implied the five planets associated with the five elements as much or more than the constellations. As the sun and moon are affiliated with yang and yin, the sun would be affiliated with heaven, the moon with earth, and the five planets and elements with the changeable conditions of human experience. This notion is supported by Wieger's explanation that Wu Chi, which in common speech has the symbol of the five elements above a square (cardinal points) and represents the self or individual human being. Mantak Chia refers to this Wu Chi as the starting point for thought as the great energy-filled void, the unnamable and the unknowable whole that precedes and underlies all appearances while it at the same time interpenetrates appearances.[40] He says that no limitation can be placed upon it. To give it a name only limits it, and this unwillingness to reduce an essence to name and form has led followers of monotheistic faiths to question the authority or integrity of such a system.

The symbol for Wu Chi then in his view refers to the cosmographical relation of the five elements with their circulating planets operating between heaven and earth and effecting the visible changes, and this cosmic identity of the awakened human being is seen as a shared human heritage.

When reduced to popular use, the same character simply identifies an individual operating between heaven and earth. Wu Chi should not be confused with the cosmic implication of *I*, the one, which in essence is unknowable and unnamable, and Wu Chi must be seen as a more complicated explanation based on the division of three combined with the concept of five elements. Even so, in this complicated character, we see Taoists using numbers properly as windows or means to understand the original one, the unknowable truth behind appearances. That is the basis of Chinese numerology and alchemy and, therefore, traditional Chinese medicine.

A cosmographical understanding of heaven, human, and earth or heaven, horizon, and earth is to see the distinction as a numeric shorthand for three different whole systems. Each system governs a specific perspective and domain of living: (1) the sun and ecliptic—that of the sun, heaven, and the plane of the ecliptic or the sun's path; (2) the moon and equator—that of the eccentric lunar orbit that circles the center of the earth and gives dimension to idea of earth as a center with its equatorial plane, prime meridian, and vertex defining locations on the earth; and (3) horizon and diurnal cycle—that of life defined by the horizon and day-to-day necessities bounded by its north, east, south, and west, its above and below unique to the individual or tribe, the center of the earth from the social or personal perspective. This horizon-based system has a whole set of conditions unique to a location and individual. The ancients were very familiar with the geometry and effects of these three planes (fig. 3).[41]

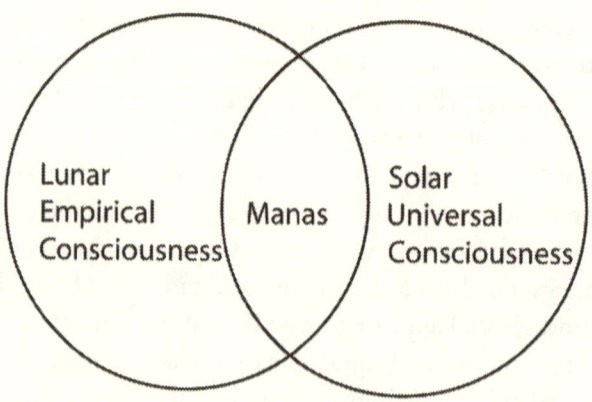

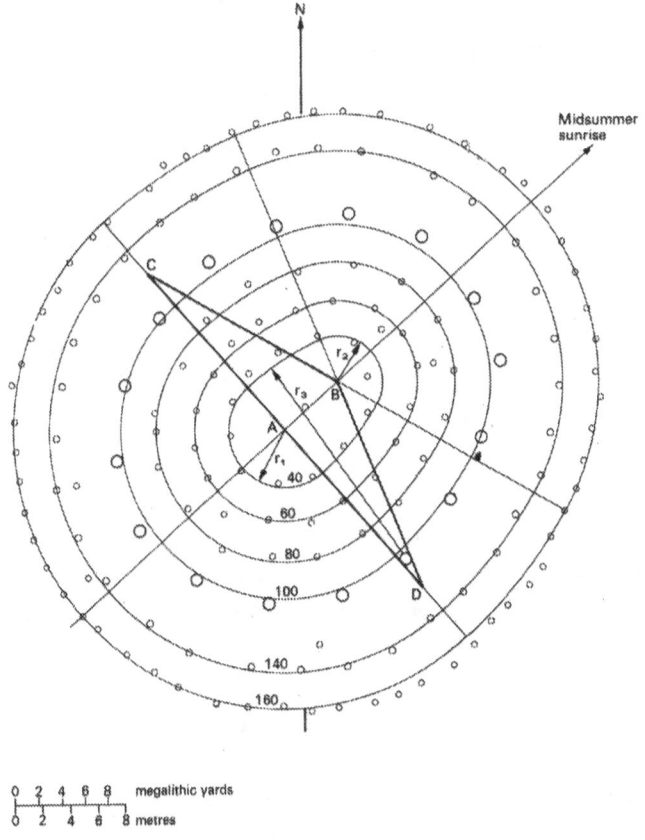

Figure 3. *A*, The Megalithic Standard of Heaven, Earth, and Horizon/Human; *B*, With Woodbridge Circle Showing the Necessary Triangular Geometrical Structures That Define Relations

The number three, as Wieger explains it, does not contribute to explaining the process or change, but it does differentiate three preconditions of life and three basic divisions of the cosmos that make action and function possible. The immobilized idea of the vertical of the heaven, horizon, and earth relationships demand the inclusion of generation and action that it evidently lacks, and it demands as well that the horizontal be explained more fully by these processes: three generates four. Whether facing north, east, south, or west, the horizon has three components—a front, a behind, and a center. This phase and location of three is not developed and won't develop until the number six allows it, but it is implicit in three. Beyond that, however, the number three, like two and one before it, must be seen as absolute preconditions, as essential defining characteristics of all things. One is a primordial unity in which a person participates, two is woman or man, and three defines the three cosmic zones of life that require human adaptation; and these are built into the vertical structures of standing, body, head (heaven), torso (horizon), legs (earth).

The vertical structure of the human is similar to that of the world and similar to that of a tree: roots, trunk, and branches. The torso has three divisions: the head, three divisions; the arm, three divisions; the legs, three divisions; and fingers and toes three divisions. These various divisions cumulatively add to nine or twelve, twenty-eight, etc.; but the whole is three: roots, trunk, and branches. The unique features of the human structure are its mobility and its verticality. The roots uniquely are above ground, retaining only symbolic affinity; the waist is the horizon, and the head and upper body is heaven. In the upper torso, the heart warms and the lungs breathe, just like the sun in heaven and the air; the dominant virtue is heat and dryness, below fluids and solid wastes recall earth and water the dominant virtue being cold and moist, and the middle has a mixture of these extremes, relatively hot or cold, relatively more moist or dry, mist or dust, rich or poor soil, moist and dry.

Three creates the context in which all dualisms can unfold in time because it creates loci for their interaction on the surface of the earth. The surface of the earth is a mediator. The line expresses the essential unity between oppositions and views the horizon as the point of intersection and change. By extension, it is the middle or equilibrium between all oppositions; and by further extension, it is the consciousness of the individual that perceives the oppositions and maintains the equilibrium needed to adapt, whether to survive or develop spiritually. Thus, the number three is the root of all moral virtue arising from conceptions of prudence, justice, equality, and the golden mean. So, although the number three has a prebirth character having to do with absolute relations and preconditions for life, it also has undeniable and important real-world implications for function and location.

I. The Number Three

In Wieger's (三) "San three, the number of heaven, earth, and humanity;" he extends this basic meaning by showing how it was used in other characters. In the symbol *wang* (2) for king, he says, the king is the one, the man who connects together the heaven, earth, and humanity.[42]

Ancient astrology was applied to the king not as an individual, but as a representative or embodiment of the whole human populace and by his intrinsic function as a mediator in a given epoch of history as defined by necessities of nature and politics between the will of heaven and terrestrial necessities.[43]

Wieger explains that when three lines are turned ninety degrees to vertical positions below two horizontal lines as in *shih* (4), it symbolizes influx coming from heaven, auspicious or inauspicious signs, by which the will of heaven is known to mankind. The three vertical lines represent what is hanging from heaven, viz. the sun, the moon, and the stars (five visible planets and agricultural stars), the mutations of which reveal to men the transcendent things."[44]

That which was implied in *erh* 一 is manifest in *san* 三. Heaven and earth now are visible with an explicit horizon (also the explicit basis of *I*, one) between represented in the three firm horizontal lines. These are the three firm lines of the *I Ching*, the lines for yang of heaven and of prebirth conditions. These three strong lines openly include the one and the two; and taken together, the three lines represent a whole or a totality (fig. 1). The totality to which it refers ultimately is that which existed before one, but the three explicit visual domains lose something in their differentiation, and they lack the dimension of the original whole. They are confined by their relationship, however useful the knowledge of relationship might be. The totality they represent does not seem to include change; it is rather a timeless, eternal set of relations or preconditions that govern all life experience. In this sense, the fact that the use of horizontal lines to represent numbers stops at three is important because the cosmographical reference they encapsulate is spatially complete.

II. Cosmographic Structures

The number three can be shown graphically in accord with the megalithic model as three vertical spheres instead of the three lines of the Chinese character. These spheres represent heaven, horizon, and earth (fig. 4).

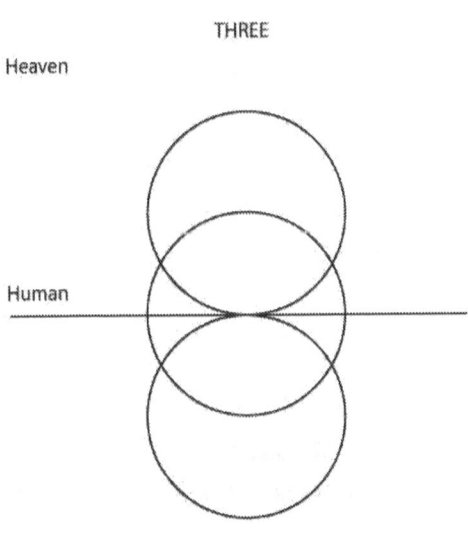

THREE

Heaven

Human

Earth

Figure 4. The Three Vertical Spheres for the Three Domains of Life

In accord with the principles of yin and yang, which divide the cosmos into heaven and earth or day and night on the one hand and dawn and dusk on the other, the principle of three clearly can be applied horizontally as well as vertically, which automatically produces an image for four, five, or six, depending on how it is used. This kind of immediate flash of numeric possibility that emerges from the distribution of early numbers two (yin) and three (yang) were what contributed to the notion that numbers are fundamentally generative in character.

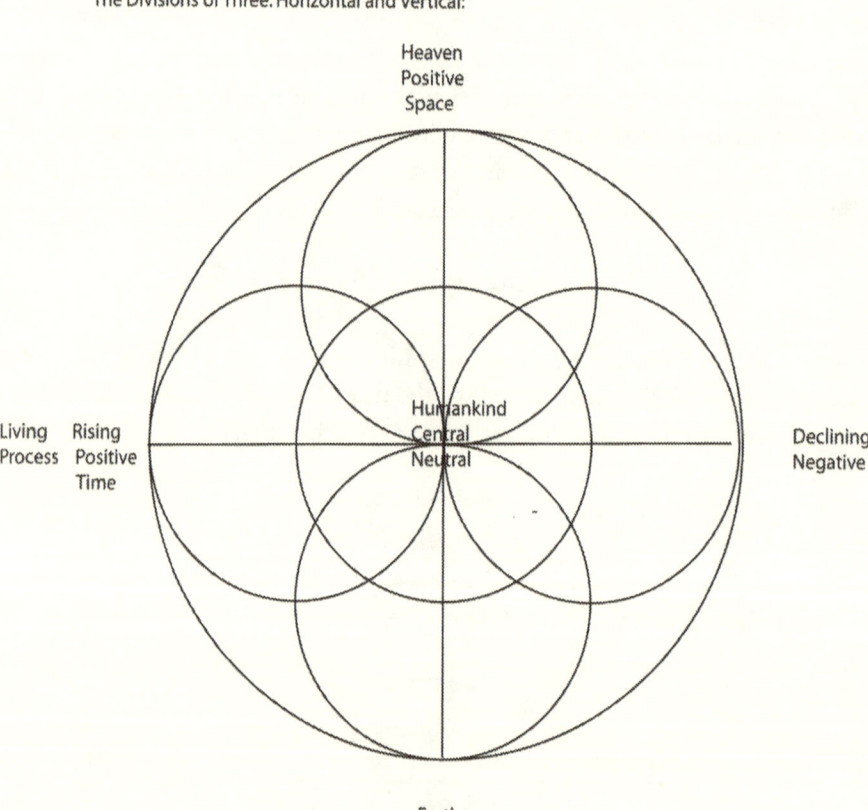

The Divisions of Three: Horizontal and Vertical:

Heaven
Positive
Space

Living   Rising
Process   Positive
Time

Humankind
Central
Neutral

Declining
Negative

Earth
Negative

Figure 5. The Horizontal and Vertical Aplication of the Number Three Based on the Conceptions of Yin and Yang. The image shows the potential evolution of the number three into a model for four directions, five elements, and six configurations.

The complex spheres referred to above can be reduced to the three basic planes that are needed to identify an object in space (fig. 6). All the implications of the number three with its triangles that form the root structures for other more complicated geometrical forms; its symbolic spatial reference to three dimensions of height, width, and depth; and its symbolic temporal references to beginning, middle, and end stem from these three planes of astronomy/anatomy.

Three

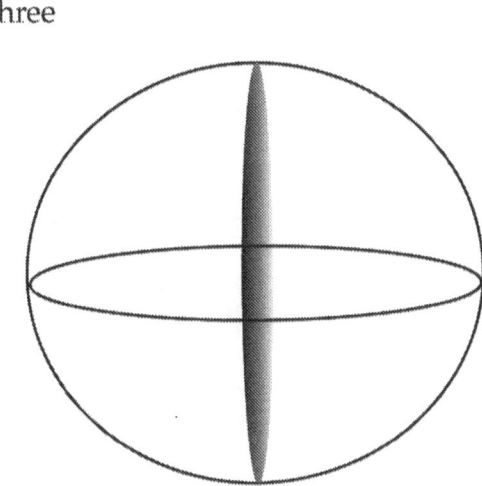

Figure 6. The Great Planes Responsible for the Ramifications of Three

The Chinese reduced the essence of this image of the three core astronomical planes to the vertical implication of these three planes. The prime vertical and the prime meridian cut through the horizon/equator/ecliptic planes defining in each case verticality. This is what is implied in the stacking of horizontal lines referring to heaven, horizon, and earth in the Chinese character for three. The plane of the horizon is the basis for this figure, and it is replicated above and below because it is from this horizon plane that all celestial observation starts and, in the end, has its experimental result. The annual and daily solar cycles and the lunar cycle can be tracked using sundials and stone circles. Nevertheless, figure 7 elaborates and reemphasizes the vertical introduced by three as opposed to the horizontal that plays out in the megalithic image because it is upon heaven, the apex of the triangle, noon, and midsummer that the whole yang construct of three depends.

Three

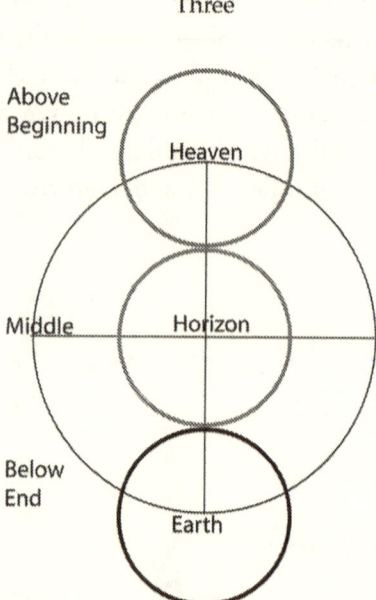

Figure 7. The Vertical Version of the Megalithic Triune Principle That Includes the Consideration of Time and the Incipient Division of Four

Figure 8 superimposes a stick figure on the basic system of planes to show the prototypical division of three, for ultimately, it is in the very structure of the human body and its divisions that the cosmic cycles can best be traced.

Three

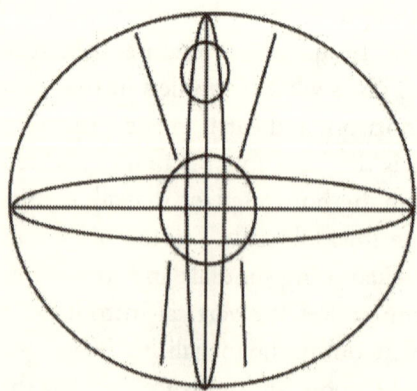

Figure 8. Stick Figure for Three

When using this model as a basis for medical treatment, the core value lies in its ability to isolate vertically on the standing patient, the zone of the body most affected by disease or most vulnerable to external pressures. These zones are fundamentally hollows on the torso defined by gross structural changes and by the large divisions of the limbs and head determined by bone articulations. In the midst of identifying the gross zones of functional disruption vertically, it should be remembered that the differentiation of left, center, and right is equally important as a structural differentiation based on processes within each vertical zone. Then again, time, the onset of disease, the progressive symptoms, and the likely outcome change depending on the location and depth of penetration in the bodily system. A cold and sniffles in the lungs (heaven) is less threatening than one that involves vomiting and digestive disruption and less threatening again than one that has gone past digestion into chronic stagnation and exhaustion that involves loss of fluid, hormonal patency, and muscle mass. All this is implied within the framework of the number three.

III. Medical Practice—Theory

Three comprehends an entire system of diagnosis and treatment in Chinese medicine in the same way that three implies three separate but integral ways of looking at experience. It describes the beginning, middle, and end of any evolutionary process and identifies the root cause, middle, or final stage in the development of a problem. Apart from trauma, all disease processes start invisibly, owing to internal or external vulnerability, and manifests with symptoms of a certain trajectory—either fever or chills, thirst or bloating, and the stages of the process and reactions of the body can be evaluated on the basis of how the balances of opposites shift.

Managing perspective is an important part of clinical practice. If a patient has a clear idea of their aims or problem, they can begin to get perspective on it. Identifying the medical or psychological issue clearly as a function of nature, society, or rationality in the widest sense can be a great help. If the patient is failing to conceive, having an affair, irrationally attacking her spouse, or refusing to care for her children or siblings, or reacting emotionally to everything, her problem may have to do with mammalian parts of life and issues having to do with reproduction, nourishment, instincts, or role models. Her perspective may be bound by repetitive reactions to the angles of lunar phases. If an active social life compels drinking, smoking, and conformity with social norms, if the person cannot adjust to a workplace, if there is excessive ambition, ceaseless activity, an unnatural craving for recognition or constant legal or political vehemence or suffering or partying till dawn day after day, then it is the mundane perspective, the day-to-day concerns that limit the perspective and lead to predictable yet unforeseen outcomes. If a person is worried, fearful, angry or depressed, or grieving and subjugated by any negative emotion or neurosis, then we can anticipate a disconnect from the ever-renewing energy of heaven.

If patients understand the zone and origin of their problems, that can be very important to their healing. Frequently, it is the time frame that we live in that dictates habitual behaviors. When consulting with patients, it can be shown that living month to month can cure the limiting perspective of the day. Trapped in a monthly cycle of passion and disappointments, the perspective of a year or goals for next year can help. If trapped in longer-term emotional or neurotic problems, the perspective of many years or a lifetime or of death can give leverage on the problem. These perspectives are built into the three light cycles and the structure of the human body. When these suggestions are reinforced by needles in appropriate symbolic locations, this can break habitual patterns and open appropriate zones of the body with amazing results.

Three provides a way to gauge the emphasis in heaven, on earth, or in between. It identifies heaven as above in the body, having to do with circulation and air, heat, and moisture, as well as the head and psychological/spiritual dimensions of human nature, and spiritual laws, if you will. Speech, identity, coherence, focus, and eye contact tell the tale.

Earth has to do with strength, stability, reproductive power, balance, familial roles and integrity, endurance, and the capacity to eliminate waste and keep fluid equilibrium.

The horizon with its equilibrium has to do with discrimination and discernment and with identifying worth, nourishment, and work that will support life. Digestion, transformation, separation of value from waste, good from evil, the social contract, the legal contract—all that has to do with the horizon where the social equilibrium develops between nature and heaven.

Spiritual and moral rules never govern entirely. The natural order and reproductive drives never govern entirely. Somewhere in between, there is equilibrium between instinct and intuition that our lives unfold, and humans learn how to function. Pathologies develop from excesses or imbalances in these three zones of experience, and both physical and psychological symptoms result.

Traditional Galenic or Roman medicine divided the three functional zones in the same way, attributed to the three vertical zones of the body: the rational (cognitive/spiritual/language functions), the animal (common sense; emotional/communal), and the vegetable (action and instinct). These were derived from Plato and Aristotle who described the mortal parts of the soul as reason, emotions, and appetites, wherein reason is located in the head, emotions in the heart, and appetites in the gut.[45] These were and remain very useful characterizations of the level of disruption, and both herbs and acupuncture can direct their healing more toward one zone or the other. The fact that there are three basic divisions of the brain—the cerebellum that deals with motor and instinctive responses, the midbrain that regulates the body and manages emotional and hormonal functions, and the cortex that handles cognitive and intellectual functions—gives further credibility to the traditional distinctions.

The cosmographical approach allows a practitioner to see in subtle signs and symptoms the early stages of imbalance that lead in time to disease, and the location of these symptoms and signs also indicates the trajectory and stage of disease.

All the dualisms stem from absolutes of light and dark, but play out in time and process. That is the difference between diagnosis and prognosis. They also play out in specific appropriate locations. If we extend the one-to-one relation between humans and the world, we immediately can take advantage of fundamental physiological and structural symmetries for diagnosis and treatment. This suggests that the hands, feet, upper torso, and forehead mutually reinforce similar principles; that the middle abdomen, forearm, leg below the knee, and middle of the face share similar functions and traits; and that the thighs lower abdomen, lower jaw, and upper arm share similar functions.

If the problem is identified as heaven, earth, or middle, the level of disruption is established. Knowing what level to treat—whether mental, emotional, or physical—is extremely important in the clinic. You can treat physical symptoms all day, but if the spirit or mind is traumatized from shock or loss, it will do little to forward the patient's healing. Using the principle of three allows you to determine the level and depth of the problem and gives a variety of body zones to treat them through. The later numbers, six and nine, further articulate these relations by further dividing each level of the body into three and making the connections just listed; but the primary idea of establishing the level and depth of a problem as physical, emotional, or spiritual is the single greatest diagnostic advantage of the principle of three.

When three is kept in mind during diagnosis, the following kinds of obvious conclusions are reached in cases with simple presentations. If all symptoms are above the diaphragm on the torso or in the head, then we must think of air, metal, and fire (lungs and heart, brain and sense organs, the world of thought and speech). These elements are hot and dry. If there is irritability, flushing, heat sensations, headaches, or dry skin, the qualities of heaven have lost their moorings, and the qualities of earth need support. If there is phlegm and moisture above, we know that these functions are weak and that some pathology is mobilized from below. Water and food generally move downward in the body, so if they make their appearance above, it is a reversal of the natural order, and above needs to be strengthened. If there is nausea or coughing, the natural movement is reversed. If the person is lethargic and tired, emotionally exhausted, or alternatively and excessively restless, irritable or verbally hyper, either can indicate a problem.

If all the symptoms are below, with weakness in the legs, reproductive disorders, lower back pain, bloating or gas, exhaustion, or instability on their feet, then there is a problem below. If sleep is difficult to achieve and if yin and rest are elusive, then nature's laws are being compromised. The inherited resources are being used up by some habit, compulsion, or circumstance.

If all the symptoms are in the digestive processes affecting the middle, then it will be an imbalance of moisture and dryness, heat and cold. The equilibrium and processes of daily living and daily social circumstances are disrupted. If the person cannot adapt and adjust, is too rigid, too controlling, or, conversely, too yielding, cannot lift a finger to support himself, eats only junk foods or compulsively one kind of food, drinks excessively, lives in jealousy, consumes or pursues properties or relationships or money then discards them without a sense of value or purpose, develops ulcers from unfulfilled desires and worry—all this has to do with the center, with the ego living a life of consumption, digestion, destruction, and elimination.

IV. Medical Practice—Treatment

The best uses of the Chinese pulse depend on making a one-to-one connection between humans and their world understood through the lens of number. The pulses have three positions that refer to heaven, center, and earth (3); and these occur bilaterally (6), and each position has three levels—heaven (superficial), center, and earth (deep) (9). This is the perfect three-dimensional meaning achieved by the number nine and played out dualistically in nature (12/18). The ramification extend to the number twelve and twelve individual positions; if the three positions are placed bilaterally, yielding six then contemplated in two depths that represent heaven and earth, then this pulse reading is the five element pulses. The standard TCM pulse has eighteen individual positions, nine on each arm. So in general, at the level of the number three, the pulse simply refers to any individual group of three pulses or the three pulses on the two arms taken simultaneously to study shared qualities of distal, medial, and proximal positions of heaven, center, earth. That tradition of assigning the distal pulse to heaven and chest above the diaphragm or head, the medial pulse to the middle of the torso or the whole torso, and the proximal pulse to the lower torso or legs confirms the notion that the hands and feet relate to heaven and so on and so forth, and the standard anatomical position in Chinese medicine is the standing human with arms raised like branches toward heaven.

The physical zones established by bilateral, distal, and superficial positions on the wrist pulse correspond exactly to the superficial outer aspect of the chest, and like an x-ray, the total pulse can be very revealing of local conditions within such physical zones as represented on the pulse. The pulses assume direct connection, communication, and correspondence precisely because of the numeric and anatomical patterns and their correspondences. The position of the arm, leg, and neck pulses are all at the mediating position between heaven and earth (divisions of three and nine): the middle of the neck, just below the hands and just above the feet. These three pulse positions represent in greater detail the heaven (neck), middle (wrist), and earth (ankle), but each contains a holographic representation of the other and implies the other in the distal, medial, and proximal positions.

Clearly, the root of all pulse analysis has to be an understanding of what the pulses are showing us about conditions in the three zones of the torso, the head, the legs, and the arms. If the distal pulses of the wrist are empty and the neck pulse is weaker than the ankle pulse over all and within the neck pulse we have indications of a particular yang or yin deficiency or obstruction that explains the deficiency, then we need go no further to find an important basis for treatment intervention. All other things being equal, then we need to nourish chi and yang, or we need to nourish yin and moisture balances in the chest.

Similarly, if we have a tense pulse with pounding and heat indications in the middle, bilaterally, and the overall pulse wrist pulse is tense or tight, then we can anticipate some kind of stagnation of chi, blood, or fluids with perhaps some history of chi or yin deficiency that predates it. We might assume that the middle position of the neck or ankle pulses would indicate some similar traits at the very least, but if they don't, we can assume that it is just a local digestive problem. And if they do, we can learn that the center and horizon social life of action is somehow frustrated or under strain. If the upper and lower pulses are significantly different from one another, either one strong and the other weak or both deficient or both showing tension, then in each case, we have clarification as to whether other factors related to heaven and earth are causing the digestive disorders.

When applying the same analysis to the proximal pulse and lower *jiao*, if distal and medial pulses are relatively healthy and the proximal pulses are feeble, we know that there has been some recent strain or cause of exhaustion. If the other pulses are weak and the proximal pulses also are weak, we can anticipate finding chronic exhaustion or chronic disease.

The anatomical division of the body into its three vertical zones and the identification of correspondences between these zones create an anatomical/topographical foundation for studying medical problems. With elaborations of in six, nine, twelve, eighteen, twenty-four, and so on, this foundation can be elaborated structurally and in time according to the kinds of divisions that project from the body in geometric space and how they ramify for the phases of the day, month, and year. It is only when physicians forget the one-to-one relationship between the body and the world that they lose their grip on diagnosis through direct perception of the common sense organization of the body and its functions.

Chapter 4 四

The Number Four and the Earth Element

The number three offers a division of a whole cycle into its structural components, and as an odd number, it is strangely timeless and structural. It defines the vertical axis of heaven, earth, and horizon, and as yet the horizontal implications for cyclic function with rising light, transition, and declining light cannot be considered. Therefore, three is not about defining process or change, but it does differentiate three preconditions of change. Three differentiates the basic divisions of the cosmos that make action and function possible. The mandate of heaven is complete with three, ready for manifestation. The immobilized idea of the vertical of the heaven, horizon, and earth relationship demands the inclusion of generation and action that it evidently lacks, and it demands as well that the horizontal be explained more fully. Three generates four, and four in turn allows for directionality that will measure heaven.

The horizon from any vantage point may be seen as an infinite plane differentiated only by the cardinal points with their seasonal and directional implications. The vertical is compressed down to a single plane, and the vertical of the midday sun is reduced to a position either to the north or to the south, depending on one's location on the surface of the earth or the backdrop selected. The movement from three to four, therefore, is a plunge from the completed vastness of a timeless prior heaven preconditioned for life into the restrictive, time-bound, earthbound processes of life itself. The infinite motion and space of heaven has been replaced conceptually with a restricted horizon with four eight or sixteen directions and a space occupied by unnumbered forms. This yin number was generated by three, but it burst from three into a context that amplifies and articulates the interaction of the two types of yin and yang that were only implied by the preceding yin number two. Early Western horoscopes show the importance of the cardinal points and the conception of the square implying four directions to all later divisions of six and twelve.

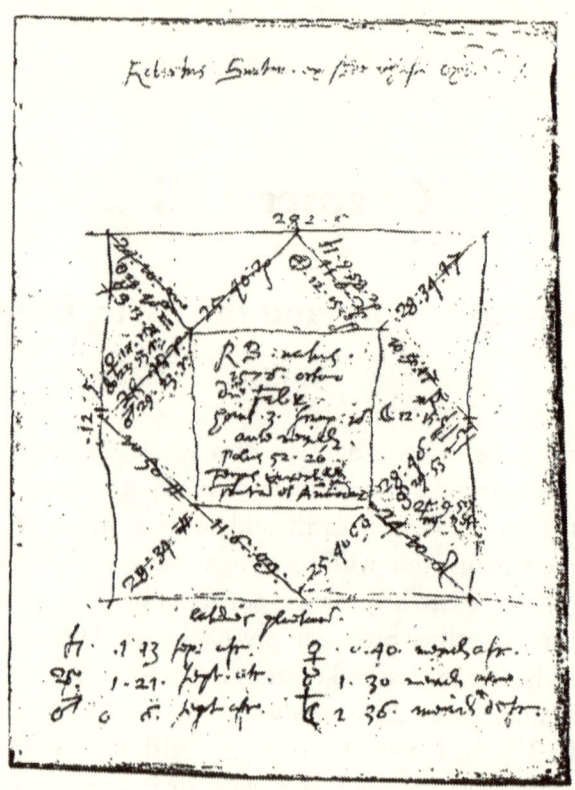

Figure 1. Robert Burton's Horoscope with the Heavenly Sphere Reduced to a Square (Squaring the Circle) with an Internal Square Showing the Cardinal Points and a Second X-Shaped Cross (8) Delineating House Divisions

I. The Number Four

These basic meanings are contained in the character *ssu* [四] for four. Wieger says, "It's a numerical sign. Even number, which is easily dividend into two halves. The old form graphically represents the division of the box into two halves."[46] The character *ssu* has two horizontal lines for heaven and earth and four vertical lines, representing four relations between heaven and earth. From any mountain or tree at the center of the earth in any traditional civilization, each of the four directions would describe different meteorological conditions and different relations to the cycles of light—such as rising and setting sun, cold when the sun is in the south, warmth when it comes farther north, and so forth. The qualities of heaven and the consequences on earth are different in each direction. The square shape implies the cross of the four cardinal points, but the roots of the character lie in the number two, which differentiates heaven from earth. It is with the number four that the absolute differentiation of heaven and earth in the number two can be reconsidered as a cohesive horizontal process. The north and south poles relect the absolute division of heaven and earth with some meteorological qualifications, while dawn and dusk or spring and autumn show the transformative processes on earth as either yin or yang. Differentiating the four seasonal and daily transitions of the cardinal points on the level of the horizon between heaven and earth is the key to this integrated reunification.

II. Cosmographic Structures

Chi Po's focus on simultaneous changes in the pure chi of heaven with its celestial bodies and on earth in the form of the elements (agents) and seasons was the first great differentiation between space and time, both linked to change and process. Used together, these division in nature could be applied to any process of increasing light/decreasing darkness and increasing darkness/decreasing light in two ways using the absolutist, vertical, spatial, or prior heaven way or the process-oriented horizontal posterior heaven way indicated in figure 1.

Four As Two Processes:

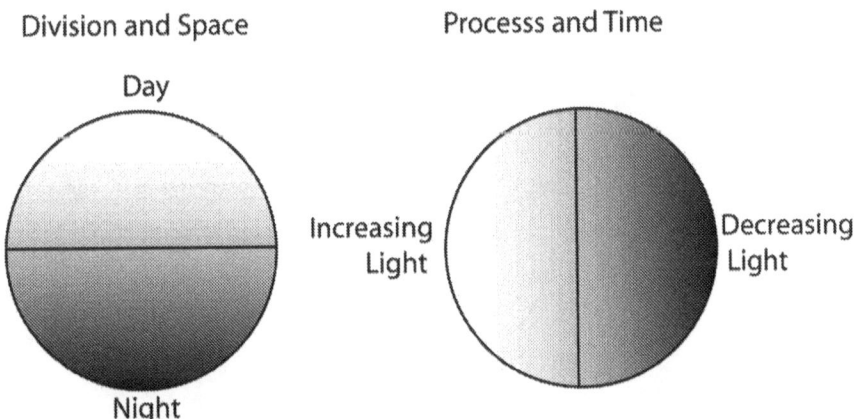

Figure 2. *A,* Division of Day and Night; *B,* Division of Midday and Midnight

The tension between these two views of nature, each with two polarized halves, will be examined here because it explains the meaning of four. The division of the great circle vertically from midday to midnight is descriptive of a relativistic conception of yin and yang that is more sympathetic to an observer than the absolute dichotomy of light and dark generated by the horizon. This is the differentiation of yin and yang through the traditional description of it as the difference between the shady side of the mountain versus the sunny side of the mountain. In this case, the mountain is the totality of all that is observable and invisible.

A composite of the two ways of perceiving yin and yang generates a sphere with a fourfold division, which can be superimposed on any complete natural cycle of light and darkness—annual, monthly, or daily.

The cardinal points—north, east, south, west—associated with the solstices and equinoxes, are directional in their meanings, and in the Northern Hemisphere, north inevitably implies the absolute source of both cold and darkness and south its opposite. The Mediterranean division of the year by four and twelve indicated the predominant agent and types of diseases associated with them.

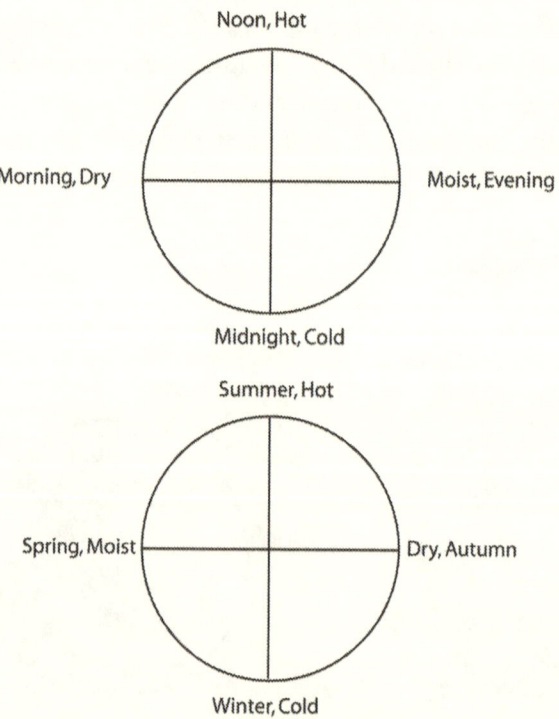

Figure 3. The Cardinal Points with Agent, Qualities, and Planetary Correlations: A, Ptolemy by Diurnal Motion; B, Ptolemy Qualities Combined by Seasonal Motion

Ptolemy links the cardinal points to daily attributes as follows:

> The eastern point . . . is chiefly dry in its nature; because, on the Sn's arrival therein, the damps occasioned by the night begin to be dried up . . . The southern point, or angle of the mid-heaven, is the most hot . . . the western point, is moist; because when the Sun is there, the moisture, which had been overpowered during the day, recommences its operation . . . The northern point, or angle of the lower heaven, is the most cold.[47]

The daily cycle stands in counterpoint with the seasonal growth cycle, where east is moist and west is dry owing to the characteristics of plants, and varies in different localities according to prevailing weather. In both cases, night/winter is moist and cold and day/summer is hot and dry. It is the transitions and their equilibrium and directionality that are in tension. The process, not the characterization, is important. The summer dries things out, completes the growth cycle, and in the autumn and evening, the cold and damp of night and winter begins to increase. At dawn and spring, moisture and damp dominate, but the sun begins to dry things out and allow for growth. It is the process and the energies behind them that needs to be understood, not the rigid concept for instance that the fall and the metal element is dry. Rather, if vegetation dries out at the completion of the growth cycle, it also is the point where cold and yin begins to encroach and dominate.

The four humors of traditional Western medicine came from the combinations of cardinal point attributions, and they symbolize the interaction of these qualities during seasonal transitions.

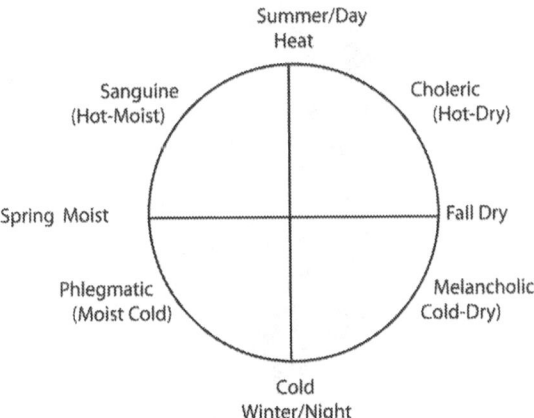

Figure 4. The Four Humors or Temperamental Types Derived from Combinations of Prime Qualities or Transitions between Seasons

In popular Chinese astrology, the element of the year, month, or hour are determined by heaven and, specifically, the position of the sun, within the larger celestial stem and branch sequences, the sun's place in the seasonal cycle (element and stem) and terrestrial phase (three-yin and three-yang phase and terrestrial branch), the lunation (month), the lunar phase and position of the moon in its constellation and house, and the phase of the sun in the diurnal cycle, whether noon, midnight, dawn, dusk, or other. The combination of these cyclic factors yields a pattern of symmetry and dissonance among the parts. The basis of Chinese astrology, like classical Mediterranean astrology, is a model based on the four cardinal points, which define the landmarks of each of these cycles.

The Chinese system has the north associated with winter and the water element, south associated with the opposite, summer and fire element, east associated with the growth of springtime and the wood element, and autumn links the decline of solar light to the brittle plants and clearing air of the metal element. Water is cold, fire is hot, wood is moist, and metal is dry. Earth was viewed as the center, the locus, and foundation of these transformations.

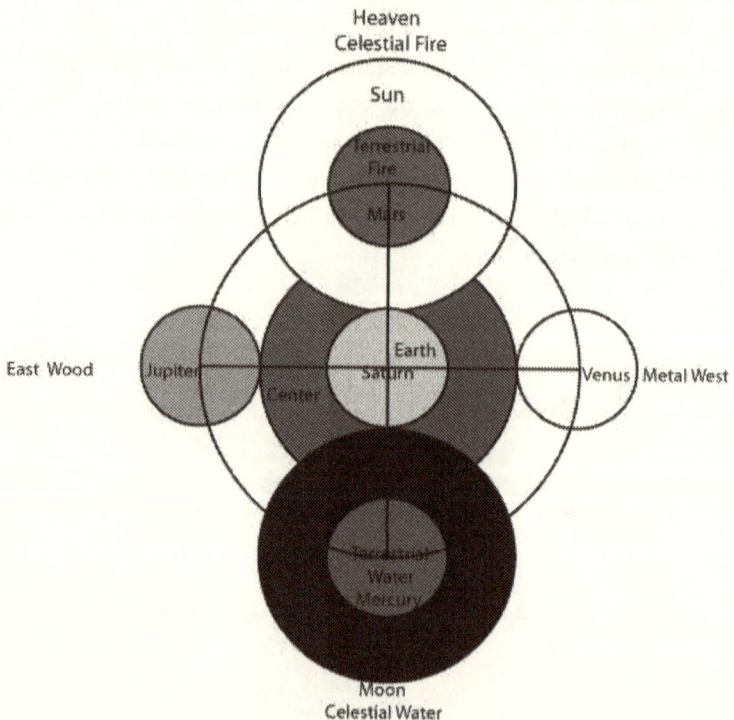

Figure 5. Earth-Centered Cardinal Points Portrait of the Elements with the Vertical Principle of Three Built In

V.—The Chinese Zodiacal Constellations, with their Corresponding Elements and Animals, the Longitude of their Determinant Stars in A. D. 1800, and their Approximate Constellations. 二十八宿（宮）.

Signs.	Element.	Animal.	Longitude.			Constellation.
1	角	木	蛟	201	3	0	Spica; ζ, θ, ι Virgo.
2	亢	金	龍	211	42	1	ι, κ, λ, β Virgo.
3	氐	土	貉	222	17	35	α, β, γ, ι Libra.
4	房	日	兔	240	8	48	β, δ, π, ν Scorpio.
5	心	月	狐	245	0	25	Antares; σ, τ Scorpio.
6	尾	火	虎	253	27	15	ε, μ, ζ, η, θ, ι, κ, λ, ν Scorpio.
7	箕	水	豹	268	28	15	γ, δ, ε, β Sagittarius.
8	斗	木	獬	277	23	6	μ, λ, ρ, σ, τ, ζ Sagittarius.
9	牛	金	牛	301	15	11	α, β, π Aries; ω, A, B Sagittarius.
10	女	土	蝠	308	55	54	ε, μ, ν, θ Aquarius.
11	虛	日	鼠	320	36	16	β, Aquarius; α Equuleus.
12	危	月	燕	330	33	45	α Aquarius; ε, θ Pegasus.
13	室	火	猪	350	41	59	α (Markab), β (Scheat) Pegasus.
14	壁	水	貐	6	22	9	γ (Algenib) Pegasus; α Andromeda.
15	奎	木	狼	17	48	12	β (Mirach) δ, ε, ζ, η, μ, ν, π Andromeda σ (2), τ, ν, φ, χ, ψ Pisces.
16	婁	金	狗	31	10	39	α, β, γ Aries.
17	胃	土	雉	44	8	47	Musca Borealis.
18	昴	日	雞	57	12	1	Pleiades.
19	畢	月	烏	65	39	58	Hyades; μ, ν Taurus.
20	觜	火	猴	80	54	47	λ, φ (2) Orion.
21	參	水	猿	79	34	6	α, β, γ, δ, ε, ζ, η, κ Orion.
22	井	木	犴	92	30	21	Gemini.
23	鬼	金	羊	122	56	24	γ, δ, η, θ Cancer.
24	柳	土	獐	127	31	4	δ, ε, ζ, η, θ, ρ, σ, ω Hydra.
25	星	日	馬	144	39	44	α, ι, τ (2), κ, ν (2) Hydra.
26	張	月	鹿	152	54	37	κ, λ, μ, ν, φ Hydra.
27	翼	火	蛇	170	56	9	22 stars in Crater and Hydra.
28	軫	水	蚓	187	56	52	β, γ, δ, ε Corvus.

VI.—Combinations of Chinese Cyclical Characters for the Years 1899-2008 inclusive.

A glance at this Table will show that it is formed by combining the Stems and Branches (天干, 地支) in regular order, beginning with the first of each. This gives the same combination every sixty years. Chinese chronology is computed by the cycles thus formed. A list of these cycles from B. C. 2637 to A. D. 1923 may be found in Mayer's Chinese Reader's Manual. The present cycle began in 1924.

己亥 1899–1959	庚子 1900–1960	辛丑 1901–1961	壬寅 1902–1962	癸卯 1903–1963
甲辰 1904–1964	乙巳 1905–1965	丙午 1906–1966	丁未 1907–1967	戊申 1908–1968
己酉 1909–1969	庚戌 1910–1970	辛亥 1911–1971	壬子 1912–1972	癸丑 1913–1973
甲寅 1914–1974	乙卯 1915–1975	丙辰 1916–1976	丁巳 1917–1977	戊午 1918–1978
己未 1919–1979	庚申 1920–1980	辛酉 1921–1981	壬戌 1922–1982	癸亥 1923–1983
甲子 1924–1984	乙丑 1925–1985	丙寅 1926–1986	丁卯 1927–1987	戊辰 1928–1988
己巳 1929–1989	庚午 1930–1990	辛未 1931–1991	壬申 1932–1992	癸酉 1933–1993
甲戌 1934–1994	乙亥 1935–1995	丙子 1936–1996	丁丑 1937–1997	戊寅 1938 1998
己卯 1939–1999	庚辰 1940–2000	辛巳 1941–2001	壬午 1942–2002	癸未 1943–2003
甲申 1944–2004	乙酉 1945–2005	丙戌 1946–2006	丁亥 1947–2007	戊子 1948–2008

Figure 6. *A*, Table of Chinese Seasons with Animal Associations and Agents;[48] *B*, Su Wen Version[49] (Needham and Liu)

It is important to recognize that Ptolemy's treatment of the prime qualities and cardinal points is similar to that of the Chinese and for the same reasons. For the most part, traditional astrologers were seeing eye to eye on real conditions and their causes despite differences in symbols and terminology. Chinese metal is identical in fact to Mediterranean air because both represent the period of decline and vegetative brittleness epitomized by the quality of dryness. That Ptolemy placed earth in the winter position, rather than water, indicates his hierarchical notion that earth generally lies below the water and in accordance, in its own way, with the Chinese notion that earth was the foundation upon which the seasonal changes occurred. His model, which was oriented toward the heavens, was more hierarchical and less introspective than the Chinese, so earth was not made the implicit central element. Nevertheless, for east and west, cold, not moisture, was the essential quality associated with winter.

Chinese models for the twelve lunar months remain true to the four-cardinal-point model by assigning elements to signs according to the cardinal point elements and qualities. The three springtime lunations and their animals are all wood, and the three autumn lunations and their animals are all metal/air. Similarly, the three winter animals are all water, and the three summer animals are all fire. The earth animals were given the quality of humid (moist and warm) to reflect positive agricultural values and to celebrate the fertility of the earth.

Table 1. Cardinal Point Model Based on Su Wen with Earth Implied at the Center and Transitions (Su Wen)[50]

Season	Element	Animal	Polarity
Winter	Water	Rat	Yang
	Earth	Ox	Yin
	Earth	Tiger	Yang
Spring	Wood	Rabbit	Yin
	Earth	Dragon	Yang
	Earth	Snake	Yin
Summer	Fire	Horse	Yang
	Earth	Monkey	Yin
	Earth	Sheep	Yang
Autumn	Metal	Rooster	Yin
	Earth	Dog	Yang
	Earth	Boar	Yin

Table 2. Cardinal Point Model with Three Animals/Agents

Season	Element	Animal	Polarity
	Water	Boar	Yin
Winter	Water	Rat	Yang
	Water	Ox	Yin
	Wood	Tiger	Yang
Spring	Wood	Rabbit	Yin
	Wood	Dragon	Yang
	Fire	Snake	Yin
Summer	Fire	Horse	yang
	Fire	Monkey	Yin
	Metal	Sheep	Yang
Autumn	Metal	Rooster	Yin
	Metal	Dog	Yang

Table 3. The Cardinal Point Model with the Associated Celestial Animal and Lunar Mansions or Constellations

Element	Heavenly Creature	Lunar constellations
Wood	Azure Dragon-East	Jiao-Horn Kang-Neck Di-Root Fang-Room Xin-Heart Wei-Tail Ji-Winnowing-Basket
Fire	Vermillion Bird-South	Jing-Well Gui-Demon Liu-Willow Xing-Star Zhang-Growth Yi Wings Zhen-Sorrow
	Yellow Quilin	
Metal	White Tiger-West	Kui-Legs Lou-Bond Wei-Stomach Mao-Hairy Head Bi-Net Zi-Turtle beak Shen-Three Stars
Water	Black Tortoise-North	Dou-Dipper Niu-Ox Xu-Emptiness Wei-Danger Shi-Room Bi-Wall

The cardinal points and prime qualities were associated with the visible planets, Sun, and Moon are as follows:

Table 2. The Visible Planets and Prime Qualities

Body

Celestial Body	Western Qualities	Western Humors	Western Element	Chinese Qualities	Chinese Element
Sun	Hot/Dry	Sanguine or Choleric	Fire	Hot/Dry	Fire
Moon	Cool/Moist	Phlegmatic or Melancholic	Water	Cold/Moist	Water
Mercury	Dry/Cold	Melancholic Mutable	Air	Cold/Moist Mutable (like water)	Water
Venus	Moist/ Warm	Phlegmatic	Air/ Metal	Cool/Dry	Metal
Mars	Hot/Dry	Choleric	Fire	Hot/Dry	Fire
Jupiter	Warm/ Moist	Sanguine	Water/ Fire	Warm/Moist	Wood
Saturn	Cold/Dry	Melancholic	Earth/Air	Damp/cool Transformation	Earth

The Roman astrologer Manilius shows the planetary relations to these dominant traits. Systems have their own weak points, so one quality was assigned to each planet, whereas the model suggests that the qualities originally were conceived as interactive.

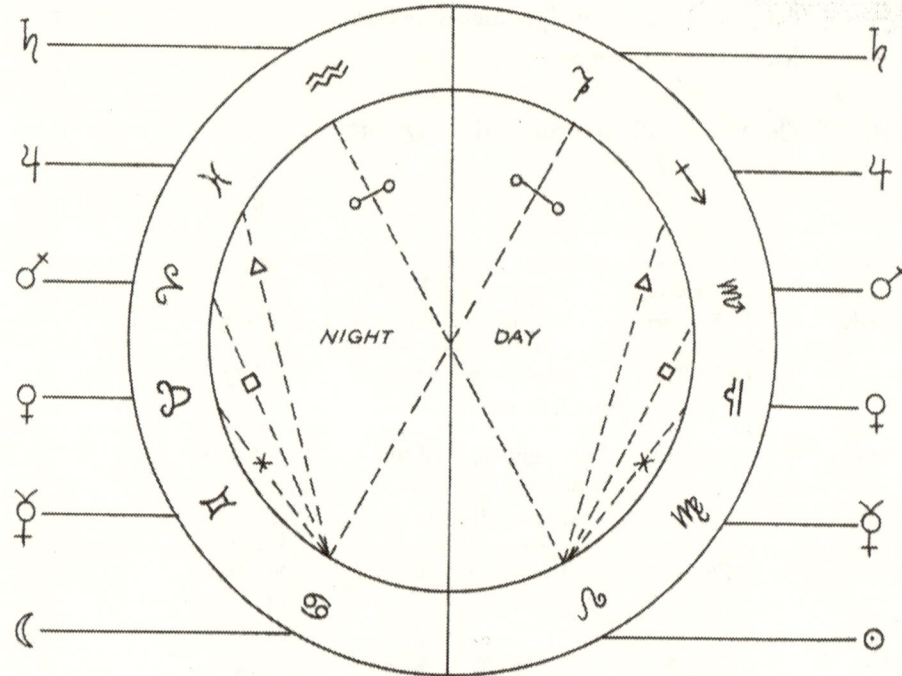

Diagram II. Zodiacal Houses of the Planets.

Figure 7. Manilius's Planetary Rulership of Zodiac Signs and Seasonal Phases

Manilius shows how correlations developed, but these abstract relations originally were connected to sense for the living process of the seasons with the oscillation of heat with cold, moisture with dryness in such a way that admixtures developed from planetary positions could be analyzed in a living relationship with local conditions and patterns. Chorography was the name of this discipline that linked astrology to epidemiology (see introduction).

Mediterranean models vary to some extent in their disposition of the qualities for different cycles, diurnal, seasonal, and annual, and they show an Egyptian flavor, and he was bound by Aristotle's rigid dichotomies. His system is more concerned with the dryness of winter than with the cold, for water in a desert is scarcer, and its lack in winter is worse than cold. If we look behind his combinations of qualities such as hot/moist or hot/dry, we can see an effort to characterize periods. Given that a day in any season may be unseasonably warm or cold, hot or dry, and seasons themselves vary, Ptolemy, like the Chinese, is here concerned with a relative emphasis of one quality or element over another rather than a categorical certainty. The Chinese's seasonal attributions are shown in figure 5 below. Thus, the aim of the ancients was to understand the bias of a season, not to say that the other qualities and tendencies were eliminated. Figure 8 shows the whole complex set of relationships.

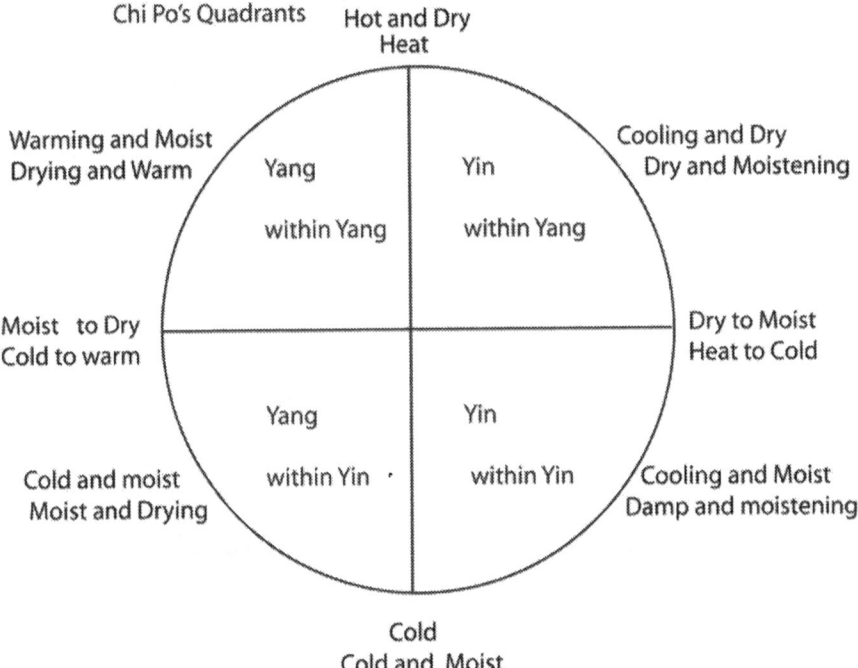

Figure 8. Dominant Single Qualities Associated with the Seasonal Cycle if Viewed Vertically and Horizontally

Chinese thinkers wrestled with the same problems of how to characterize movement in a cycle when divided by four into a spatial configuration. Chi Po describes the relations within the Chinese division when it is superimposed on the diurnal cycle as follows:

> Therefore, it is maintained that there is yin within yin and there is yang within yang. From dawn until noon is hot the period of yang of Heaven and it is yang within yang; from noon until sunset is the period of yang of Heaven and it is yin within yang; from sunset until crowing of the cock is the period of yin of Heaven and it is yin within yin; from crowing of the cock until dawn is the period of yin of Heaven and it is yang within yin.[51]

Chi Po focuses here not on the absolute poles but on the interaction of them in the processes contained in the quadrants between the cardinal points. Each quadrant of the diurnal cycle has its own tone, its on set of balances derived from the rotation of the earth, and its changing relation to the sun. The peak points have transitional significance for Chi Po, but implicitly the point of transition, which approximates the cardinal points, elicit from nature a peak expression of a particular combination of qualities. By shifting the quadrant model of the circle forty-five degrees, the cardinal points can be viewed as the stable center of a seasonal process or the point of greatest manifestation or directional mobility of yin or yang.

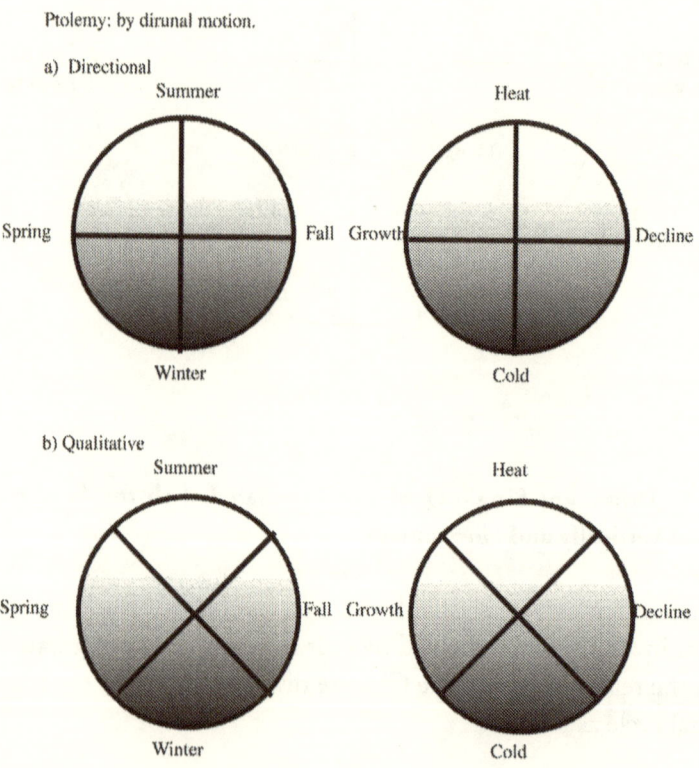

Figure 9. Quadrant Division: *A*, Directional and *B*, Qualitative

Both views discriminate separate useful understandings of the functional character of any cycle. But together they show that the point of greatest extension or emphasis is the point of greatest transformation. These points have to do structurally with the plane of the horizon, the plane of the prime meridian that divides east from west, and the plane of the prime vertical that divides north and south. These invisible geometrical concepts have real meaning when applied to human anatomy. Su Wen asks, "Why would one want to know about yin in yin and yang in yang?"

> This is because
> In winter, diseases are in the yin [sections];
> In summer, disease are in the yang [sections]:
> In spring, diseases are in the yin [sections];
> In autumn, diseases are in the yang [sections].
> In all cases one must look for their location to apply needles and [pointed] stones.[52]

What is not commonly appreciated in the simple opposition of the prime qualities—hot/cold, moist/dry—is their interaction and the relation of that interaction to the great planes and implicit anatomical structures. If hot and cold form an axis that aligns with the prime meridian, then this defines the structural absolute as the vertical plane. This is a spatial precondition to change. If hot and cold are the constants, then moisture and dryness define the effects or process. Whenever it becomes colder and rain may result, whenever it becomes warmer on the oppositional gradient, the equilibrium between moisture and dryness is affected; but the long-term affect is that dryness results.

If on the other hand moisture and dryness become the constant, that is the prior heaven precondition associated with the horizon. Then, the more rain falls, the cooler it gets; and the more dry things become, the warmer it gets. These meteorological relationships are the reason prime qualities, were selected from an infinite number of opposite traits to occupy the cardinal points. The cardinal points taken together define the golden mean, the equilibrium of qualities in nature and human behavior. A complete cycle must have equilibrium for life to flourish.

The number four and the cardinal points then display the relations between meteorology and the transformations in nature, which occur on or near the horizon. This is a subordinate process to the greater actions and preconditions of heaven and earth, and indeed heaven and earth are implicit behind the screen of weather and seasonal changes. But the number four provides a comprehensive standard for evaluating the equally fundamental relations between constants (chi) and substance (transformations) that Chi Po defines as primary differentiations between yang and yin. The real insight here is that each of the polarities (hot/cold, moist/dry) has a yang (prior heaven spatial) and yin (posterior heaven temporal) implication, which yields four directional poles for each condition when it is the defining condition (fig. 10).

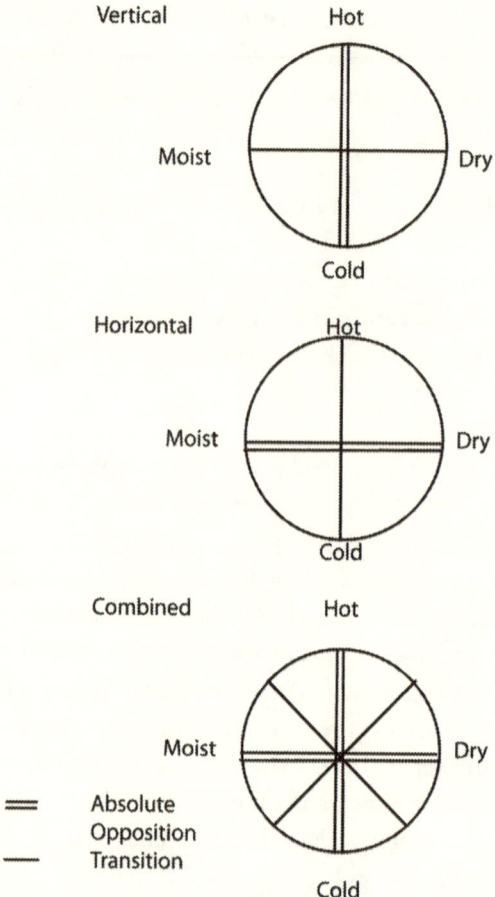

Figure 10. *A*, **Four Directions Showing Hot and Cold as Absolutes, the Other Suppressed and Implicit;** *B*, **Four Directions Showing Moist and Dry as Absolutes;** *C*, **Recombined in an Image That Establishes the Ruling Polarity**

The combined image merely establishes which dichotomy of the interdependent polarity is setting the standard, which sets the other in dynamic motion.

To further flummox the mind, the standards are reversible and indeed reversed in nature, where the space of prior heaven is the time of posterior heaven; and the time of prior heaven is the space of posterior heaven. Specifically, the cardinal points define processes in nature (time) relative to the number three (space, heaven, horizon, earth) but define space (the four directions) relative to the number five, which focuses on processes. This reversal that plays out in number theory and in the double quadrature (eight) implied by the functional interplay, offers the seed of understanding of why the five elements—which derive from the number ten, the Sun, and the heavenly stems—have direct association with the cardinal points and seasonal transformations (space) and the yin substances of wood, fire, metal, etc., on earth while the reverse also is true that the six terrestrial configurations that derive from the twelve lunar terrestrial branches and form the basis of the three-yin and three-yang configurations describe the phases (time) of the Chinese clock that defines and conditions the circulation of heaven and chi. The conundrum is simple only when the root cosmographical standards are understood. The invisible prime vertical, meridian, and horizon define space so that time can be recorded; and visible transitory objects define time so that space can be understood.

The simplest way to understand the cardinal points is to understand that for the horizons midday and midnight, located on the prime meridian, are conditions defined by it, while dawn and dusk on the horizon are conditions defined by the prime meridian.

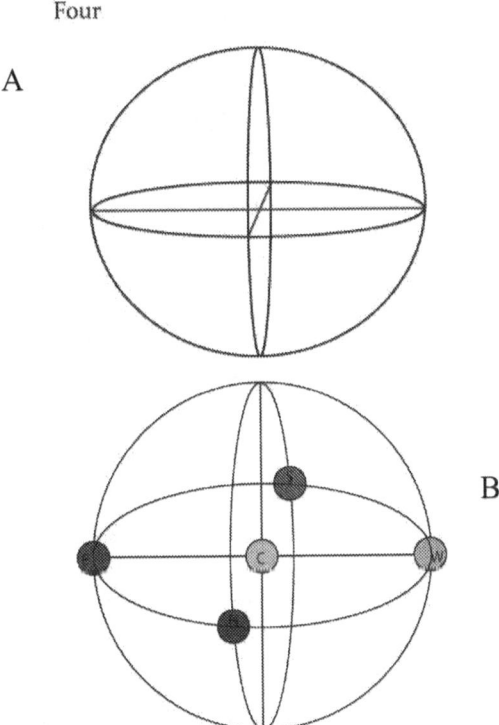

Four

A

B

Figure 11. *A*, Graphic for the Number Four Showing Constituent Great Planes; *B*, Graphic for the Number Four Showing the Cardinal Point Consequences on the Horizon with Element Associations

Four really describes the location when plotted on the horizon of these vertical planes, which give the four directions.

Four

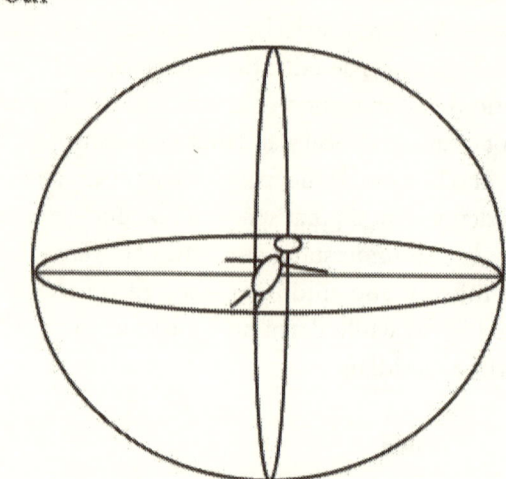

Figure 12. Graphic with Stick Figure for the Number Four

The stick figure shows a human form lying down, either facing heaven or earth with legs together or head to the south and arms at ninety degrees. This figure emphasizes the human relation to the four directions and the connection between the four limbs and the four directions. It is noteworthy that the arms can lay ninety degrees, but the legs cannot; but they do extend the main trunk of the body along the north-south axis with its implied vertical connection between heaven and earth. There is a rational upper body understanding of the geometry on the plane of the horizon. The prone human body unites the plane of the prime vertical (of the individual) with the plane of the horizon. The emphasis is on the mundane directions and material orientation of life. The compression results in a human experience in space, not time; and four is the first basic clarification of how an individual can be located in space. As far as the body goes, there is back and front, which are assumed in the number four; but what is made explicit on the body in this position is above and below and left and right, which are the first really basic internalized understanding of orientation. When this physical relationship is extended from the prone position, it gives the four directions when facing south; and standing north and south or back and front are assumed, and the basic divisions describe the heaven and earth, the prime vertical, and horizon east and west. I have chosen the former because four always has been about the cardinal points north, east, south, and west with the standard associations with prime qualities and from their element and seasonal affiliations.

The number four, then, has each of two oppositions on a plane that relate to a third invisible function. This is both the residual implication of the number three that emerged from two for the same reasons and now anticipates five. From both the perspective of the horizon and the midheaven, the third or missing component is defined by the intersection of the horizon, the prime meridian, and the prime vertical. The locus where this takes place is on the horizon at an imaginary point at the center of the earth. The genesis of the number five is from the cosmic conception of the intersection of all the basic planes for identifying objects in space where they intersect the horizon, the four cardinal points with the implicit center. It is not really earth that is at the center but the planes that intersect at the geometric plane of the horizon, the meeting point of the invisible heaven with its geometrical laws and the tangible earth. This earth is a process in motion.

III. Medical Theory

The *Nei Jing* makes a basic link between seasons and health in book 1, chapter 2. Therefore, yin, yang, and the four seasons are the beginnings and the ends of everything; they are the roots of life and death. To live against them will bring about destruction of life, and to live in harmony with them will prevent disease . . .[53] Traditional medicine typically associates the elements with the prime qualities as in figures 4 and 5 with fire in the south associated with heat and, therefore, with heaven, sun, and Mars. Water in the north is associated with cold, associated with the Moon and the planet Mercury. Wood in the east is associated with springtime moisture and the planet Jupiter, while metal to the west is associated with the hardening and drying of plants in the autumn and the planet Venus. Based on the above, we can see that there is a basic opposition between these elements based on the prime qualities and their planetary corollaries.

Fire and water form a vertical pole that mediates moisture, growth, and decline. Moisture and dryness form a horizontal opposition in nature that mediates between the peaks of summer and winter, midday and midnight, controlling the goals of growth and decline.

Remembering the roots of the five elements in these seasonal constructs is vital to clinical practice. Each element has an opponent and mediates those at ninety degrees.

Wood

Cosmic structure tells a practitioner that problems with moisture—i.e., water accumulation, dampness, weight gain, sugar problems, and appetite issues—fundamentally are imbalances in wood, the rising light cycle, and the liver. If the liver is not processing sugars, metabolizing chemicals, breaking down fats, and purifying blood, the body will allow water to accumulate around cells to diffuse toxins, which results in damp problems. Chi deficiency is a problem fundamentally of the rising chi being insufficient to overcome inertia (winter, rigidity, habit, death) in the forms of toxins, metabolic wastes, yin damp, and cold. Healthy liver and gallbladders together support and regulated growth, smooth flow of chi, and the regulated increase of warm and activity.

Wood works with cold to overcome dryness and death and to keep cells nourished. When wood flourishes, it protects the earth, contributes to oxygenation of air, and supports chi. Rain and moisture in the body control excess heat and keep rising fire useful by chaining it to the job of moving fluids around the body and, therefore, supporting the heart. *Shao yang* wood is aggressive and accelerates the light cycle in surges of heat. *Jue yin* liver controls these surges with rainfall, which cools things down. Rain occurs when moisture evaporates from the earth and rises into cooler air or when cool air intrudes on a region of localized heat. This process is very much like the fluid metabolism in the body. When the body is active, it produces heat; and to disperse the heat, the pores open and people perspire. Similarly, in cold, damp weather, the only remedy is to keep active and to generate heat; otherwise, exterior cold and rain can rapidly overwhelm the body, leading to cold conditions. The wood element is the key to these meteorological and physiological processes. The wood element regulates our level of activity and growth adapting to environmental changes to keep equilibrium between heat and cold while counteracting the principle of inertia and death enshrined in the metal element.

Metal

The metal element originated as a way to characterize the hardening and withering phases of the growth cycle. The movement of light is decline; the principle movement is descent. Form diminishes liberating the spirit; the two separate. This is the principle of breath, to inhale, then to exhale. Metal is the necessary exhale; wood is the inhale of the annual cycle. Our last breath is an exhale. Metal is the corollary of air, but the focus of the four directions is on the horizon; and therefore, some choices were made to discover what material traits would best express the action of air. Air is drying as it blows across surfaces. This capacity to dry things often is a result of the air being heated. The late summer heat yellows plants, which later as the fall progresses harden. This single process shows how heat and air dry; but as the days get colder, the plants do not recover but harden further until in the middle of winter. Even water, the softest of substances, hardens into ice. Things become more brittle with cold, and things become softer and more pliant with heat. Metal is the mediator in this process of moving from softness to hardness, from life to death, from kinetic energy to inertia. It is a singe process.

The fact that wood/wind is the inhale and gentle ambient air/metal is the exhale of any cycle gives us a glimpse of the interdependency of the two functions. While lung is attributed to metal along with the colon, which also exhales waste air/gas to move matter, the interdependency is important to keep in mind. Each occupies one side of the diaphragm. If the liver is swollen or there is an accumulation of fluids, the inhale will be difficult. If people are chi deficient from lack of yang and light, lung problems will develop. If there is fluid and yin accumulation in the lung, you need to consider the liver; but more importantly, the lung is vulnerable to dryness, then to cough. Frequently, it is the liver's ability to manage moisture in the equation of moisture and dryness that results in the lung having problems. If the lung is dry, it is important to treat the liver to reestablish the capacity for rain and moisture.

Even visually and anatomically, the open lung looks like a downward hanging tree that is white and decaying; the liver is shaped like a healthy deeply colored tree with the gallbladder as its trunk facing upward. These physiological shapes and colors were not lost to early Chinese anatomists, and they symbolize the kinds of equilibrium we see between wood and metal. Wood sets the conditions for equilibrium of moisture and growth that allow for the accumulation of chi even from oxygen; the lungs use the chi and circulate moisture, keeping moisture in check with dryness by distributing and drying it when it accumulates. The opposition affects heat and cold. If it's a rainy year, it will be cooler than normal. If it's dry, it will be hotter in the summer but colder and more barren in winter. It is the static oppositions with their process relation to the plane at ninety degrees that account for change and equilibrium. It is these fundamentals that establish the meanings for the elements when applied to the five-element scheme.

Fire

Long days, short nights, work and play, pleasure, excess, abundance and war, strife, decorum, activity, intensity, joy and sorrow—these are the trappings of heat. Heat is a summer shower; and intimate pleasures with a healthy liver are travel, war parties, exploration, honors, respect, and celebration with a healthy lung. Heat? How does heat affect us? Heat always is associated with the sun and in a secondary way with bellicose red-colored Mars. The sun brings life and warmth. Mars brings activity and strife. Both exhibit the peak of a process of increasing light symbolized by the summer solstice, full moon, and noon. Yet clearly some summers are subjectively moist, some dry, some warmer, some cooler;

and it is this equilibrium between heat, moisture, and dryness that modulates the temperature. This also is the basis of health. Good blood circulation (heart, sun, heaven) results in the stable management of heat and energy from the wood and dispersing and cooling from the metal. The lungs ventilate excess heat, and the liver keeps the metabolism and fluids in equilibrium. Yet these together are not enough to manage environmental extremes or to keep the tendencies toward exuberance in check.

The oceans, rivers, and lakes—the accumulated bodies of water—also must be there to create a reservoir of yin to compensate for prolonged disequilibrium either by storing or eliminating excesses of fluid or compensating for excesses of heat and dryness. Summer heat also is stored in water, helping to get through the winter, while winter cold creates a resource for slowing the flaring heat of spring and summer. The same operations take place. It is calcium and minerals in the bone that help humans survive periods of prolonged stress, excessive heat, and starvation. It is the bladder that eliminates excess fluids and disburses the metabolic wastes from excessive activity.

Water

Water is stored below the surface of the earth, moves on the surface, or appears as rain and vapors in a way that corresponds to the three zones of the body where the bladder stores water below the waist, the stomach gurgles above the waist, and vapor emerge from the lungs on a cold day. Froth in urine and stagnant water in the tissues are audible only when the abdomen is percussed, or fluids streaming from above all indicate clinical imbalances in the appropriate distribution of water in the body. Water and its cooling functions like heat and its warming functions affect the whole body and that vertically.

Clinical observations suggest that each level has its need for activation and calming, heating and cooling to maintain equilibrium. Equilibrium of water depends on its reservoirs and the hormonal and mineral storage in the deepest levels of the body, the lower glands and the bones. Traditional physicians considered the premature exhaustion of these resources from overwork, overactivity, and overindulgence—common to summer heat and fire to be so harmful as to make recovery almost impossible. Water is conserving and conservative in this sense. There always is a limit that must not be crossed without endangering all the other elements and functions; and fear is the emotion that insists on safeguards, protections, caution, and basal standards that protect the integrity of the gene pool, the family, and the tribe. Water has a psychological and a physical ramification, as does its opposite fire, and the two together establish the capacity and equilibrium of an individual; wood and metal simply express how this equilibrium plays out in experience. Leverage on this equilibrium of water and fire is innate on the one hand but is supported by hydration (swimming, standing in a rain shower, and exercise.

The interaction of the prime qualities around the four cardinal points is the conceptual root of both the five elements and the eight principles, because it is the same cycle and equilibrium described by each model; but the oppositions that result in the ko cycle relations described by the five elements and the dynamic oppositions generate the model for eight that underlies most Chinese medical theory. All systems describe the same cyclic structures and processes, but the number four sets the standard for associative meanings where it counts on the horizon in the manifestations of annual growth cycle. The importance of this perspective, and particularly this perspective, applied to the middle of the human body where digestive processes rule is vital. At this level, both the spleen and the liver must be taken together as a process of equilibrium that is counterbalanced by the lungs, whose vertical relationship in the body and world (air being above verdure) is brought down to the level of the horizon in the model of four. This compression to two dimensions highlights the basic functional balance between invisible chi and visible substance on the horizon. At that level, chi and air are fully integrated functionally into equilibrium with fluids.

When applied to the human body, this cardinal-point model is the human form lying flat on the earth: (a) If when lying prone the body faces upward, it is receptive to heaven and acts as its yin, earth counterpart. With this orientation the head and lungs/heart is to the south, the liver sits to the east, the spleen is to the west, and the kidney, bladder, and feet lie to the north. The arms are pointing east west at ninety degrees to the straight relation of the spine with feet together completes the cosmic symmetry and shows the alignment between the legs as an extension of the earth aspect of the shao yin, kidney, tai yang bladder centerline, (b) If the body faces down with the yang back conforming with the yang of heaven, then the head faces north, the liver is to the east and right, the spleen to the west and left, and the UB/kidney and legs to the south. Facing up the body displays the changing functions and forms of nature in relation to the four directions; facing down the body displays the energetic action of heaven that generates life and cause these changes. Clearly, these positions are symbolic and static, and the bilateral topography of the body suggests that humans are designed uniquely to adapt to every position and configuration while retaining their equilibrium.

The clinical importance of these anatomical positions is that we can view the prone patient cosmographically. When the patient faces up, the organs are laid out on a functional grid that mirrors the functions of nature, and they stand in a one-to-one relationship with nature. The prone position allows us to see the literal processes that generate symptoms laid out geographically as it is in nature. The lungs and liver are on the same plane operating in a rhythmic relationship on either side of the diaphragm, the dealing with fluid metabolism the other with oxygen, which together create a systemwide equilibrium between hot and cold, moisture, and dryness.

Thus, we should be able to note the effects of meteorological trends in nature in our patients and advise accordingly. If the weather is dry and hot for prolonged periods, the acupuncturist or herbalist can recommend eating less, drinking fluids, resting, eating moist foods and fruits, etc. From the personal perspective, the functions of the body that show dry cough, dry lips, cracked or flaky skin, and fatigue can be seen as a problem of dryness that may need the treatment either of (1) treating the related organ directly, focusing on the equilibrium in the lungs and air that are responsible for distribution and diffusion of moisture and dryness, (2) the liver to improve the distribution of fluids and rain, (3) dipping into the reservoir of the kidney or the yin aspect of the lung and large intestine to hydrate (dipping into the kidney abates the heat of the fire, which slows the drying process in air, but it also adds cold, which is the precondition for generating the hydrating rain and activating the liver), (4) improving the heart and circulation to mediate between moisture and dryness, emphasizing the yin aspect of the heart to ameliorate the ultimate cause (prime meridian to the horizon) of the dryness in the long-term relative excess of heat (if that is insufficient, again, drawing on the kidney to support that action).

This one-to-one relationship between nature and humans clearly forms a theoretical foundation for medical practice. It is not enough to treat a particular organ; the whole system needs to be understood, and treatments prioritized according to the depth of the problem from relatively superficial to deep as above to (1) treat the equilibrium of the system, whose symptoms are exhibited, that is, moisture = liver/GB dryness = lung/LI, heat = heart/SI, cold = kidney/Bl; (2) treat the opposite principle: if dryness is the problem, treat the liver/GB to moisten; (3) get leverage on both poles of opposition by treating the controlling cause found on the plane at ninety degrees (treat a problem either of moisture or dryness by treating its cause heat for dryness, cold for moisture); or (4) if the opposite, treat heat to cause dryness in the patient, cold to increase rain or moisture; and (5) draw on reserves of moisture or chi to create equilibrium between functions and break the local cycle of imbalance, i.e., treat GB, TH, and si DU to activate reserves of chi to replenish chi or move fluids.

This is a basic and profound clinical approach that is often overlooked because standard clinical approaches tend to lack finesse in diagnosing the level of a problem. Clinicians too frequently will bring out the heavy guns at level 2 or 3 by automatically treating the opposite or using the reserves to treat a local problem. It is all too common to treat dampness by treating spleen/stomach and lung points, and damp conditions are automatically assigned to spleen when orthopathic dampness in the Chinese system originally was associated with spring and the equilibrium between *shao yang*, GB/TH, and Li/PC.

Many damp conditions can and should be seen to develop from imbalance in these orbs caused by hyperactivity (GB/TH) and consequent heat or emotionally induced hypoactivity (Li/PC) and consequent lethargy and damp. The tendency in Chinese medicine is to leap to opposites, to the point that the original model for healthy equilibrium is lost, and the source of the problems and solutions are misidentified in these very opposites.

Dryness is a condition that results from the failure of a whole process, starting with a disrupted equilibrium in the distribution function of the lung and air, which keep clouds and moisture and temperatures in some kind of equilibrium. An exhausted marathon runner who has destroyed the equilibrium in the lungs by drying out the system, must hydrate, rest, and repair; and this provides a good example of what happens more slowly to people who are too active or sedentary in a way that disrupts the equilibrium between liver and lung. The person who overworks all the time will dry out the lung because the excessive catabolism and heat will dry up yin and accelerate the drying out as the lungs try to keep every shrinking a supply of moisture moving around the body to put out the flame. Thus, in long term, the lung suffers when the activity of the spring is out of balance. The person who is too sedentary also fails to distribute moisture and nourishment to the tissues of the body, and ingested moisture sits and accumulates, creating localized dampness and stagnation. The dampness isn't dispersed and distributed. This leads to mixed symptoms of dryness on the surface and dampness internally. By looking at the prone patient, we can see these dynamics playing out literally before our eyes. When we lose this capacity for literal observation, we get into systems and lose immediacy in our clinical actions.

The absence of earth as a manifest element at the level of the number four anticipates its emergence in five because it is everywhere assumed but not explicit. The whole locus is on the horizon. Physically, if we assign the head to heaven and the legs to earth, the entire circuit of the torso aligns as earth, the locus of transformation in the cardinal-point model. Abdominal diagnosis stems from and depends on this one-to-one directional correlation. The yin number four with its directional emphasis is the perfect organizer for the otherwise complicated contours of the yin side of the body. Heart is south, water is north, liver is like a tree with the roots facing north but growing upward toward light, and lung is like a tree with the roots facing south but growing downward toward cavernous darkness. Structurally speaking, the diaphragm is where these two processes meet. This structural relationship is the basis of clinical diagnosis, and the body zones link directly to areas where elements predominate; and the way the elements interact with each other spatially also are clear.

This means for treatment that to treat water, you treat below, fire above, wood in the abdomen, lung in the chest, and these throughout their zones, not exclusively on meridians. Moreover, if one element dominates another, the opposite can be treated to bring it into balance. If heart is depleted from excessive use, water can supplement and control it at the same time. If lung is depleted from excessive use, wood can supplement it and control it. These treatments have less to do with arm and leg meridians than with structural zones on the body. In this context, leg meridians are mere extensions and support of water and wood below the diaphragm. Arm meridians are mere extensions and support of fire and metal.

As usual, this straightforward notion is supported and contradicted at the same time, because the arms extend to ninety degrees, suggesting their relationship with wood and metal, while the legs extend vertically suggesting connection with water and fire. The reason for this partial overlap is that all these diagrammatic approaches attempt to describe three-dimensional and functional relationships in two dimensions and absolute terms.

Another feature of the symmetry of four is that it shows that heaven always is one; Earth always is two. The head above is contrasted with the two legs below. The heart (yang) is one, and the lung (yin) is two. The liver is one; the kidney below is two. The reversal here is that the legs act together as one, and the arms above act independently as two. Once again, there is a pattern and compensation that allows for multidimensional functioning. Where the body reversed, the arms can be extended vertically together just like the legs. The legs can be extended to forty-five degrees to accommodate the number eight. The critical thing is to distinguish in each case what symmetrical structural relations apply uniquely to the number that forms the basis for analysis. For the number four and the four-directional model, the symmetries described above form a very basic and persistent set of structures that relate to function, practically and symbolically.

In a way, the functional levels of treatment established by three and heaven, earth, and human are played out structurally and given a pragmatic set of element correlations. Fire and water create an outer surround of yang and yin; metal and wood represent a lesser more intimate contrast that focuses on the breath as a mediator (fig. 9). The arms at ninety degrees point in a way to the seasonal monthly and daily east-west mediators and trends of spring and fall, which are having to do with the horizon, while the legs torso and head recollect heaven and earth with the basic core polarity of heat and cold. The tension between metal (lung) and wood (Liver/digestion) is a more complete explanation of the interaction of yin and yang at the level of the middle than was available in the number three, where the yin and yang of the middle were taken together as the horizon or middle zone. Four is an important articulation of the number three that clarifies the dynamic equilibrium that three generates at the level of the middle and horizon while it points to the seasonal and bodily structures that cause the oscillation.

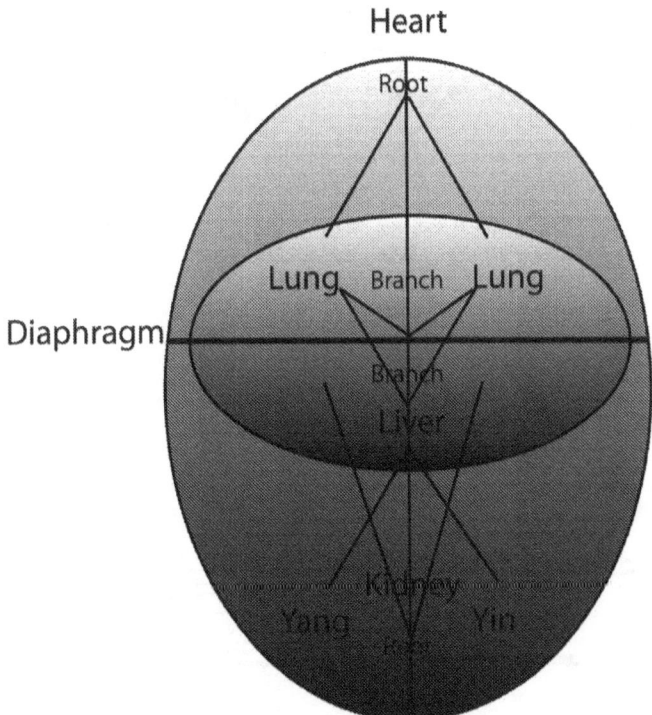

Figure 13. The Basic Structural and Functional Extension of Four with Its Focus on the Abdomen

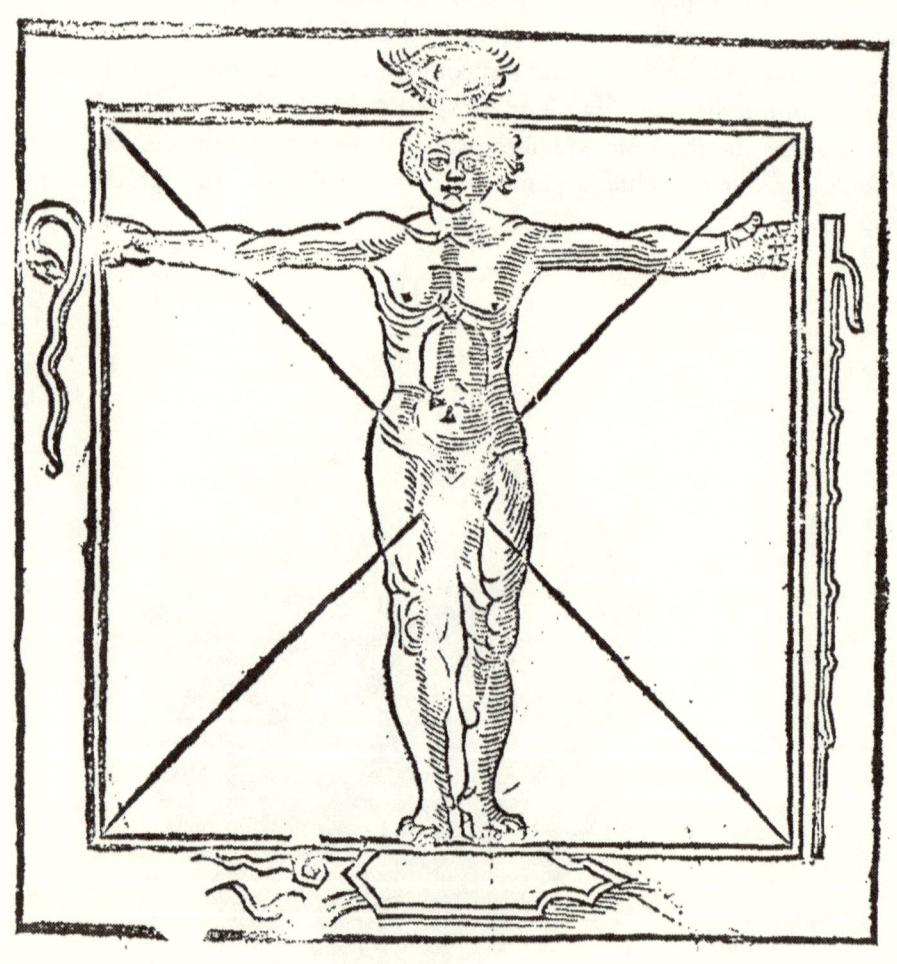

Figure 14. Cornelius Agrippa's Geometrical Representation for the Number Four

Chapter 5 五

The Five Agents and Their Origin, Correspondences, and Meaning

Chapter 5 shows the progression from the number four to the number five and explains how the elements that are associated with the cardinal points were adapted to a sophisticated model based more on process than on structure or space. The last chapter's description of an implicit process in the two oppositions of water/fire and wood/metal now are explicitly differentiated into a web of five interdependent phase relations (fig. 1). The number five explains the process implicit in four and leads to a dialectic with the number six that then explains the structural implications of the number four in even greater detail. The five-element system is, fundamentally, a numerical extension of four that makes explicit in five phases, certain processes that could not be fully described in the fourfold division of the seasonal, monthly, and daily cycle. However, the number five with its traditional associations to the five agents receives nearly iconic treatment in Chinese tradition because of the clinical importance of the insight it has into human character and physiology.

This iconic status is well justified owing to the compression and immediacy of the associations, associations that blend functions of heaven and earth recognizably onto the plane of the horizon and yet deals uniquely both with pure function and pure movement. Manfred Porkert extracts from *Huang Di Nei Jing* is a model for the five-element circulation that shows the standard sequence of circulation on the inner circle and on the outer a pattern based on the ten stems organized in five pairs, which represents the variable emphasis

Five

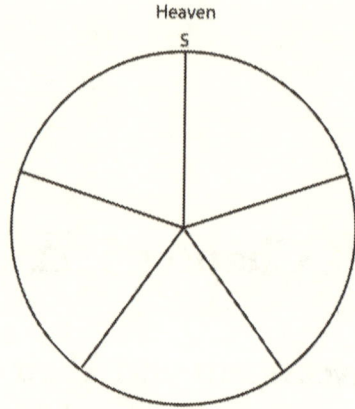

Figure 1. Basic Image of the Five Phases on the Plane of the Horizon

that may develop from one year to the next.[54] The sequence of elements in the outer circle is the same; but the distribution shows (1) that the dominant character of the year may be governed by a bias toward one or another of the elements and (2) the duration of the individual seasonal phases may vary from year to year (fig. 1).

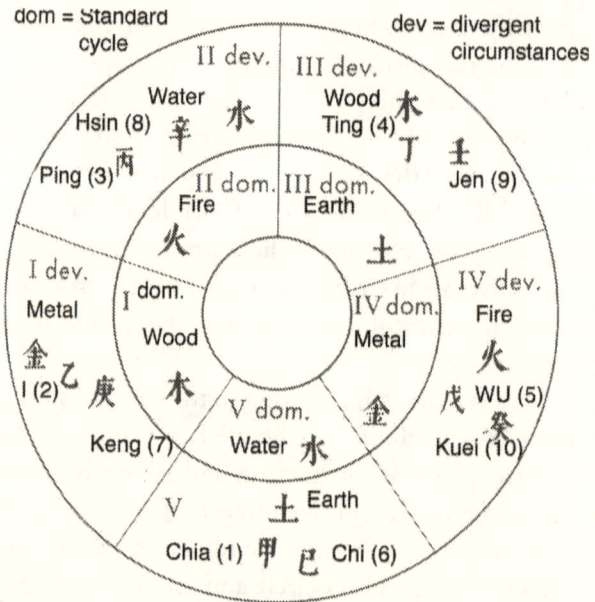

Figure 2. Standard and Divergent Circulation of the Five Agents and the Stem Pairs That Define the Phases from Su Wen and Porkert[55]

Nevertheless, the aim of this chapter is to introduce and justify a slightly different, more useful model for the five elements that are symbolized in the five divisions of figures 1 and 3.

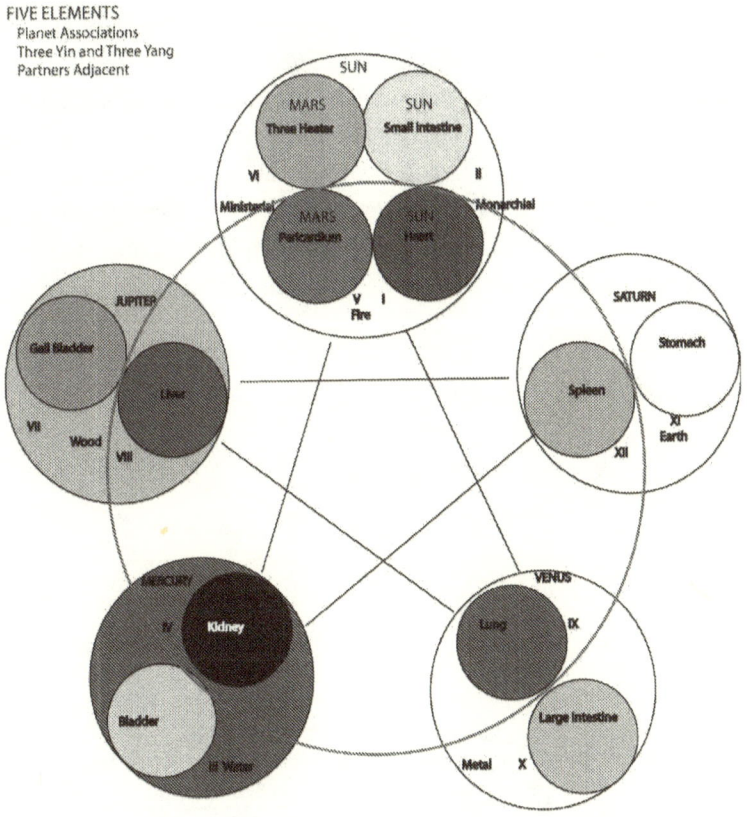

Figure 3. Proposed Five Element Model with Planetary and Meridian Associations

The five phases describe first and foremost the physical changes in nature as the annual growth cycle unfolds. Five phases of seventy-two degrees are asymmetrical with the cardinal points but because of the element associations, they retain a close association with real conditions. Placing earth in late summer matches earth with the yellow drying fields—with their ripening harvest and abundance, with the period where light begins to decline but not to the point where cold, hard, frost, and the frozen white ground of the metal element appear. The story of the light cycles as told by five elements is a family narrative that explains the cyclic process in human terms of mother to son, grandparents, generation, and restraint. The whole story has makes explicit in its agricultural and social reference, the conditions of life on the horizon.

Yet these earthly conditions are described in terms of time and an uninterrupted flow of transformations, whose root in Chinese thought is the celestial stems and the five visible mobile planets. Literally, the five elements embroider the planetary motions, describing the whirling circular motion around the heavenly pivot anchored by the North Star. Thus, within the five-element system is a fundamental pressure to return to the vertical. Five can barely be contained in the two-dimensional portrait of the horizon. Figure 4 shows in a three-dimensional model the relationship of five elements above to the next in sequence six of the terrestrial configurations below. This vertical relationship is implied in five and realized fully in six. It is implied in the earthly idea of elements themselves even when five derives from celestial stems and is compressed to circumstances at the horizon by the fact that the ten stems are combined in pairs to form the basis of the five-element series.

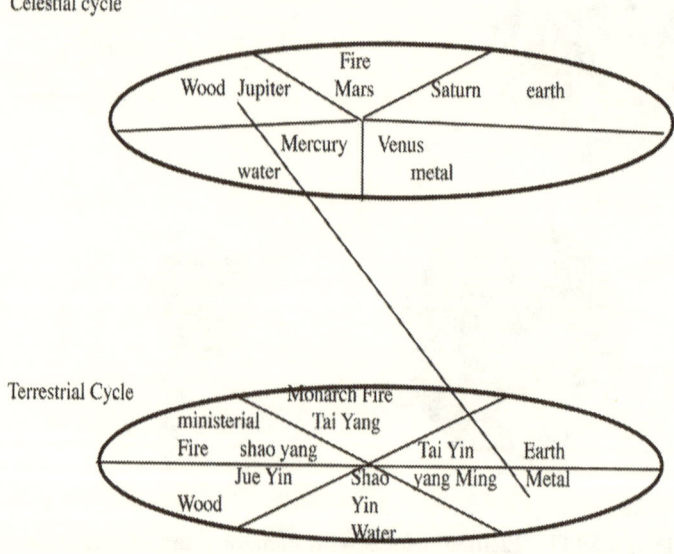

Figure 4. Five Elements Derived from the Ten Stems Above with the Six Terrestrial Configurations Derived from the Twelve Terrestrial Branches Below

This drawing shows that the standard five-element sequence is followed in both cycles (five and six), and it shows how conditions above in heaven such as spring and wood affect earthly conditions below both in the wood configuration and its opposite metal configuration. This model shows how five-element transformations translate into symmetrical structures in four seasons or six configurations (four plus up and down) and into the symmetrical structures of the body in terms of the three-yin and three-yang meridian structures that correspond to the six cosmographical phases and configurations. This will be covered in detail in chapter 6.

In this chapter, it is with the number five that numerology first provides a model sophisticated enough to explain human behavior and temperamental types. The story of the five elements is a story of family relations—mother to son, grandparent to child, etc. Five phases describe five root types and twenty-five subordinate combinations that define behavioral traits in sufficient detail and accuracy to be useful. The possibility of describing a living human being and in all their capacities is fully realized with the number ten, but the number five is particularly important in defining the macrocosmic circumstances that allow this. Mantak Chia in his *Fusion of the Five Elements* organizes meditation around these basic types.[56] When recovering the family narrative, it is important to remember the heritage of the ten stems and twelve branches mentioned in the introduction and the background of the *I Ching*, which in its turn will be described fully in chapter 8 for their contribution to five-element thinking. The ten stems contain a completion that explains the meaning and value of the five divisions, and the five divisions anticipate reciprocally the completion of ten. Both stem and branch systems alternately differentiate time and space, heaven and earth, and prior heaven and posterior heaven in ways that help to position the five-element system as a yang time-based explanation of the light cycles and, therefore, a model of pure process.

The cosmographical and geometric structure of the four cardinal points divides visible space into the four directions. Although cold seems to come with the north wind and heat from the south as the sun creeps north in its annual cycle, east is associated with dawn and west with dusk; and certain links between experience and direction emerge from these perceptions. The model begs for a counterpart that shows the seasonal phases in time. There is a need to explain the process of agricultural transformation. The five-element system mechanistically shows how the prime qualities of the cardinal points transform into each other with the march of time. The cardinal point planes and square structures are assumed, but now the location of the horizon on the earth is made explicit; and the five transformations influenced by time play out principally in the shapes and hues of nature that comprise the list of element associations.

The Greek insight that the dodecahedron with its ten pentagon facets of all geometric solids most nearly reflects the spherical nature of the cosmos helps to understand the Chinese conception of nearly seamless sequence in the generative sheng cycle of the agents. The links to the five planets reinforce the cosmic structural linkage and defines that with five outer transformations; there will be a corresponding five inner responses, and both inner and outer motions will be activated by the planets under different conditions. Similarly, it is in the detailed divisions of ten that the decan of the zodiac that the heliacal risings and settings of fixed star can be measured with some accuracy and relevance to the calendar in support of agricultural decision making. Two of these decans constitute an element phase, so it is within the seasonal sequence of the elements that decisions about planting and harvesting could be made with some accuracy.

I. The Number Five

The secret to this mechanism has to do with identifying and expanding further on the in the five-element scheme. Wieger confirms this basic generative sequence in his treatment of *wu* 五.

> The character is a stylization of the x a numerical sign. It represents says Glose the five-elements as four sides and the centre; compare cross lesson 24, later on two strokes were added, to represent heaven and earth, thus was formed wu (2) Five the two principles yin and yang, begetting the five agents, between heaven and earth it forms Wu the individual between heaven and earth as an embodiment of the five agents.[57] [ ]Wu (3) Lesson 39, Five

Clearly, the evolution of the character for five from an *x* to the present symbol underlines a perception in early Chinese history that the horizon, the zone between heaven and earth, was a place of dynamic change. The x points to the transitional points between the cardinal points as the focal point for change and process (fig. 1-5).

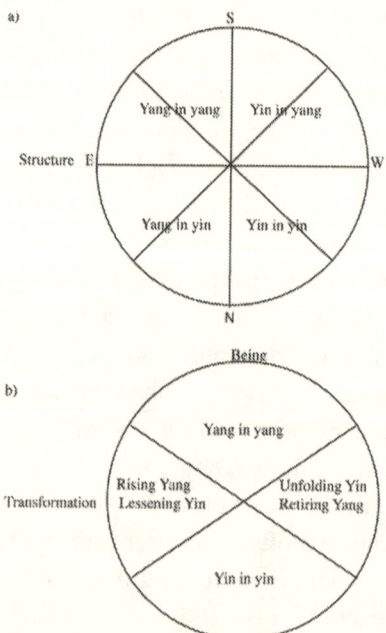

Figure 5. *A*, **The Cardinal Points by Direction and with Midpoints: the Cross and the X;** *B*, **The Cardinal Point Emphasizing the Function the X**

The cross of the four directions point directionally to the center points for summer, winter, spring, and fall with the *X* pointing directionally to transitions between the seasons. The cross standing alone is purely directional; the X standing alone provides the outline for each season and, therefore, is a purely functional model as opposed to a directional model. The X divides the zones of heat from cold, moist from dry. Of great importance is the notion that the cross and X imply a center that in effect represents five, not four. "Four" automatically generates five, just as four limbs on the body require the torso. This anatomically translates as the cross with legs together and arms at ninety degrees, and the X with legs and arms at forty-five degrees. Both versions of four assume the presence of an essential and implicit fifth center, torso, or function. This reemphasizes the generative notion behind numbers and behind the Chinese anthropomorphic conception of the universe. The cosmos generates life; numbers generate each other. In this living world, conditions are not fixed by principles, especially abstract numeric ones; rather, a principle always implies a consequence, and any manifest form always has an unstated, unmanifest component bursting to emerge.

The idea of a center to the character of the cross, a center—which implies home, an individual or a tribe as an observer—includes the observer in the system of four. Five then uniquely expresses the contemplative identity and role of the human being in life. The individual can only identify him or herself as a separate object by location. Humans never see themselves whole, only their context. Using the prime meridian, prime vertical, and horizon as reference points, the context is fully rendered. These planes are implicit in the four directions, identifying the quadrants of the horizon. The four points of the cross, often as in the number four, result in a square or box; the box is the torso, and the content of that box is the individual.

Chinese sages clearly saw the nonperception of a separate self or emptiness as a virtue, a tendency of mind closer to the childlike original one out of which all appearances emerge. An implied human observer, the enlightened human, occupied the emptiness at the center of the box (4) and represents in the emptiness the unmanifest self. Placing the X between heaven and earth manifests the individual that is implied in the square. The same implication is evident in the traditional square Western horoscope, but it is difficult to say whether it had the same understandings of self behind it. The fact that the observer is detached rather than central in philosophical models for modern Western science suggests that the outlooks were different.

II. Esoteric Aspects of Chinese Cosmology
Related to Numbers and Agents

Mantak and Maneewant Chia describe how the original one or great emptiness, which he calls Wu Chi, divides into two as yin and yang, then three, and how the principle of three unfolds as heaven, earth, and human spheres of activity. Chia applies these numeric concepts as well to an abstracted prebirth spiritual body where they are recognized as the three pure ones or three emperors.[58] These emperors contain the principle behind three subtle manifestations of energy, namely, Shen (individualized spiritual consciousness), Qi or Zheng Qi (the healthy activated chi or life force of the body that is derived from a combination of breath, nourishment, and original genetic resource), and Jing (the cellular and hormonal root energies of the body that organize and trigger its capacities and functions). The three divisions of the head, torso, and limbs are reflections of this original prebirth template. It is out of the dynamic action of these first almost conceptual divisions of two and three which when consolidated and extended by the agents generate; for Chia and the Taoists, the whole universe was first as whirling energy, then consolidating as elements, then heavenly bodies, and in particular as Polaris with the five visible planets, the Sun, and Moon of the solar system.[59]

Heaven and the pattern of heavenly order manifests first; and it is important to recognize that in esoteric traditions, humans, who achieve awareness of the primeval one, also directly operate in harmony with the operations and phases of the celestial bodies and their cycles. These manifestations, the planets, the earth, and forms of all kinds are viewed as residual effects of the movements of the elements, not the reverse.

The meditations may use symbolic animals from Chinese mythology to represent the spirits of organs and the qualities of the five energies, but these helpful visualizations, ultimately are merged back into the primordial one. These visualizations, like the story of the five-element phase relations, may be seen as a kind of temporary scaffolding by which the human being can reorient their internal perspective and moral consciousness. The return journey is guided literally in the Taoist world by the North Star, which symbolized the fixed, still point in the center of heaven and the brain. Chinese astrology is the root of these meditations and clarifies the need for an integrated understanding of human experience in the widest possible context.

In astrology, the relative prominence of a planet in the heavens, combined with its resonance with constellations and seasonal conditions, explained the cause of weather and medical conditions. A prominent Mars would indicate strife, heat, heart, or liver fire. Clearly, any of the above five categories could have a larger number of discernable qualities; there could be seven or twelve emotions, and indeed that is the case. The need to emphasize associations that conformed to the planets (Mars is red, heat and masculine; Venus is silver, cool andfeminine, etc.) dictated to some extent the limits for medical description within the system. This fact clarifies that when we assert these particular associations that we are viewing, the human being through a particular controlled lens, whose associations make sense only within that context. Links between emotions—a sense organ, a planet prominent in the sky, and a functional problem—identify a pattern. Similarly, different lenses can be used. The five emotions, five planets, and five senses yield one set of relations, where seven emotions might better describe another.

III. The Standard Associations Between Five Elements and the World

Following are the basic chain of associations for the five phases, which extend from heaven to the Earth and include the body as follows:

Table 1. Five Element Correspondences Cosmological Correlations

Element	Season	Direction	Planet	I Ching King Wen	Musical Note
Wood	Spring	East	Jupiter	Mountain, Wind	Jue (Mi)
Fire	Summer	South	Mars	Fire, Thunder	Zhi (sol)
Earth	Late Summer/ Transition	Center	Saturn	Earth	Gong
Metal	Autumn	West	Venus	Sky, lake	Shang
Water	Winter	North	Mercury	Water	Yu

Virtues and Toxic Emotional States from Lonny Jarrett[60]

Element	Emotion	Virtue	Toxic	Behavior	Display
Wood	Anger	Benevolence Discernment Flexibility Esteem Patience	Resentment	Passive Aggression	Seething
Fire	Joy/Sadness	Propriety Sage Insight Intimacy Wu wei Mastery	Bitterness	Lying	Sarcastic/ Teasing
Earth	Sympathy, Worry	Integrity Altruism Reciprocity Engagement Ingratiation Adaptability Transformation Contemplation	Disgust	Integration	Complaining
Metal	Grief, Melancholy, Inflexibility	Righteousness Balance Non attachment Purity Inspiration Self-worth Receptivity	Disdain	Pontification	Snide
Water	Fear/courage	Wisdom Contemplation Cleverness Concentration Courage, Adaptability	Paranoia	Secrecy	Intimidation

The table above shows how essential these points are to balancing *zhang* and *fu* orbs and for clearing meridians of stagnation, moving things either inward or outward. Luo points along with *zi* cleft points clear pathogens from meridians. Jarrett has a great chart about emotional correlations to pathogens that can be affected by these points:

> Cold Fear Wind Confusion, outrage Heat Bitterness, angst Fire Mania, compulsion, carelessness Damp Burden, Worry Dry Grief, Longing, acerbic Trauma Shock, dislocation

Table 2. Acupuncture Correlations

Element	Finger	Zhang (yin)	Fu Yang	Sense	Color	Sound	Odor	Taste
Wood	Ring Finger	Liver	Gall Bladder	Sight	Green or blue	Shout/ Non-shout	Rancid	Sour
Fire	Middle Finger	Heart	Small Intestine	Sound	Red/ashen	Laugh	Scorched	Bitter
Earth	Index Finger	spleen	Stomach	Smell	Yellow	Sing	Fragrant	Sweet
Metal	Thumb	Lung	Large Intestine	Taste	White	Weeping/ Cutting Off	Rotten	Spicy
Water	Little finger	Kidney	Urinary Bladder	Touch	Black/Blue	Groan/ Silence	Putrid	Salty

In addition to the standard correspondences in table 2 here, it is important to add the simple recollection that the five fingers and toes imply bodily systems of both energy and structure that are only revealed partially by the acupuncture meridian system and division by five. The ten pathways (including above and below) implied in human structures need to be researched, and the energetic harmonics should have direct relations to the five seasonal phases in their yang and yin aspects. There is every reason to expect that a fully evolved acupuncture will achieve congruent meanings for many more acupuncture points than have been expressed in the numerological structures of present day meridians that devolved from the anatomical divisions of three with the three-yin and three-yang divisions and a meridian system based on the number six.

These five manifestations also are reflected in the five senses. Close study of deep pathways in the Chinese meridian system show conclusively that meridians connect directly to associated sense organs, which in fact either explains or expresses established beliefs about their connections with five flavors, five colors, five sounds, five odors, and five solid organs with their emotional counterparts. Five-yang and five-yin meridians taken together would yield ten full-body meridians governed by the ten stems and a vastly expanded and more detailed system of treatment.

IV. The Five Planets

The too frequently forgotten medical component of this five-element chain of associations is the traditional connection to the five visible planets, whose everyday motions resonate with the sensory motor activity of the body and are projected in the meridian and meteorological associations. They, as much as any other single factor, gave the five-element model its medical and astrological affinities and familiarity. In astrology, the five planets are designated in characters showing their element affinities (table 1 and 2), which connects their celestial motions with seasonal meanings. When these five qualities are explained in their daytime and nighttime manifestations, this yields the ten stems or ten basic qualitative values associated with heaven. This, in a nutshell, is how the five-element system developed from the ten stems. That is another reason why the wu yun wheel is associated with the stems and heaven rather than with the temporal periods that they so clearly describe.

Planets have clearly defined cyclic periods that provide benchmarks to individual human lives. Saturn cycles are twenty-eight years and are comprised of four distinctive seven-year periods. In the traditional mind, this long Saturnian epoch matched and extended the symmetrical twenty-eight days and four seven-day periods (weeks) of the lunations. That is why the Chinese associated Saturn with the earth element. The number two showed that where the sun ruled heaven, the moon ruled earth, and in this cosmographical association, the moon and Saturn were linked by their connection to change, earth, and terrestrial branches generally. A natural projection from such associations would be three lunations per season, three Saturn cycles per life, one Saturn cycle per generation, etc.

Jupiter with its twelve-year cycle has always been linked with the authority of the sun and its regular twelve-month annual cycle. The moon separates and returns to conjunction with the sun, but it always is the sun in its relation to solstices and equinoxes that establishes the law of the sheng cycle, not the moon. Jupiter has a brilliant blue-green light—which links it to verdure, growth, and periods of increasing light. It is considered to be moist and warm, and this is why it affiliates with the wood element, where it mediates between and shares attributes both of water and fire.

Mars has a two-year period, Jupiter twelve, while Venus and Mercury are tied to the sun's annual orbit. The bright red light of Mars and its regular biannual circuit associates it with the summer aspect of the solar cycle. In ten years, Mars completes five complete circuits; and this also links Mars by the same logic as the five planets taken together to the five elements and ten stems and explains its subsidiary role as the governor of ministerial fire.

Venus is associated with balance because sometimes it appears ahead of the sun, sometimes behind it. With the sun in the center, its annual motion forms a balance like that of the scales, so it became associated with balance, with scales, with harvest, with coin, and with value. Mercury has a cool light, and it wanders rapidly like water in its movements, which are tied to the sun's motion. Its cool light and qualities associated it with the winter Sun and water more than the summer sun and fire.

These regular planetary periods, along with the fixed stars, were very important in their outline of the agricultural year and phases of human life. They were the logical foundation for naturalistic diagnosis and medicine, and on these simple foundations the large chain of associations in five elements were built. In the *Nei Jing*, Su Wen asks Lei Gong, "The categories of yin and yang, and the Way of the conduits and vessels, that is what is ruled by the five inside. Which depot is the most precious?" This question is something of a testing question because all the phases are coequal and necessary; but Lei Gong responded, "Spring, [that is] jia and yi, [that is] virid. Inside it rules the liver. It governs for seventy-two days, this is the main reason for the vessels."[61] The yellow emperor challenges him, and he goes away for seven days to think; but the simple point of replicating this passage here is to underline its emphasis on time and the certainty that the five periods of seventy-two days determined by geometrical division of the year were used.

The planets established standards for analysis and synthesis in time and space, but more than that, they do affect human character and behavior by lending a unique bias to the otherwise standard progression of the seasons. Much of the variation from year to year in the annual cycles depend not on numerology so much as on changing planetary relationships and their relative prominence in the sky overhead. Of great significance are the number of planets and character of the planets that occupy the sky in quadrants or phases that represent the critical phases or elements in the diurnal, monthly, or annual cycles. Thus, if Saturn, the planet associated most with change and transformation (earth), is prominent in a fire phase of the year, the fire phase of the lunation (the full moon), and midday—all other factors being equal—fire will be a very important element to consider in any evaluation of the patient. Study of these relationships of inhibition or excitation by planets can be very helpful both in diagnosis of a particular condition or in assessment of the relative yin or yang balance in a personality. Appendix 1 gives numerical weights to the medical stress imposed by individual planets in different element configurations and summarizes the relative significance of planets with their element affiliations in an individual natal chart. These meanings can be extended to transits of planets over birth positions or to simple planetary emphasis in a given season.

V. Sheng, Ko, and Counteractive Cycles in Five Elements, and Why the Fire Element Is Given Special Emphasis

The relations between phases are generally viewed as regulated by three different kinds of force: the sheng cycle, the ko cycle, and the counteractive cycle. Each describes a different kind of bias in human physiology and nature imposed by different conditions. The central point of these cycles is to give a vocabulary for describing the pattern of element relations causing patient symptoms. Each cycle describes the typical compensations that happen when one element is suppressed or deficient. Recognizing the pattern is a terrific confirmation of causative factor and coequal to it in importance because the pattern explains the total display of signs and symptoms in a way that causative factor alone cannot.

The sheng cycle is the relentless ongoing cycle of transformation that goes on from one season to the next. It depends fundamentally on the dynamic tension caused by yin and yang in all aspects. Moreover, it shows the natural progression of any cycle of thought, feeling or action, from its inception in emptiness to its taking on of potential, arrival at form, fruition, disintegration, and reabsorption or conclusion. Natural change with it's transformations of verdure and animals—its birth, life, and death cycles—and its regular cycles of light offer a rich resource for understanding this cycle, both directly and by analogy.

The five-element cycle is a yang-, solar-, celestial-, and time-driven portrait of change that is governed by celestial stems (fig. 1). This contrasts with the four directional models that precede it and the six configurations that follow in sequence. It is more yang, and the combined functions of heaven and earth are compressed into a two-dimensional cycle associated with the horizon just as the cardinal points were similarly compressed. Nevertheless, the five-element cycle is continuous with only an implicit, relationship to four, and tenuous connection with the four cardinal directions. The sheng cycle epitomizes this relentless forward movement in time. Though the importance of the oppositions between water and fire and wood and metal in defining the meaning of elements should be retained, the main purpose in the number five seems to be to show the signs of how invisible chi or energy (the invisible driving force behind life) changes predictably on the level of the horizon and the body. There is not only a lack of directionality; there also is a lack of verticality in this model of five elements. The five elements are all about process and time.

The presence of the earth element on the sequential wheel of seasonal change extends directly from its status as an implicit element in the cardinal-point model. As discussed above, this earth is not the earth of heaven, horizon, earth; it is the symbol for the interaction of the prime meridian, prime vertical, and horizon to produce change on the surface of the earth. Therefore, derivative earth could have been placed as a symbol for change at any point between the cardinal points representing transition or change as it was in some models (chap. 4, fig. 7, 12). Below in figure 4, four terrestrial branches are assigned to the four transitional quadrants in the center (earth phases) between the cardinal points (fig. 6; Porkert, fig. 11).[62] This standard yin model based on the number four is significantly changed in the yang models associated with the stems and the number five. It is in the ko cycle and counteractive cycle that these regulatory relations between former oppositions are maintained and extended.

Figure 6. Manfred Porkert's Placement of the Twelve Terrestrial Branches in a Configuration with Four in the Center Acting as Transitional Phases in Sequence between the Cardinal Points

The ko cycle that is depicted in the middle of figure 5 has the element sequence of wood, earth, water, fire, and metal shows a braking action exercised on one phase by the phase before its predecessor. The grandparent keeps the grandchild in the circle, keeps it regulated, because its qualities generally are opposite. Practically speaking, winter cold and rain (water) creates a reservoir of both that keeps the heat of summer from spiraling into extreme heat and drought. Each ko cycle relation abounds in such justifications. The ko cycle relations are one way. The predecessor on the sheng cycle acts on the subsequent phases, the child by giving birth (shen cycle), and the grandchild by regulation (ko cycle). If a particular phase is weak, the grandparent can nourish it—provided that its system was well regulated in the previous phase and was able to store resources. The grandparent also regulates the grandchild if its qualities and functions become deranged or excessively pronounced. However, a weak grandparent will have little support to give; and if the grandparent function is deranged or shows evidence of pathology, this pathology can be transmitted to the grandchild via the ko cycle.

Figure 7 shows the pentagon formed by the fivefold division of the year with its seventy-two-degree angles and days. The two sides that radiate from any given point connect that point to the prior and next in sequence phase. This gives a sense for the structural connection between grandparent and grandchild as a coequal equilibrium and triangle that forms owing to the five divisions. This stable structure requires that if one moves the other, two are immediately affected.

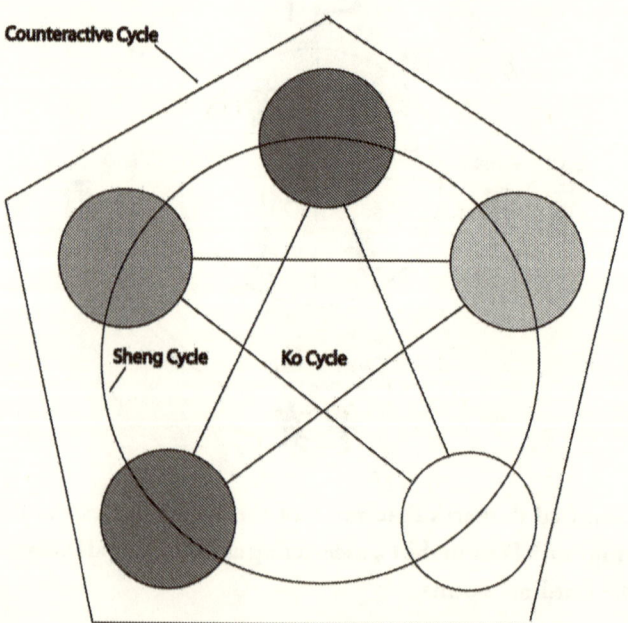

Figure 7. The Pentagon Formed by the Five Elements and Its Structural Implications

Each pole of the pentagon is equal and exerts equivalent force on the whole structure; however, if directional precedence is given to the apex of fire as an intermediary, we see that metal and wood become the focus of two equivalent triangles, one for the rising cycle and one for the declining cycle. If the apex itself is the focus of a triangle that has a structural relation to the world, we see that on the plane of the horizon, it is no different from any other apex; but facing south or as an emissary of the vertical, it creates a triangle of three above (heaven) and two (yin and yang, earth) below. From both vantage points, the pentagon laid out on a plane shows the basis for two different aspects to fire and seeds for the generation of the number six.

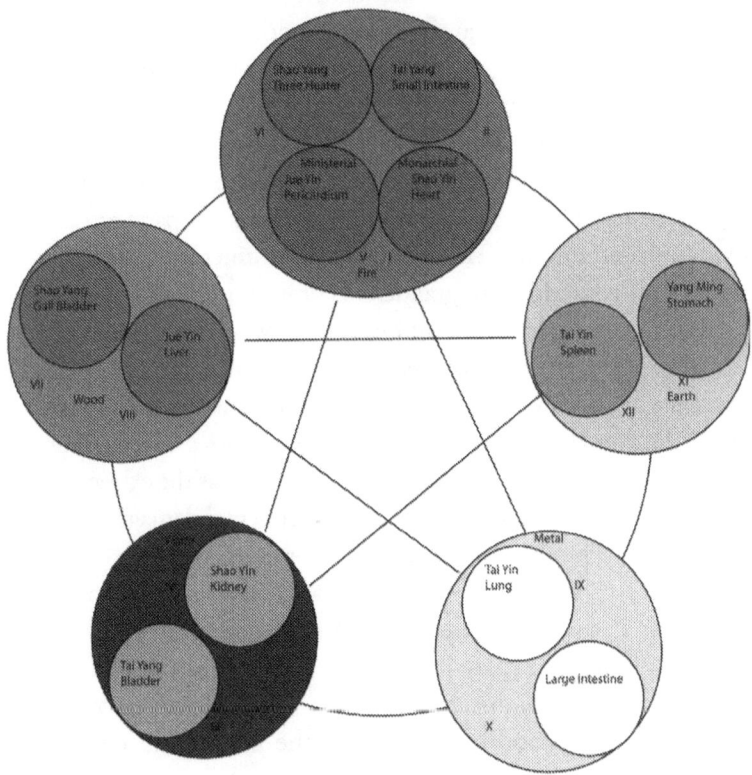

Figure 8. The Standard Five-Element Sequence with Meridian Arrangement— Four Meridians in Fire

The fact that the four terrestrial branches are assigned to fire rather than to earth in most five-element models is significant (fig. 8). It may merely express the importance to the total growth cycle as a process of warmth, light, and heat. Water is a given, the balance of moisture, and dryness is just about equilibrium;

heat is the essential driving force behind change, something more evident in the springtime surge than at any other time. With the number six, one element or another derived from ten stems has to receive the extra two terrestrial branches and their allied meridians. Precedent suggests earth would be the recipient, but traditional cosmologists chose fire. Doubling fire also allows wood/fire as a rising light pair to eventually oppose earth/metal the declining light pair in the six configurations. In the ko cycle, wood controls earth (metal), maintaining the oppositional relationship between wood and metal. Similarly, metal controls wood. The new ko configuration in the number five shown in figures 2 and 5 above of fire-controlling metal extends the wood (fire, as *jue yin, shao yang*) opposite to the metal of the cardinal-point model and extends into fire (wood) opposite the metal in the six configurations. Thus, model 2 maintains all oppositions in four and extends them into the three yins and three yangs of six in their proper alignment. The counteracting cycles maintain similar balance.

The portrait of sheng and ko are not adequate by themselves to describe the relations between phases because in any system, all parts are connected and communicate by some means. The third cycle, the counteractive, describes in its family narrative the means by which the grandchild communicates with the grandparent. In traditional medicine, such communication occurs when the grandchild is in distress. If, for instance, the heart (fire) is overheated or agitated for a prolonged period of time, this can exhaust the kidneys (water), whose function is to regulate the heart and to keep the thoughts and feelings within boundaries that contribute to survival. This physiological situation would project onto nature as the extreme drought from summer heat of the fire phase that causes environmental damage. The parched land and dead plants will not retain and absorb water in the winter.

The opposite extreme would be a cold, rainy summer—a weak summer—that is overcontrolled by the strong grandparent winter, which counteracts on the grandparent as an excessive cold or damp condition in the subsequent winter or for humans as a difficult winter owing to lack of food and resource. In bodily physiology, the acting and counteracting is a simultaneous present dynamic that can be biased or augmented by seasonal conditions. Clearly, the way traditional physicians and astrologers thought was by analogies, but these analogies were derived from the simple observation that humans respond to cycles in much the same way as nature does.

The reason for the continuity of balance between sheng, ko, and counteractive functions of five and the oppositions of four and six is that the basic solstice arrangement around the center pole of fire and water is maintained symmetrically by adding two branches to fire. Had they been added to earth or wood, the models would have become asymmetrical, and the ko and counteractive cycles would not have made sense.

Cosmology offers an additional insight into this emphasis on fire within the model derived from five. Five is a yang number. Adding weight to fire emphasizes what heaven is bringing to the horizon, all the growth enhancing light and verticality leading to and contained in midday and midsummer. The implicit vertical in the five-element system is made explicit in the four fire meridians and their associated terrestrial stems when the division of six is manifest.

It is uniquely ironic to the five elements with its root in the celestial stems as elements acting on substance that it should be restricted somewhat to the two dimensions of the horizon as the basis for describing transformations in nature. This compression results from a fall from grace that occurred after the prebirth number three, where the preconditions for life were established; and it is within the constraint imposed by the number four and on the horizon plane now that human life and consciousness has to unfold. Each number, then, can be seen as an evolutionary restraining feature on the original chi—forcing it and compressing it into the restraints of body, of form, and of nature but also as an evolutionary option, a window into the one that adds insight and comprehension to the whole. "Five" fully examines the dynamic flux of time and circumstance and in so doing, adds a profound immediacy and individuality to the perception of life. Five elements can only be understood by contemplating its relations to the sequence of compression and to the adjacent numbers four and six. The cardinal points remain the interior essential source of relations for the more dynamic five-phase display. The proper sequence of six configurations with the reintroduction of the vertical dimension must derive directly from the proper preconditions in five.

Figure 3 above shows the five phases with elements in synch with branch configurations and fire meridians appropriately sequenced according to their phase and orb compatibilities. This is markedly different from the standard sequence shown in figure 8.

VI. Cosmographic Structures

The five-phase portrait of the annual and daily cycle is a yang portrait, whose sequence is associated with the waxing and waning light cycles of heaven and the combinations of ten celestial stems in pairs and to heavenly standards that are regular and unaffected by earthly circumstance. Yet each year has its own variations as described in figure 1 at the beginning of this chapter. Figure 6 shows the five-element cycle as five equal divisions of the plane of the horizon in blue. Only one spoke of the wheel aligns with the plane of the prime meridian and that has been attributed to fire, south. The northern direction is implicit but not explicit until the division of six that allows for the full differentiation of the planes of the prime vertical and prime meridian in relation to the horizon, without reducing them to two dimensions.

Five

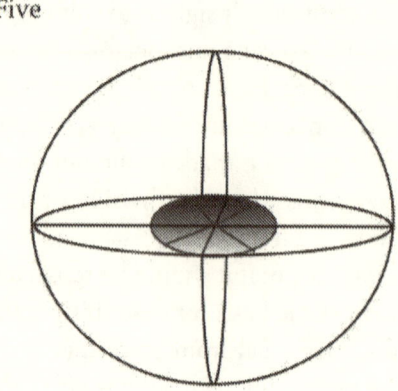

Figure 9. The Cosmological Drawing in the Number Five

The stick figure imposed on the framework of the five elements differentiates the human head with the four limbs (fig. 10). This as much as any single factor gives the number five a uniquely human touch, and it differentiates it from the rather impersonal cardinal points, which only emphasize the limbs in various configurations. The head, as the seat of consciousness affiliated with heaven, reestablishes on the anatomical level the distinct importance of *tai yang* fire and the associated concept of monarchial fire, which is directly associated with the celestial stems. The asymmetry of the limbs in relation to the cardinal points emphasizes movement and change rather than fixed spatial relationships; and it is change or the will of heaven that is emphasized in the five-element theory. It is in the single act, the flash of color, the smell, the momentary exchange that the five-element diagnosis is realized. Action—not form, movement, not diagnosis—is the key to using the number five in the clinic.

Five

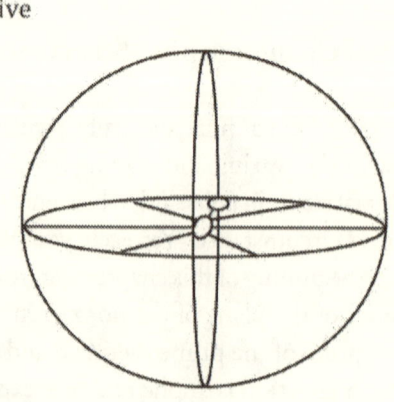

Figure 10. Stick Figure for Five

The five elements are endlessly adaptive and the symbolism of the principal strake of the five divisions being associated with the head and a cardinal point suggests that the head could be associated with any cardinal point or any element and that treatments should be organized around that. People usually show primary disruptions in one element that creates a pattern of disruption in the others, and this confirms the importance of diagnostically identifying the principal element causing trouble in the system, as this will be the key to any five-element treatment.

The immediacy and rapidity of change between the five elements in any situation or cycle and celestial stems contrasts to the twelve branches—which are concerned more with space, form, and circumstance. Thus, the five phases may be seen as an unchanging heavenly substrate that regularly generates the annual changes below on earth.

The same cycle of the year viewed from the point of view of the branch sequence is seen through the window of the six terrestrial energetic configurations treated in the next chapter. Figure 4 showed the two levels of heaven and earth; five and six are operating simultaneously with the element positions synchronized above and below. The drawing shows that the dominant element in heaven affects not only the configuration below but also its opposite as the dynamic compensations in nature occur holistically.

VII. Medical Practice—Theory

The great clinical opportunity of the five division's year or cycle derives from the ability to link cause and effect, diagnosis to prognosis, and the ability to project ideal outcomes based on clinical reality. This time-based approach allows the practitioner to follow a thread through the labyrinth of pattern, signs, and symptoms. In the final analysis, it is people whom we treat, and we have to know them and treat them in a way uniquely adapted to who they are. People have life-changing moments, and these leave imprints on function and behavior. These imprints can be noticed in five elements and through careful history taking, traced through time and circumstances into a chain of causation. The eventual outcomes are as clear as the present stage of the process. By visualizing a reasonable outcome for treatment, given a patient's basic condition, a five-element practitioner can plan a sequence of treatments to get the patient to where they could or should be rather than simply alleviating pain or recovering a comfort level.

Very frequently, patients get some relief from symptoms and then their problems recur. Very often, this is because the root is being neglected. Unless the root of a problem is treated, a trend cannot be established in treatment. A very common condition is spleen chi deficiency with dampness, but a good percentage of people with this condition are fire causative factors with weakness, sadness or discouragement, and self-neglect—which then has the effect of allowing damp to accumulate. The causes fall in the functional orbit of the fire element, relationship issues, social expectations, and the lack of spiritual connection. Without strengthening the mother fire, no real progress can be made by treating stomach 36 and 40, spleen 6, UB 20, etc. From the cardinal-point model, this perfectly illustrates the leverage that can be achieved on damp accumulations by mobilizing fire, yang and chi, for it is exhaustion from heat and emotion that is leading to the inability to mobilize fluids:

1. The ability to see actions as part of a process and in their very nature as expressions of the general phases and laws symbolized. This allows for the use of every gesture, sound, expression, symptom, and sign to reveal general tendencies useful to clinical practice.
2. The focus on individuality and social expression implied by the emphasis on the fire element and the number five itself, which avoids prescriptive treatments, categorical analysis, and requires immediacy and relationship between patient and practitioner. This approach harnesses placebo and self-understanding in the healing process.

People will decline and die. Therefore, the aim of treatment is not merely cure and symptom relief but something personal, something to do with quality of life and awareness. Within the context of the one, humans are viewed in the widest context as part of a continuum, and the five phases identify phases in the development of that continuum. If an individual is trapped and identified in a particular phase, they lose adaptability and resonance with the aims of the larger whole and the complete cycle. By identifying the locus of agitation, weakness and compensation acupuncturists can help a person recover a wider view and more options. Many benefits can happen from this, but what about when the syndromes resolve themselves?

3. The compression of heaven and earth to the plane of the horizon allows for the assessment of psyche and material simultaneously as actions and manifestations reveal them in traits.

The great weakness of the five-element divisions is their lack of anatomical coherence and the limitations of the number five to locate anatomically relevant points on the body for given dynamic expressions. It is only by exploiting the three-yin and three-yang structural relations that the five-element treatments translate into coherent structural wholes.

VIII. Medical Practice—Five-Element Treatments

The principal insight of the cosmographical and anatomical number five is the ability to perceive and treat patterns of dynamic equilibrium between and within agents.

These patterns are treated by using acupuncture points through those points with assigned elements on the four limbs.

The symbolism of the stick figure (fig. 10) emphasizes that the treatment of these patterns require the identification of one element as the key to treatment.

Five elements diagnosis and treatment is very immediate and needle insertions or moxibustion should register immediately on the pulses and in the patient's demeanor. The practitioner needs to have an intimate understanding of how the patient functions and be able to notice minute changes in their process as treatment progresses.

Within Elements

Individual elements are supported by using the following strategies:

1. *Yuan* (source) points in either the yin or yang parts of the element orb. Yuan points are the key points for five-element treatment because of their unique ability to focus the patient's sense of identity and purpose toward their original purpose. This kind of effect is very close to the overall aim of yang five-element treatments, which tend to encourage stability and purposes of hun and shen spirits as a vehicle for healing.
2. Any disequilibrium between the yin and yang orbs can be corrected by using the luo point of the deficient yin or yang function.
3. Source and luo points can be used together to support and create equilibrium at the same time.
4. The element point on the meridian of the element being treated can strongly reinforce a meridian, more so if used in the appropriate season, moon phase, or time of day.

Between Elements

Transfers are the key to treating patterns. Simple transfers should never be underestimated in their power to affect whole system. It is better to spend time and be certain of the diagnosis then use the person's strengths to create a better equilibrium than to force the energy in a particular direction. The basic transfers include:

1. Tonification is using the mother to treat the child. This can be done strongly by treating the mother point on the child, such as heart 9 the wood point if heart is weaker than wood. Treating the fire point on the wood meridian (the sedation point of the mother) also can do it more gently. Both points can be used together if the mother and child are disconnected as indicated in the pulse or behavior. Alternatively, if there is a good connection but both are weak, for example, the heart and fire relatively more than the wood, the wood element point in the mother (liver 1) can support the wood and heart 9 the wood point on the fire meridian will affect the transfer without depleting the liver.

2. Sometimes, a few moxa on the tonification point, particularly in the case of the fire element, cold season, or deficient chi, can have a dramatic effect where needles might not cause much change. Thus, it is clear that even in the simplest of transfers from mother to child the state of the mother and child must be evaluated and the means of intervention must be adapted to the situation.

   The needle technique used is a short-duration insertion on the exhale, turn and removal on the inhale.

3. Sedation can be accomplished by longer duration insertions on the child point in any meridian with a relative excess. I only use this treatment as a gentle means of transfer to the child to establish equilibrium, or as a means to break a persistent stagnation of chi in an organ system.

4. Resonance. Five-element treatments fundamentally are designed to keep patients in alignment with nature. Whatever the primary imbalance in the individual, every patient can be treated by seasonal and horary means so long as their bias is considered. This means that in the spring season wood points are good to do, in the summer fire points, late summer earth, etc. It also means that fire points can be augmented at mid-day, water points by midnight, wood by dawn, and metal by sunset treatments. The same element applications can be applied to dark of the moon, new moon, quarter moon, and full moon positions.

Naturally, juggling these various cycles takes some judgment and requires a clear understanding of what the treatment aims are. If your aim is to support yin, you can choose afternoon treatments, if to support yang morning treatments, and then select appropriate resonant points appropriate to the season and time of day. You can work with the lunations generally to support yang and yin, yang on the waxing cycle, yin on the declining cycle. If a patient is exhibiting heat signs you can choose relatively more yin times of day, phases of the moon (toward the dark of the moon) and work with seasonal energies when possible. The treatment aims such as supporting yin and reducing heat signs come first, whether it is summer or winter, but obviously there will be more support for this in the winter months provided there is some lifestyle adaptation.

If the element of a patient's causative factor is well known, then the seasonal treatments can consist simply of doing the element point of the season or the element point for the two hourly period of the treatment on the meridian of the causative factor. This will be the key to the treatment.

Alternatively, the meridian for the season and a point on it for the two-hourly period can be treated simultaneously with the point selected for the causative factor. That is with the fire causative factor in a springtime treatment at noon: (1) The seasonal and horary point would be liver 2 or gallbladder 38, the fire points on the wood meridians, which open the body to alignment with the cosmic standard. (2) The practitioner could then treat the causative factor with a source point heart 7 or element point which also is the horary point for noon, heart 8, or the tonification point, heart 9 depending on symptoms and body zones.

These selections can be refined further based on the moon phase. This process will be discussed in detail in chapter 7. A glimpse ahead would recommend using a corresponding point on a partner meridian below the waist on a foot meridian for a waning or new moon and above the waist and on a hand meridian with the moon prominent in the sky. That is, for the springtime treatments with a waning and new moon, wood points on liver or gallbladder would be selected, taking into account prevailing weather or physical signs and symptoms just has they have been above. However, with an appointment close to the full moon, treatment would be preferred on the *jue yin/shao yang* partner fire meridians, using the wood points on the pericardium or three heater. Such a selection would allow the practitioner to economize in the selection of fire points, probably using pericardium 8 as the principal point and not using heart and Small intestine points.

5. In every pulse taking a basic discrimination can be made as to whether a patient is chi deficient or blood deficient, yang or yin deficient. If both are yang and yin are deficient which is relatively more deficient. If the pulse is tight and wiry its yin deficient. If it's soft its chi deficient. Knowing this, the normal approach when treating a causative factor is to support the yin by using metal, earth, or water points or to support the yang by using wood, yuan, and fire points. Similarly, among points

on the limbs, the hands are the most yang, the segments of the arm or leg closest to the torso are relatively yin. These general guidelines can be used to select primary and auxiliary points on a meridian. Thus, with a fire causative factor and a tense pulse as in the above example, it would be natural as an alternative to tonifying the heart meridian or sedating the wood, to treat the metal or water point on the heart meridian. The metal point would help to control stress in the wood orb, which very frequently is the root of a tense pulse, and it would have a calming, restricting and sedating effect on the heart. The water would calm and control the excesses of the fire meridian helping to control yang while supporting yin.

There are no absolute rules as to how a person will compensate for a problem. If a patient presents with fire and metal imbalances together, the metal point on the fire meridian or the fire point on the metal meridian can be considered as a harmonic treatment. This may be viewed as an element within and element treatment or the treatment of a pattern of disruption related to fire and metal. Fire is the ko cycle controller of metal, and metal counteracts on fire when under stress or when it is relatively in excess and fire functions are disrupted. These relationships need not be involved directly in the selection of the resonant points for treatment but they may inform the choice.

Similarly, a fire and metal imbalance may in many instances drop out of consideration from the perspective of five elements and may indicate a disruption of heaven and the upper torso in the window of the number three.

Using the Grandparents

The ko cycle and counteractive cycle can be treated by strategies that strengthen the weaker orb of the grandparent/grandchild pair and restore the flow of chi to the natural sheng cycle patterns of distribution.

Ko imbalances can be repaired by: a) strengthening the grandchild in our example of a fire causative factor, metal, which is being acted upon by the fire. b) Treat the metal point on the fire meridian to reestablish equilibrium, lending some chi to the metal from the fire, which is stronger, or if the metal is relatively stronger and unregulated, to strengthen the fire meridian. Generally the aim is to have an indirect effect on metal and in particular to use the metal point on the fire meridian to accomplish this. b) By selecting the fire point on the metal meridian to regulate or inhibit the disruptive grandchild directly.

To treat the counteractive imbalance the principal aim is to: a) strengthen the grandparent fire meridian directly by source, element, or horary point and/or b) transfer extra chi from the metal around the sheng cycle to eventually tonify fire, c) use Korean four needle technique to tonify the fire while sedating the metal. The Korean technique is powerful and effective, but I prefer the simpler transfers that either tonify or sedate rather than do both in the same treatment. By using the "laws" of nature symbolized in the sheng cycle and delineated by the patients own relative strengths and weaknesses, excess or deficiency, no imposed force is needed.

The principal strength of the five-element system when viewed as a part of the numbered cosmographical sequence is its ability to identify a primary element as the key to a pattern of imbalance. Five is utterly dependent on the structural and functional implications of earlier divisions of the circle and congruent with them. The five phases have yin and yang aspects (ten). They still refer to imbalances in the three zones of the head, torso, and legs and their associations have both meteorological and physical repercussions, heaven and earth. The elements are dynamic counterparts to the elements and qualities of the cardinal points.

The selection of the causative factor has ramifications for all subsequent numbered divisions. The number six, by restoring the vertical, will allow the five-element causative factor to find a network of anatomical structures that can support treatment. Seven allows the three-yin and three-yang pattern of anatomical correspondences to be refined to accommodate the lunar cycles. Eight allows the five-element diagnosis to be treated by using all the systems within the dominant quadrant affected. The number nine allows for resonant treatments that use an elaboration of the principal of three into three dimensions, which apply to the whole body. Thus, if you have a metal causative factor with lung symptoms, treatment can be augmented by points on the top of the head, the hands, the feet, and effects can be counterbalanced back-to-front, side-to-side, heaven-to-earth according to the phase, beginning, middle, or end of the process. The number ten allows for the full development of the five elements into its implicit five yin and five yang aspects, which symbolically are represented by the five fingers and five toes, which extended bilaterally represent the complete three-dimensional human, and the full extension of the meaning of the ten stems, now ready for action on earth. Each of these windows will allow a more refined selection of points on the meridian of the causative factor and of counterbalancing points on other symmetrical planes of the body that either will augment or regulate the causative factor to strengthen treatment effects.

Summary

The key point in using the window of five elements as a basis for treatment is that you can work very closely to the core of identity in a person owing to the celestial stem and mid-heaven emphasis of the system. Key points used in the system emphasize heaven as tonification points on the hand and feet, the transitions between heaven and horizon (source points, windows of the sky). The ability to identify general associative patterns in normal human behavior that link to distinct traits in nature allows for a good first cut at assessing underlying tendencies in the personality of a patient. This helps with advising and treating.

The general approach used in five elements is that of less is more. Treating the least necessary, clarifies the aim of treatment and allows for clear feedback as to what is working and what is not. When using many needles it sometimes is more difficult to discern what is working.

The general approach outlined above for treating a five elements diagnosis is (1) to treat the source/*yuan* point. It is self-equilibrating, safe, and resonates with a person's core inherited chi and will reestablish order and a basic connection with self. (2) Then tonify the causative factor by treating the point associated with the prior element, if earth treat the fire point, if water, treat the metal point, etc. ; this treatment can be augmented by strengthening the mother by treating its source or element point when doing the transfer. The transfer should not be made if the pulse shows a relative weakness in the mother when compared to the target element. If the mother is weak a transfer can be directed through it from another stronger element to the target causative factor. (show examples) (3) Alternatively, if the response has been good to the source point treat the element point next. This will give a powerful boost to the causative factor, which usually is weak. (4) Then next in sequence treat the *shu* or *mu* point of the element or organ involved to strengthen a treatment. (5) If the meridian is unresponsive or stays weak, check the entry-exit points. (6) Establish the level of the causative factor imbalance, is it body mind or spirit? To treat the spirit try to select point names and functions that aligns with the spiritual, mental or emotional problems. Windows of the sky can give perspective, outer *shu* points can treat the spirit, ghost points can treat compulsions and desires, the mental aspect of particular meridians, and the names of points can in some cases indicate their spiritual potential, as "great esteem", or spirit seal, or spirit burial ground. When treating an element or orb think of the three-yin three-yang partner meridian to get symmetrical leverage on the problem, and to create balance above to below. When treating the five elements you are treating the root.

Circuitus sanguinis & Spirituum devehentium humidum radi-
cale & calorem primigenium per venas & vias duodecim membrorum spatio viginti
quatuor horarum, quo tempore quinquagies circumvolvuntur, secundum cœlo-
rum per quinquaginta domos conversionem. (Literarum maju-
scularum ordo indicat ordinem circuitus
seu circulationis.)

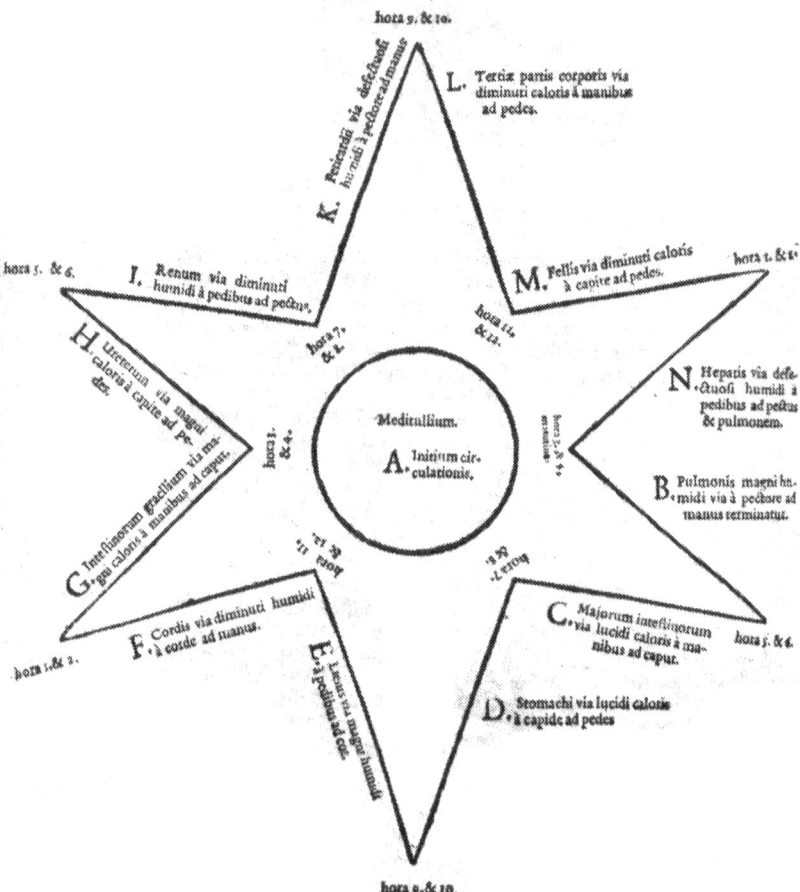

Figure 10. Andreas Cleyer's diagram (+ 1682) of the double-hour dominance of
the successive acu-tracts in the Circulation Order.[63]

Chapter 6 六

The Three Yins and Three Yangs

Chapter 6 is devoted to displaying the cosmographical structures and medical implications of the number six. The chapter is somewhat lengthy because the division of the annual cycle by six is much neglected but essential counterpart to the five-element system. The system of six is the basis for the standard twelve meridians, and it is with the number six that verticality is fully developed in anatomical description. This is a very sophisticated conception that integrates all the knowledge of the horizon from four and five division systems but that integrates notions of physiological process with physiological structure in a new way. The distinction between the cardinal point conception of four seasons, four elements and four directions and the six directions are important yet seldom examined in depth. This chapter shows how it is the number six more than the number five that links physiological process to anatomy.

The rationale behind the pattern of the three-yin and three-yang meridians is cosmological and geometrical, which means that these meridians express proportions of relation between the horizon and the prime meridian. We will show that the *jue yin* and *shao yang* meridians are relatively more tied to the vertical while the *yang ming* and *tai yin* are relatively more tied to the horizontal planes. These spatial distinctions have a lot to do with the meaning of the meridians and how they function. The very notion that this is true destroys the idea that the meridians are physical structures that can be traced, and it suggests rather that they may have been conceived originally as mathematical structures that conform with emprical conditions.

On one very basic level the number five generates the number six and it is the division of six that provides the physical structure and definitions that show how best to apply the five elements to the human body. This goes beyond transfers, tonification, sedation and various typical strategies to creat balance, because it gives structural correlations that can be balanced directy and automatically on the basis of location alone. The wood meridians have functionally a mother son relationship with "fire" in five elements, but in the three yins and three yangs they have an explicit below to above relationship with pericardium and three heater that is unique and differentated from the relationship to heart and small intestine. Treating above to below or across or bilaterally greatly strengthens the leverage applied to any such treatment target.

The fundamental questions, where do we place needles based on five elements or how do the ko cycle relations of five element theory tanslate into structural relations in the human body and the distribution of the twelve meridians can only be answered by studying the division of the daily and seasonal cycle by six. Six contains implicitly all previous whole numbered divisions, and its descriptions must emerge naturally from them. "One" implied an observer, two; "two" implied something between, three. Three vertical zones and the planes that describe them implied four directions. Four cardinal points and their associated seasons implied an un-manifest center (earth) and a causative sequence generating five. Therefore, when that center moves into appearances and five is fully manifest, i.e., as a wheel of five Wu yun, the center is left vacant again requiring six. Similarly, the corresponding five seasonal periods of the wu yun in their turn imply that this "yang" movement have a structural consequence or at least occurs in space (6). The context is the four directions with up or heaven (stems and five heavenly agents) and down or earth (branches and four terrestrial directions).

In traditional thinking the number six is generated from five. The odd and therefore masculine numbered five-fold division of the year or day was in Chinese theory associated with pairings of the ten celestial stems. What is implicit but not expressed in a system based on heaven, time, and movement, is earth, space, and structure. Thus, the next in sequence even number must represent these necessary but unexpressed facets of the number five. The traditional six-pointed star above displays the source in heaven of the six sided geometrical form. What is implied by the image is the great circle of heaven divided by the points of the star into six equal division.

The even numbered division of the cosmos by six is spatial and determined by six yin and yang pairs of the twelve terrestrial branches (See appendix three for lists and meanings of stems and branches). The divisions are not identical but they are symmetrical in relationship to the midheaven and seasons. The light cycle they describe is the same, but the periods of the six divisions are sixty degrees/ days in duration not seventy-two. Unlike the number five, which is not divisible except by itself, six can be symmetrically divided at the horizon into three-yin and three-yang components, and it is, therefore, a direct elaboration of the principles defined by the number three. However, each phase of five or six implies a whole comprised of seasonal phases with predictable attributes, that when divided into yin and yang aspects will describe the same whole in greater detail, either as ten stems (five yin five yang) or twelve branches (six yin and six yang). The combination of the stems and branches expresses something preliminary about what will be fully expressed when each stem or branch is articulated in the numbers ten and twelve.

Six produces three-yang and three-yin divisions of the cosmos. As with the numbers three, four, and five, the horizon line is the defining essential that distinguishes above from below or conditions on the plane of the horizon, but it is in the number six that the vertical dimension is preeminent for the first time in relation to the complete circle of the horizon with its four directions. This cosmographical conception is embedded in the meaning of the number six itself.

I. The Number Six

According to Wieger, the number six is [六]Liu (4). He stated, "The even number, also easily divisible that comes from four. The symbol for four is marked with a dot on top."[64]

Liu, the number six, implies the division of four with a dot above a line, below which is the symbol for division. The single line above is the horizon line of one, I, but it contains an implicit memory in the dot of the division of heaven and earth (2). The dot above is unitive; the two legs below refer to the yin and yang of earth. The dot creates three levels, which supercede the horizon-based system of the cardinal points (4) and the five elements. The three levels recollect the verticality, established by *san* (3) and the heaven, earth, and horizon. Liu (6) economically communicates a further division of the three by two, yielding six, whose cosmographical meaning is tied to the four directions on the horizon line with heaven above and earth below. The character derives from the two stroke figures for eight [八], which communicates that six is an even-numbered transitional link between four and eight. The single line with the dot above can be conceived as the line of the horizon with the sun above it in the noon position. This establishes the emergence of the structures of six from the celestial stems and the solar cycle of five elements, but it ties both of these systems to the organizing yang principle of midday and midheaven. The four fire acupuncture meridians were shown in the last chapter to be tied to the solar character of the five elements. Six always has been considered in numerology to be a perfect number because it is divisible by even and odd numbers and comprehensively includes the wisdom of the dynamic transformations associated with four directions on the horizon with a survey of the timeless vertical structures of life.

The legendary mystical advisor to Su Wen in his *Nei Jing*, Chi Po, provides an essential link between number theory, cosmic cycles, and human anatomy in his portrayal of the generative character of the number six. In the following quote, he linked the six divisions directly to the five elements and applies the six functions to heaven and earth (12), then as extensions of the principle of three to heaven earth and human functions (9). As in the conditions of nature, so too in number theory; six emerges from five with many ramifications.

> Therefore, the ten celestial stems generate the five—elements which, in turn, generate three-yin energies and three-yang energies; the three-yin energies and the three-yang energies combine to generate six energies of the heaven (wind, cold, summer heat, dampness, dryness, fire) and six energies of the earth (metal, wood, water, earth, monarch fire, minister fire) and six energies of man (three yin and three yang). Three times three equals nine, and nine separates into nine distant areas of China to which the nine organs of man correspond.[65]

The particular language of Chi Po profoundly and economically explains that the vertical dimension that is only implied in the seasonal manifestations of the five phases becomes manifest as heaven and earth, yang and yin in the number six. Yang and yin, when in combination with the rotating plane of the horizon with its cardinal points and five elements together, form cosmographically the number three, represented by three lines of the Chinese number three, which represents heaven, horizon, and earth (fig. 1).

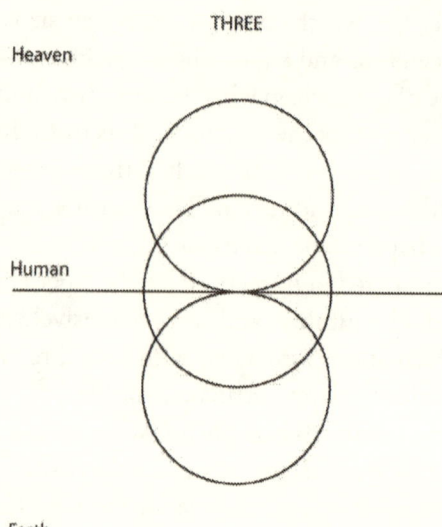

Figure 1. Heaven, Horizon, and Earth

With the lessons learned about process and structure from the cardinal points and five elements, it is easy to apply the principle of three to the horizon as well as to the vertical—which yields figure two with six zones, three for process (on the horizon), and three for structure (vertical). Each defines the other.

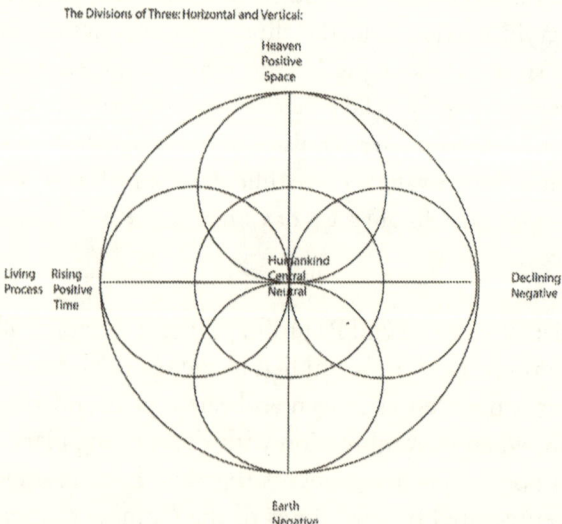

Figure 2. Six as the Principle of Three, Applied to Process (Horizon) and Structure (Vertical)

Figure 2 may also be viewed as in Chi Po's heaven to earth division of three as three yins function below and three yangs function above with the horizon implicit (fig. 3).

Division of Six

Three Yang

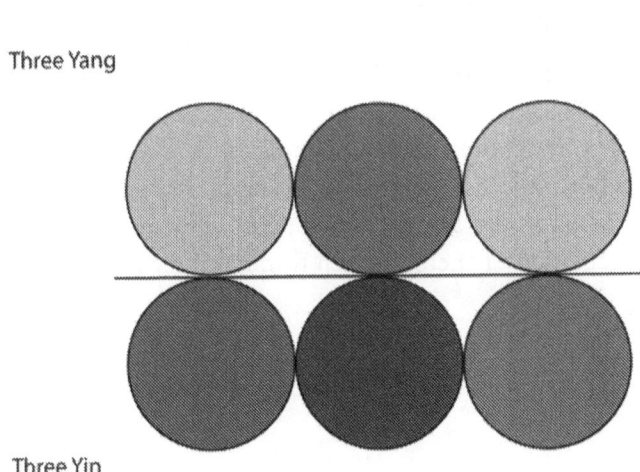

Three Yin

Figure 3. Three Functions above and below the Horizon

This image reduces to a two-dimensional proposed configuration for the three yins and three yangs conceived as six divisions of any cycle of light (fig. 4).

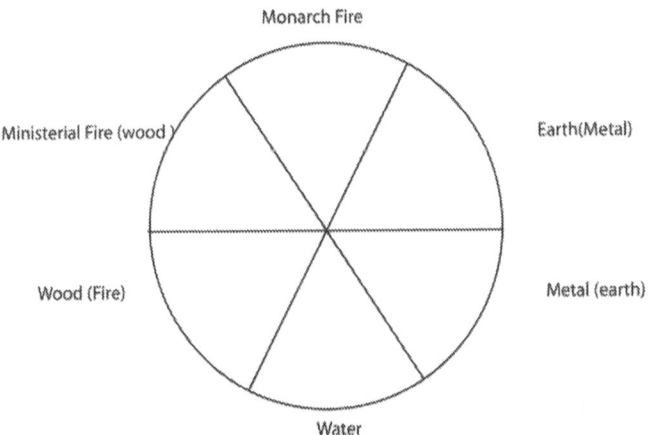

Figure 4. Proposed Three Yins and Three Yangs

Chi Po's shorthand numerology leaps to nine, bypassing numbers seven and eight because the natural extension of the principle of three yangs of heaven and three yins of earth; or if treated functionally, the six meteorological conditions and six earthly elements—each containing the complete three-yin and three-yang sequence—is to reapply the same process to the center or horizon. This was required by the quadrant differentiation established in figure 2 and the principles of the number four. In Chi Po's description, nine is simultaneously geographical (the nine divisions of China) and relates to nine specific organs in the body and, therefore, in Chi Po's mind, has to do with a full three-dimensional rendering of the cosmos and body. Figure 4 shows the application of the process of three to all vertical dimensions, and this complete version emphasizes that the reintroduction of the vertical in six is just that the development of some basic principles that apply to the circulation of energy as seen through the window of six.

Nine

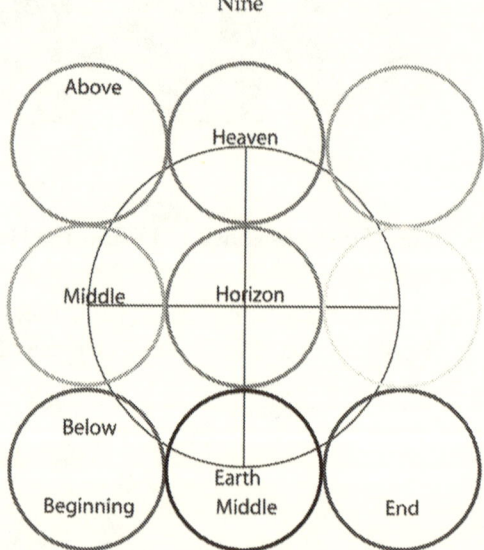

Figure 5. The Process of Three Applied to Heaven, Horizon, and Earth—Nine

The last few images have elaborated some of the geometrical structures implied in Chi Po's statement. Before examining more deeply the structures of six divisions, it is important to place Chi Po's comments in the wider syncratist perspective. It is Su Wen's portrayal of a close evolutionary development from five to six parts of a wider system that use numbers to connect outer structures to human experience. Later in Su Wen, we read, "In the sky there are the Five Sounds; human beings have their five yin visceral systems. In the sky there are the Six Pitches; human beings have their six yang visceral systems,"[66] and, "What is meant by 'six fu?' The Throat, the palace where [goods] are measured and entered. The stomach, the palace of the five grains. The large intestine, the palace of transportiation. The small intestine, the palace of acceptance and abundance. The gallbladder, the palace of accumulated essence/semen. The urinary bladder, the place of the body liquids."[67]

In this description, the six hollow organs are first hollow spaces in accord with the spatial character of the six divisions; and in geometrical code, they are distributed spatially in this description, verticallly with throat above, bladder below, and the stomach as center with five grains (reference to circulating elements), and horizontally with gallbladder (rising cycle, jing) on one side, large intestine (declining cycle, chi) on the other, and small intestine (neutral center, shen) balancing the center. The *fu* are empty holders for the positions of functions. These kinds of portrayals show how important the number six and other numbers conceived as geometrical structures were to early medical philosophy for providing windows into the organization of the body.

II. Cosmographic Structures

In chapter 4, the cardinal points were defined by the intersection of the planes of the prime meridian and the prime vertical. These planes are vertical and, therefore, stand at a ninety-degree angle to the plane of the horizon (fig. 6). Therefore, a line can be drawn vertically at ninety degrees to the horizon from any location to where they intersect above and below as in figure 4.

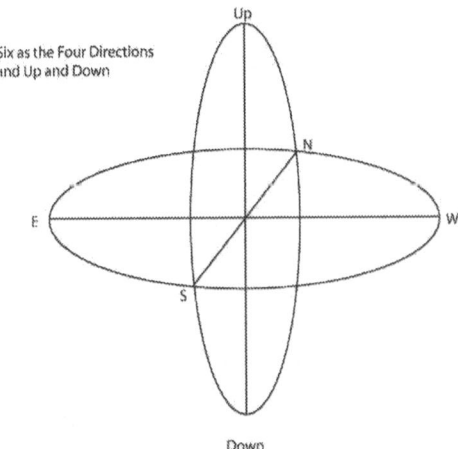

Figure 6. The Planes That Form the Cardinal Points with Matching Four Divisions on the Prime Meridian Showing the Vertical Dimension

Figure 6 shows the cardinal points (4) with up and down (2), which together forms the essence of the cosmographical idea of six.

The three yins of heaven and the three yangs of earth are experienced in time as three phases of daylight or any natural cycle represents the rising daylight of morning, the transitional peak at noon, and the declining light of afternoon. The three phases of yin are the descent into darkness of the evening, the pitch darkness of midnight, and the slow awakening toward dawn. With the human facing south, the phases can divide space into three equal zones above and below that are traversed by the sun, daily and seasonally, and the moon monthly. These three phases can be represented two dimensionally as in figure 5, where the names of the three yins and three yangs are identified in the inner circle, and the element affiliations are identified in the outer circle.

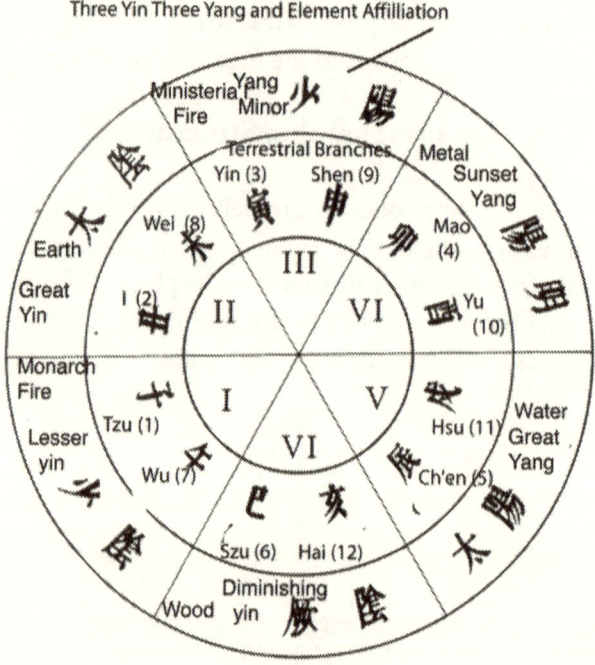

Figure 7. The Three Yins and Three Yangs (Inner Circle) and Their Element Affiliations (Outer Circle) Representing the Three Phases of Day and Three Phases of Night[68]

The horizontal line represents the horizon, which divides day from night. The element sequence is the same as that of the five elements; but fire has been divided into monarch fire, which is associated with the prime meridian at noon and ministerial fire, which in the cardinal-point model and the later model of twelve is associated with the south direction on the horizon and the rising heat of morning and springtime.

Since the cardinal points and five phases occupy the same plane, they manifest in three dimensions and are portrayed by Chi Po in a cardinal-point model reduced to two dimensions as three-yang phases above the horizon and three-yin phases below the horizon (fig. 4). However, once the vertical dimension is added to the horizon with its four directions and cyclic motions, the divisions of three may either be viewed from the perspective of above and below the horizon or left and right on the horizon line (or in any of four directions on the plane, though the vertical component in the number six implies the left-right, east-west dimension because the north-south dimension is folded into the vertical in a two-dimensional drawing) as three rising-yang or three descending-yin phases. Figure 4 really results from two planes formed by the midpoints between the prime vertical and the horizon. This relationship is shown in the cosmographical, figure 8, that shows the planes at forty-five degrees along with the great circles that determine them.

Six

Three Yang

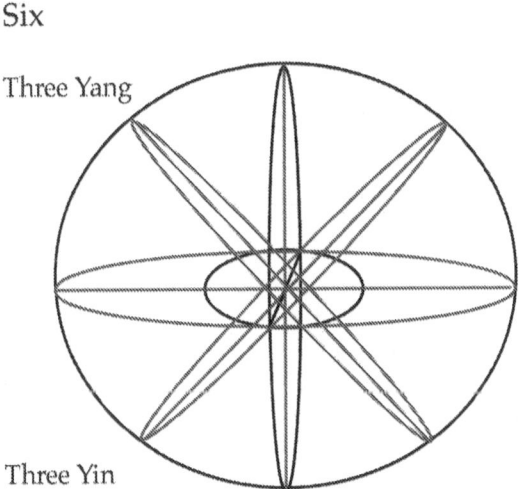

Three Yin

Figure 8. Cosmological 3-D Drawing of Great Planes and Divisions That Define the Three Yins and Three Yangs

That, for Chi Po, the three yins and three yangs generate six energies of heaven, man (horizon and overall movement), and earth, instead of six energies alone, reinforces the fact that the number six as a child of the terrestrial branches has to do uniquely and transitionally with what is close to the horizon but just above and below it, not merely the undifferentiated whole from that perspective as the cardinal points and five phases do. For Chi Po, the horizon is implicit in his six meteorological energies above and physical transformations in nature below (fig. 7). These pathological traditions have persisted in acupuncture mystegogery as the four external evil qi—wind, cold, heat, and moisture (associated with the cardinal points)—and two added factors of dryness and fire to achieve six.[69] These were present, and together these result in the twelve displays of the nonhuman observable world, above and below.

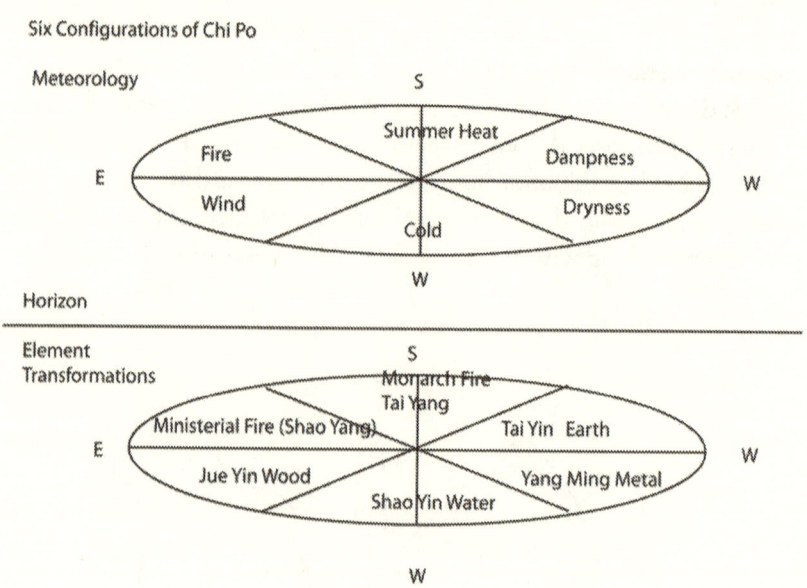

Figure 9. Six Meteorological Conditions Above, Six Corresponding Material Transformations Below—Driven by Seasonal Structures, the Celestial Stems, and Five Elements

Chi Po clearly views the six configurations simultaneously vertically as the three yangs above and three yins below and as six energies that circulate horizontally affecting weather, seasonal changes, and the body. He carries a three-dimensional notion with an interpenetrating nondimensional perception of energy operating in emptiness, which has material effects on the level of the appearances and structures of objects. Figure 8 reverts back to the human ramifications by showing the three-yin phases above and the three-yang phases below; but based on Chi Po's outlook, we have to look on this as both an immediate set of direct correlations and as a sequence that extends the five-element concept into physiological functions and structural relations anatomical relations.

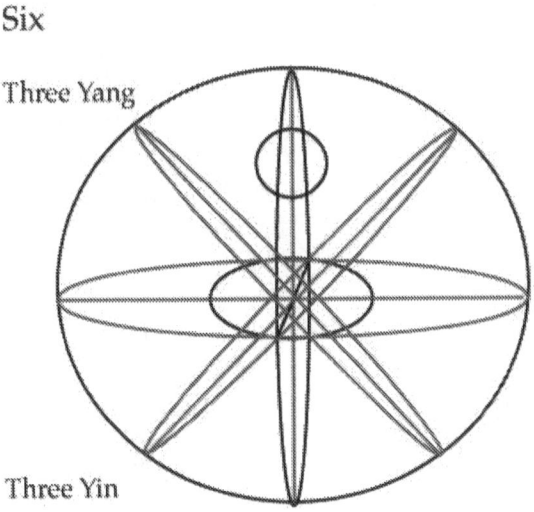

Six

Three Yang

Three Yin

Figure 10. The Six Divisions Applied to the Vertical Human Body

The three-dimensional portrait of the three yins and three yangs functions with the human body superimposed shows clearly that the arms and legs are at a forty-five-degree angle from the centerline. The prime meridian and centerline are implicit and the arms and legs from an explicit line of demarcation, an X shape which was one of the essential possibilities in the number four when viewed as a symbol for movement and four limbs. The arms and legs in this position signify transition and movement, which explains how the six configurations evolve from the yang five elements as further extensions or explanations of change. The structure is implicit except for the line and plane of the horizon, which again refers to the torso as it did in the number four. This body display sets the stage for the full realization of structure and function seen in the number eight, where the cross and the X, both are fully manifested, together are containing images for space and time, the horizontal/the vertical, and the planes of the functional transitions. Six leaves some facets of the structure implicit and is more dynamic as a differentiation of the effects of the five elements in specific planes of relationship on the structures of the body. These relationships are expressed seamlessly as direct correlations with anatomical structures and physiological functions.

III. Medical Practice—Theory

The three yins and three yangs function as energy and structures applied uniquely to human physiology and anatomy (both topography and structure) respectively. The human being is the only vertical animal, and six begins to deal with this becoming the basis for the six meridians of acupuncture that aligns vertically on the body. The human body is literally the union of the six principles applied above and below or side to side, to be realized more comprehensively in the number twelve and with spatial completion in the number nine where the principle of three that governs both six and nine achieves a full physiological representation.

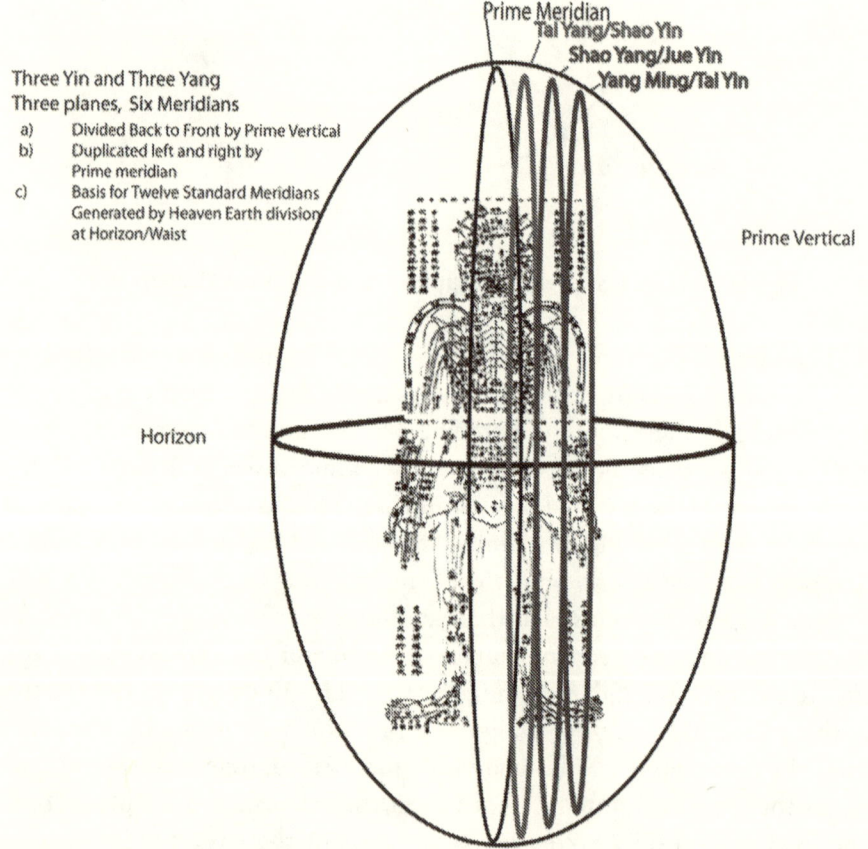

Three Yin and Three Yang
Three planes, Six Meridians
  a) Divided Back to Front by Prime Vertical
  b) Duplicated left and right by
      Prime meridian
  c) Basis for Twelve Standard Meridians
      Generated by Heaven Earth division
      at Horizon/Waist

Prime Meridian
Tai Yang/Shao Yin
Shao Yang/Jue Yin
Yang Ming/Tai Yin

Prime Vertical

Horizon

Figure 11. The Six Body-Length Meridians, Which Are Three Planes That Vertically Divide the Body in Sequence from the Midline Outward

Meridians have a back-and-front yang and yin differentiation and are duplicated left and right. When the planes are bisected at the waist and differentiated back from front, they generate the twelve standard acupuncture meridians of the body.

Chi Po's conception of number is a solid basis for evaluating any historical versions of the six configurations. From his unique vantage point, the six configurations must devolve from the traditional four cardinal points and seasonal values of water/winter, wood/spring, fire/summer/metal/ air/autumn. In derivative models, the distinctly yang features of the day or season must be prominent above in any anatomical or cosmographical diagram, and yin features of the day or season must be prominent below. Both the five elements and six configurations have the same object and element circulation in view and must correspond in their facts and theory in fundamental ways.

Links between the six configurations and the topography and functions of the human body too must conform. Everything in Chi Po's cosmos is linked directly. Body zones and their functions must link physiologically and structurally to concepts behind energies at the appropriate levels. Wind, fire, and summer heat for instance should be associated with yang configurations above the horizon and waist simultaneously and with elements and organs that display heat or wind; therefore, liver and heart and lung—either as ministerial fire or monarch fire above—are related to the rising cycle and midday, or *yang ming* or *tai yin* metal are related to afternoon sunlight. The one-to-one relation between the body and the world must be retained for each numerical increment.

The most direct diagram for six configurations does not establish a vertical centerline to distinguish east from west or the symmetrical sides of the body. This emphasizes that the functions are bilaterally undifferentiated, though the differentiation is implicit. The twelve functions of the division by six concern the whole body in two aspects—heaven and earth. This cosmic pattern functions the same way as the five phases by circulating freely through four seasons and directions.

As with the numbers, for four and five the horizon line is the principle locus for the functions of three yins and three yangs. This means that the horizon line, not the vertical, is the correct, fixed anatomical base for the drawing that depicts the three yins and three yangs. The model for six (three zones above and three below) with three implied zones at the waist yields a total of nine functional zones (fig. 3). The ultimate root for the nine zones was the three zones of the head, the three zones of the torso, and the three divisions of the limbs.

With the body facing south, the whole east, center, west divisions above and below the waist of the three yins and three yangs treat the bilateral symmetry of the body as a positive rising sun on the left, a neutral center (midline of body), and a negative center (setting sun to the west). Here, the four quadrants of the year are applied to the body with the implied center and a vertical component. The central vertical axis acts as a functional transition, pivot, or balance point but is viewed as a zone of activity. If the body faces north toward the pole star, the positive (east) is on the right, not left; and negative (west) is now on the left. The north view places the liver to the right toward the rising sun and spleen on the left—which anatomically accords perfectly with the *shao yang/jue yin* east, rising light correspondence, and *tai yin/yang ming* west, declining light correspondence. This suggests that the basis for zang/fu (solid organ versus hollow organ distinctions) and anatomical correspondences is with the north view, the view of night and water. The standard anatomical disposition of traditional medicine toward the south, which is the view of sun and fire, is in a way the view of outer life, of illusion, of disruption of function, and of cyclic redundancy.

The views north and south have an implied moral attitude. The south view is one of disapprobation toward an implied reversal of the natural order, where the body is vulnerable to external conditions that generate change, strife, and disease in contrast to the integrity of true chi facing introspectively north toward the unmoving pole start, which aligns a person with original nature and healing.

Regardless of moral interpretation, the reversibility of the human orientation, introspective, or extrovert means that the principle of six has to apply to both sides of the body independently so that the wholeness can be explained. Thus, bilateral anatomical symmetry would imply adaptability and require symmetrical meridian patterns on either side of the body, patterns whose functions are mirrors of each other. This allows the unprecedented freedom when compared to plants for instance to reverse position while maintaining the adaptation of the human body to its environment.

This insight offers an explanation of why the complete pattern of three yins and three yangs are applied bilaterally and completely to each side of the body. The meridians of the three yins and three yangs apply to the central, medial, and lateral aspect of each quadrant.

It designates the attributive names of the six full-body meridians, which reflect the three basic kinds of macrocosmic cycle microcosmically as great yang, medial, and lesser and great yin, medial, and lesser. The meridian correspondences are placed on literal bodily planes at the edge, medial, and edge of arms and legs either on the outside (yang) or the inside (yin).

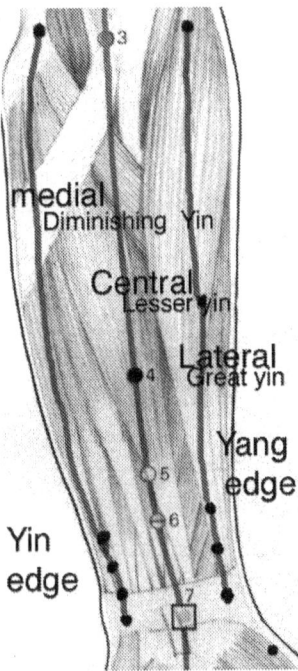

Figure 12. Placement of the Three-Yin Meridians of the Arm to Show Their Medial, Central, and Lateral Placement

The great yin meridian is the most lateral, near the yang side when in lowered position; and the shao (diminishing) yin nears the medial, interior, and relatively yin side. This organization reflects the position of the foot meridian and the adjacent torso meridian placements with *shao yin* nearer the midline. When hands are over the head in Chinese anatomical position with palms forward, the position reverses, which shows profound relations between yin meridians, which allow them to retain valuable meanings in two different diagrams and sets of relations between phases: an anatomical—and temporal-phase model. This will be examined later in discussions about the twelve-phase Chinese clock, where these discriminations are more exact. The meridian placements, at this point, confirms the idea that the body expresses the function of six bilaterally in a total of twelve functions that are fully integrated side to side but also vertically up and down.

The hand position changes indicate in a practical way the conceptions behind the three yins and three yangs when applied both to phases and cosmology. Hand position above or below is a literal indicator of the context for interpreting interior to exterior relations. In particular, the *shao yin* shares proximity to the centerline with *tai yang*, its yin-yang partner, and this alone resolves the questions about which of the three-yin meridians/orbs macrocosmically is to be viewed as the most quintessentially yin or interior of the three. Great or *tai yang* may be seen to be great because it is the most exterior and, therefore, visible function of yin is (when hands are above the head) great because its signs and symptoms may be seen at work in the daytime albeit in the afternoon in the declining portion of the day.

The development of three bilateral functions back and front and above and below the waist yields twelve zones/meridians on each side of the body, each side being a complete expression of the three yins and three yangs. Whichever way the side is oriented, it will operate either as yang or yin, as all functions are present. The symmetry of the body requires that the same complete range of functions be mirrored on the other side of the functional center, owing to the fixed east-west orientation of the waist/horizon.

Another basic anatomical approach to the three yins and three yangs in body topography is to divide the waist into back and front using the prime vertical along the east and west. This is the basic distinction between the three-yin and three-yang meridians as they are laid out—yang in back and yin in front. Then dividing the three yins and yangs using the plane of the horizon into an upper and lower zones, four gross spatial zones that conform to heaven and Earth, back and front, are created. The direct application of the three-yin and three-yang principle to the four zones yields three-yang upper zones and three lower zones below the waist, back and front, or twelve zones. The real meaning of Chi Po's theory comes into play here. For him, the three yins and three yangs implied a division at the horizon with a jump from three to six functional meteorological conditions above the horizon and six (three yins and three yangs) conditions at the horizon and the six configurations below the horizon.

The consequences of adding three more functions and areas—central, medial, and lateral—to the four basic quadrants give considerable refinement to geometrical differentiations within the quadrants. Symptoms having to do with the lateral plane *tai yin* and *yang ming* can be treated symmetrically back to front, upper to lower, and bilaterally along the various planes involved. This becomes the basis for a whole method or approach to intervention. It implies that if the meridian system works at all, it works because of precise geometrical symmetry, and compensating stimulation works clinically. In so far as symmetry and function go together, the six configurations and their related planes show how the five elements and their associated organs translate into structure. Here, the six orbs of the three yins and three yangs focus on the above (heart, lung), middle (liver, spleen), lower (two kidneys), and their corollaries in the head and limbs.

The heart and lung meridians have a medial-to-lateral relationship on the arm, which can be used for differential diagnosis; and if one or the other is implicated, the medial plane of the front of the arm can be treated symmetrically using the medial plane of the back of the arm or the medial front or back of the leg.

Differentiations about level in the body/mind have to do with the level of the organs and their functions, so heaven, earth and central considerations can contribute to a diagnosis that will translate into central, medial, or lateral treatments in the zones. Similarly, tactile differentiation between lateral, central, and medial will indicate which organs are suffering. Processes having to do with rising heat, moisture, light, and purpose or declining heat, moisture, light, and purpose translate into three-yin and three-yang meridians and their structural zones. This means that the five-element processes and disruptions of function can be differentiated directly into zones for diagnosis and treatment. The twelve meridians are merely a more detailed differentiation of these functions above and below the waist.

IV. The Evolution of the Proposed Three-Yin and Three-Yang Concept

As indicated in the writings of Eberhard and Porkert, neither tradition nor modern scholarship is uniform in the assigning of the sheng cycle element sequence of wood, fire, earth, metal, and water or starting points to the terrestrial sequence of the three yins and three yangs.[70] Needham, with the help of Western scholarship, establishes some firm links between the astrological conceptions of the Babylonians and the Chinese of the first millennium, suggesting that some communication between ancient cultures was likely.[71] He sites Eisler by saying, "In his recent survey of the astrological element in ancient astronomy, the outstanding characteristic of the oldest astrology is that it was never concerned with individual human beings (unless they were of royal blood), and always with prognostications concerning affairs of State, the chances of war, the prospects of the harvest, and so on."[72] He explains that only starting in the third century AD did Chinese astrologers begin to attempt individual horoscopes. He also traces some of the gradual evolution in Chinese cosmographical conceptions, and these developments in turn had a bearing on medical theory. The application of terrestrial branches and animals to time of day rather than season gained prevalence only after the ninth century AD. The theory of the Wu Hsing (five elements) has ancient components in oral tradition and the characters of numbers themselves but was first systematically expressed by Tsou Wun, who lived comparatively recently approximately on 350-270 BC. The Wu Yun theory was only fully established in Chinese writings by the third century AD and matured further through the ninth century.[73] The dynamic development and elaboration of number systems involved in the evolution of the Wu Yun explain the many family tradition anomalies in the cosmic models from one scholar to another and from one region of practice to another. The following discussion attempts, late in the game, to reconstruct a wide view of the evolving cosmic number sequence based on the implicit unity of the seasonal, monthly, and daily cycles. The facts of the light cycles require congruence between descriptions based on different whole numbers.

I. Comparing Uses of Terrestrial Branches in Cardinal-Point and Six-Division Models

Manfred Porkert shows the terrestrial branches in a cardinal point distribution with four branches in the center symbolizing the transitions between the seasonal attributions of the cardinal points (fig. 13 and fig. 4-6 above).[74] The branches are applied in yang and yin pairs at the cardinal points, and they preserve their seasonal sequence. The cardinal-point image is the standard image with four directions and earth, which is viewed as a process of change more than a location, at the center.

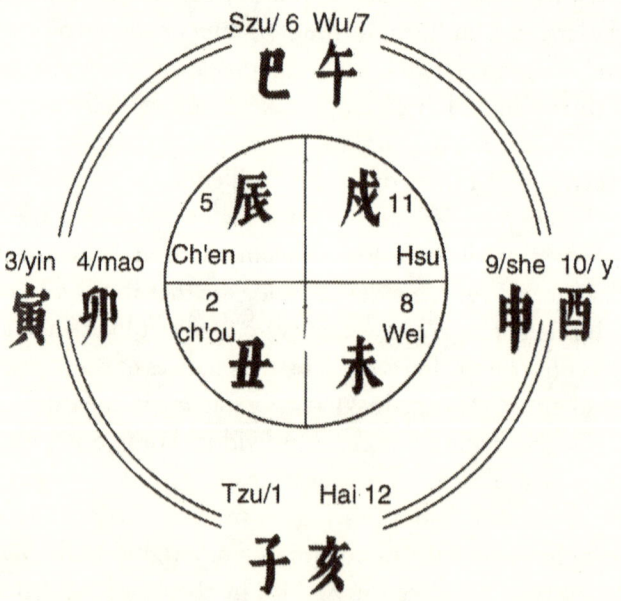

Figure 13. Twelve Branches Applied to the Cardinal Points and Earth-Centered Model of the Cosmos and the Number Four

We have to assume that the assignment of terrestrial branches to even-numbered divisions of the year is significant. Moreover, the sequence of branches in model of twelve monthly divisions would give each branch a connection to the seasonal attributes implicit in the structure of its symbol. Figure 13 preserves this sequence but in a King Wen, postnatal version that anticipates the five-element pattern with its terminal sequence of earth, metal, and water. Figure 14 reestablishes the terrestrial branch sequence in its original four seasonal template, which is a *fu*, his prenatal, structural model.

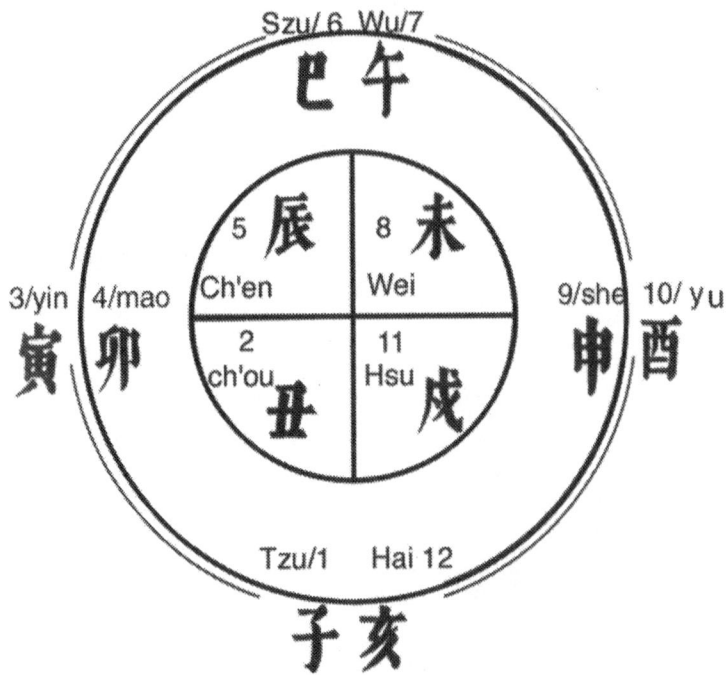

Figure 14. Modified Application of the Twelve Terrestrial Branches to the Four Cardinal Points (Position of *Wei* and *Hsu* Switched for Cardinal Point Symmetry)

Scholarship confirms that tradition does assign terrestrial branches to the three-yin and three-yang configurations and according to the pattern established by the elements and four branches to fire identifying two with monarch fire (*tai yang* and *shao yin*) and two with ministerial fire (*shao yang* and *jue yin*) as in table 1.

Table 1. Combinations of Twelve Terrestrial Branches into Six Pairs with Element Affiliations[75]

Branch order	Liu ch'i	Three Yin/yang	Organ	Quality	Element phase
1/7 Tzu/Wu	2	lesser yin	heart (kidney)	heat	monarch fire
2/8 Ch'ou/Wei	4	great yin	spleen (lung)	humid	earth
3/9 Yin/Shen	3	lesser yang	Triple Htr. (Gall bl)	Summer heat	ministerial fire
4/10 Mao/yu	5	Splendor yang	colon (stomach)	dry	metal
5/11 Ch'en/Hsu	6	Great Yang	bladder (SI)	cold/moist (Summer heat)	water (fire)
6/12 Szu/Hai	1	Diminishing yin	liver (PC)	wind (Heat)	wood (Fire)

( ) = Subordinated expression of the functional orb in phase.

The table describes the most common traditional sequence for the annual cycle displayed in figure 11 (fig. 15).[76]

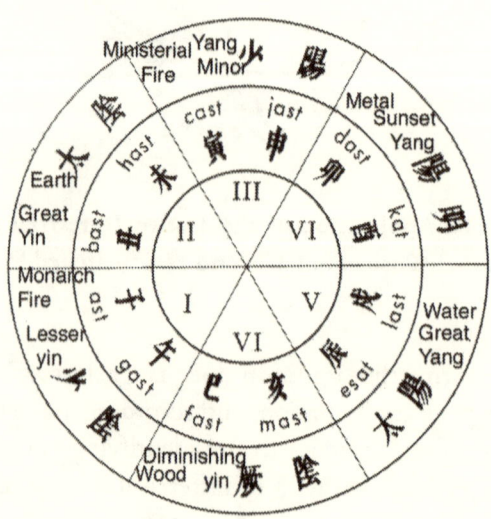

Figure 15. Traditional Version of the Three Yins and Three Yangs

This figure and table shows a sequence of elements and transformations that do not conform with the standard element sequence of five elements nor the matching sequence nor even the reverse sequence that might be the consequence of a lunar model of circulation. Monarch fire is the fire element at its maximum expression at and just after midsummer. Ministerial fire is the rising heat that emerges after the late spring, after the equinox. If these well-established links between qualities, elements, and seasons persist in delineations of six, then why is earth sitting between the two fires? Even the oppositional root of the combinations of branches one to seven and two to eight, should produce a more standard sequence. It is true that the branches have a purely numerical and symbolic function on one level, but their whole identification with yin structures and conditions that derived from the lunations suggest that a complete gap between branches and circumstances is unlikely. Figure 16 displays in three dimensions the element and structural disharmony that results from the most commonly used sequence selected by Porkert.

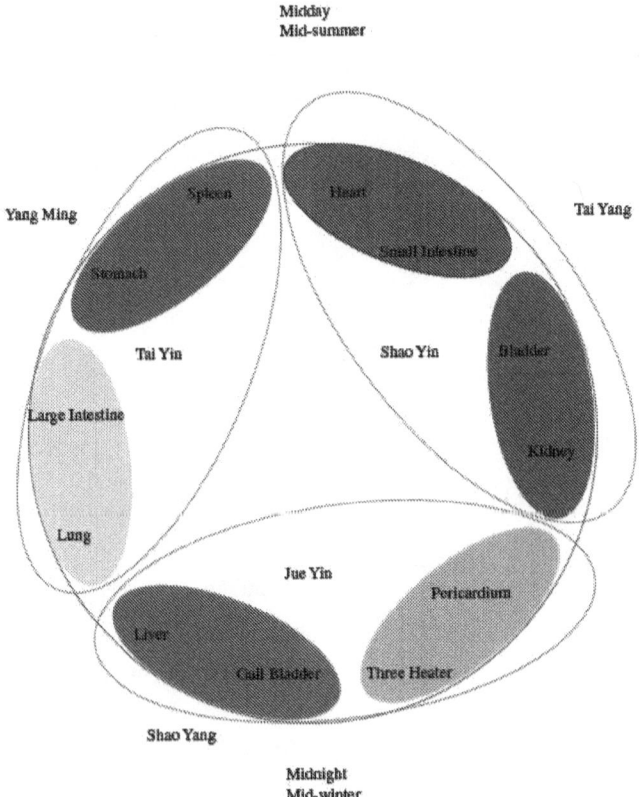

Figure 16. The Forced Sequence of the Traditional Arrangement of Blocks of Entry-Exit Points

If six configurations are an extension of five elements as suggested by Chi Po, then greater congruity between models for phases representing the same basic seasonal cycle should be expected. What is arbitrary and binding on such decisions of placement is the horizon and midheaven and demarcations like the cardinal points, along with evidence from seasonal change. The phase changes may happen to different degrees earlier or later, depending on the lens applied to the cycle; but generally, it happens in a regular way. Figures 5 and 7 above transplants the regular established cycle of the five elements to its six configurations counterpart and in so doin,g realigns the regular cycle of six configurations in a pattern that conforms with the seasons as they commonly are understood. All the variations of the sixty-year cycle and consequent divergent versions of the stem and branch cycles can still be applied to this foundation, but it becomes a foundation that conforms with nature and with long-determined meridian physiology.

Table 2 is the counterpart to table 1 and shows how the combinations of branches apply to figure 5, the proposed six-configuration model.

Table 2. The Element, Meteorological Condition, Orb and Meridian, and Three-Yin and Three-Yang Correspondences for the Paired Branches

Branch order	Liu ch'i	Three Yin/yang	Organ	Quality	Element phase
Tzu(1)/Wu(7)	1	Shao Yin	KIdney (heart)	Winter cold Summer heat	water (Monarch fire)
Ch'ou(2)/Wei(8)	2	Jue yin (PC.)	liver Moist warm Wind	Moist cool (Ministerial Fire)	wood
Yin(3)/Shen(9)	3	Shao yang	Triple (G.B.)	Heat Moist heater Warm/moist	Ministerial fire (wood)
Mao(4)/yu(10)	4	Tai yang	S.I. (Bladder)	Summer Heat Winter Cold	Monarch fire (water)
Ch'en(5)/Hsu(11)	5	Tai yin (lung)	spleen Dry	Residual damp (metal)	earth
Szu(6)/Hai(12)	6	Yang Ming	L.I. (stomach)	dry Damp	metal (earth)

The principal advantage of this table and its combination of branches linked to elements is that certain fundamental links are preserved with the earlier numbered models and the directional model of the cardinal points. Each dominant branch implies in brackets the presence of its yin or yang partner. Clearly, in models with different purposes, one or the other pairs can be emphasized as in the case of the differentiation between figures 16 and 17 for the three-yin and three-yang sequence.

Branch Distribution By Element Sequence

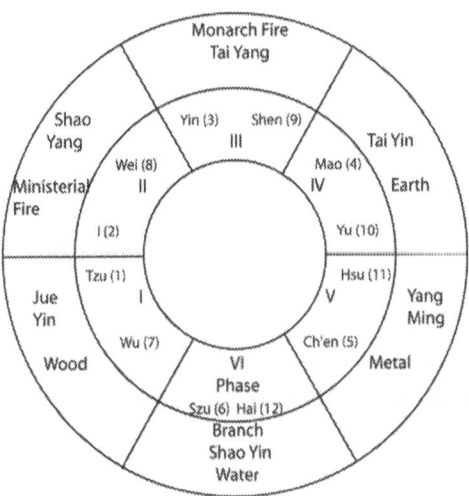

Figure 17. *A*, Cardinal-Point Terrestrial Branches in Comparison; *B*, Structural Three-Yin and Three-Yang Distribution

Branch Distribution By Three Yin Three Yang
Structural Arrangement

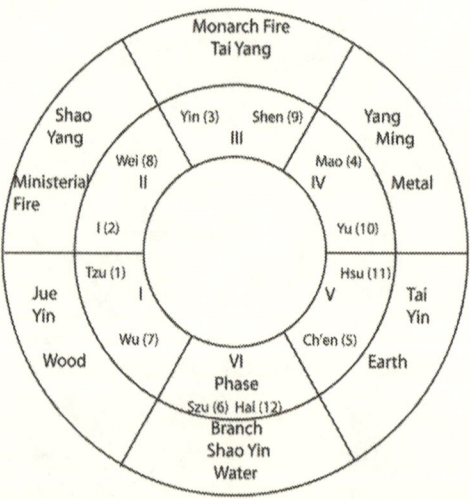

Figure 18. *A*, **Cardinal-Point Terrestrial Branches in Comparison with Phase Distribution;** *B*, **with Five-Element Phase Emphasis in Distribution of Branches and Elements**

Analysis of the transition between cardinal point and six-division distribution of the terrestrial pairs highlight some important understandings about the meaning of branch relations as the cosmographical portraits build to the complete system of twelve divisions. The paired branches Tzu (1) and Wu (7) and Szu (6) and Hai (12) of the three yins and three yangs were formerly opposed across the vertical in cardinal points; similarly, Ch'ou (2) and wei (8) govern yin transition periods, while Ch'en (5) and Hsu (11) govern yang transitions, and Yin (3) and She (9) and Mao (4) and Yu (10) are oppositions across the horizon.

In the four-directional models, six and seven, were paired together in order to represent the yin and yang aspect of fire while one and twelve represented the yin and yang aspect of water. This means that in figure 5, the new model for the division of six that the combination of Tzu (1) and Szu (6) links the yang (Tzu Wu being yang numbers) fire and yang water together in a single vertical trajectory. Similarly, the combination Hai (12) with Szu (6) links the yin fire and yin water together. These are the important *tai yang* and *shao yin* fire and water combinations. This pair occupies the medial position near the centerline of the body. Once again, the assumed vertical in north and south of the compressed cardinal-point model (four/horizon) is unpacked into an openly vertical combination that combines branches associated with water and fire and the heaven-to-earth opposition. The number combinations caused by different divisions of the cycle drive the interpretive meaning of the system that results.

The position near the centerline rather than on the centerline is explained here because the centerline of the north-south axis in the cardinal-point drawing is visually absent in the vertical three-yin and three-yang distributions. The important thing to realize is that the function occupies the zone that includes both the north and south on the horizon and vertical prime meridian centerline, and therefore, the medical and conceptual functions are the same. The ren and du meridians occupy the plane of the prime meridian at noon with the figure facing south, but their meanings are nearly identical within the context of the number eight, and the *tai yang* and *shao yin* are in the context of the number six.

The unique introduction of the vertical in six introduces the so-called special organs into the standard mix of *zhang* and *fu* organs. The brain occupying the top of the body and the uterus and male generative organs at the bottom form a linked heaven-to-earth reflection on the centerline; and their functions, each on their own levels, unify all other functions during certain kinds of activity and are responsible for physical, social, and ultimately spiritual creation.

In the combinations of Ch'ou (2) and Wei (8) in the modified structural version of the three yins and three yangs, the opposing principle of wood and metal in the cardinal-point model are combined. Similarly, with the combination Ch'en (5) and Hsu (11), the wood-to-metal oppositions in the cardinal-point opposition are united. They are united in the structural three-yin and three-yang models as an odd-numbered pair as *shao yang* ministerial fire to *yang ming* Earth. This perfectly symbolizes the differentiation at the horizon between the lower and upper halves of the X-shaped planes for the transitions that define the three yangs and three yins in figures 10, 14 and 17.

The sequenced pairing seems to reemphasize the heaven-to-earth balance combined with the east-west balance, which is in keeping with the number six. As transitions, these relations suggest equilibrium between hot and cold from the vertical opposition and moisture and dryness from the horizontal opposition.

In the combination of yin (3) and she (9), the yang aspects are of wood (ministerial fire) and metal (earth) elements and the moist and dry qualities, while mao (4) and yu (10) has their their yin aspect. Ministerial fire and earth are new explicit extensions in the six divisions that were functionally implicit in the vertical and center respectively of the four cardinal-point model. The main emphasis in the 3/9 and 4/10 combinations is on the quality of moist linking with the opposite quality of dry with the memory from the east-west cardinal points of wood to metal. However, the pattern plays out structurally as well as functionally. These combinations place opposites in yoke in such a way that the center/horizon combination (yin branch, structural earth reference) is lateral on the body—representing the east-to-west or left-to-right extremes of the horizon. The more vertical, as opposed to horizontal the combinations, the closer to the centerline of the body they become.

The relation of the plane, ministerial fire and wood, is medial on the body compared to *tai yin/yang ming* plane. This is because the *tai yin* values are more tied to the earth, the horizon values and east/west focus of the prime vertical while the *jue yin/shao yang* leans more to the north-south direction of the prime meridian, although it too has links to the horizon. *Jue yin/shao yang* has an equilibrium similar to that suggested by ch'ou (2) and ch'en (5). Therefore, *jue yin* and *shao yang* in their characters mediate between the vertical and the horizontal with an equilibrium between structure and function that is less evident in the other three-yin and three-yang pairs, which emphasize either vertical structure (heart/kidney) on the one hand and horizontal process (digestion) on the other.

The regular oppositions of the terrestrial stems in the cardinal-point model become combinations in the model of six, and the combinations match the element and spatial arrangements in the proposed three-yin and three-yang meridians. This symmetry is nearly perfect, because the horizontal oppositions are preserved with the vertical; and the combination produces the central, medial, and lateral structural distribution as the functional principles of the horizon (2). And the four directions (4) are brought into play with heaven, earth, and horizon (3) divisions to produce a graduated equilibrium between form expressed uniquely in the chi and corollaries of the three planes, three-yin and three-yang meridians, and twelve terrestrial principles.

In Porkert's traditional cardinal-point image in figure 12, Ch'ou (2) and Wei (8) occupy the two yin transitions even though Wei (8) is earlier in sequence than yu (11). This pair, then, is more expressive of cold and moisture, *ch'en* and *hsu* of heat and dryness. These pairs seem to define functional zones more than spatial zones, which means that any arrangement of meridians in an above-to-below arrangement must conform to the qualities and functional conditions of the zones. However, in each pair, one partner dominates; and the position of the prior number establishes the tone of the pair and its location in the circle. Ch'ou is associated with the *jue yin* wood, Wei with *tai yin* earth, Ch'en (a yang symbol) with *shao yang* ministerial fire, Hsu with *yang ming* metal. Similarly, in Porkert's version, with the upper combination Ch'en (5) and Hsu (11), they act as celestial and seasonal counterpoints to Ch'ou and Wei. Indeed, their meridian relationships suggest that Ch'ou and Ch'en constitute the real underpinnings of *jue yin* and Wei, and Hsu form the underpinnings of *tai yin*.

The pattern of distribution of terrestrial branches in the center suggests this distribution is a postnatal King Wen version with Wei representing metal: the lungs or large intestine. The natural cosmographical structure of the seasons according to four and cardinal points have Wei (8) above exchanging position with hsu (11) (fig. 13). The aim of this configuration seems to be to bring ch'ou (2) and wei (8) together to emphasize and anticipate the five-element sequence of earth, metal, and water rather than the structural arrangement of the quadrants, which moves from metal to earth to water.

Whether the cardinal-point representation of the twelve branches follows Fu Hsi, structure, or King Wen, function; they both reflect the unique combination of relations in a graded sequence between horizontal and vertical principles, which yields horizontal, vertical, and in between. The outcome of including day and night, summer and winter, in the seasonal rotation into this structure is three-yin and three-yang phases in any cycle. At every point of a light cycle, the central, medial, or lateral topography or function of the body is activated according to the phase.

II. Differentiating Levels of Function

The tension between five elements and six configurations phase sequence has to do with the sequence of elements and the asymmetric relation between function and form; given two options for the element sequence and this needs some analysis and explanation. A sound hypothesis about the movement of chi (function) would be that it is determined by the basic structure of the body and world, but that it is activated in a regular sequence that first builds yang then intensifies yin in an endless rhythmic relationship. Such a sequence is free of theoretical absolutes. Though simple in concept, the rhythm plays out in three dimensions, vertically and horizontally. In traditional theory, the light and vitality of heaven descends, penetrates the earth, and from the earth all the multifarious forms of life develop. This is the simple explanation of how yin and yang generate the third and of how heaven, earth, and horizon form the basis of life. However, this vertical dimension plays out in the anatomical and topographical order of the body and the world. This interaction describes a movement from the formless universal toward form and a return movement from form toward the formless.

Practically speaking, this constant interplay defined by interactions between the upper and lower, outer and inner functions in the human body are amplified by the horizontal cycle of the seasons—where a similar oscillation plays out as a progressive sequence in which first there is heaven (sun, summer); then earth (moon, winter) starts small, emerge, then decline relative to each other. In nature, something is relatively small at its inception, and then it grows to a maximum point of extension then declines. In the agricultural cycle, this occurs in space as the form expands and contracts and in time, in accordance with the seasons. The two actions together form a spiral or an evolutionary development, not just a mindlessly repeated cycle. This model suggests that both the anatomical (vertical) and the phase relationships can be accommodated by a twelve-phase model of the year that includes the theories of five and six together, provided that the levels of function are differentiated.

Certain tensions develop when the models of the human relation to the cosmos based on time (*wu yun*) are coordinated with those based on space (six divisions). Human anatomy is organized by space, and its functional capacities unfold in time. The above-to-below relations and functions dictate human physiology. All the passive movements of fluids and heat in the body depend on considerations of proximity and structure, while the volume of vessels and nerve zones similarly are structured by spatial relations, which condition the sequence of their responses to environmental conditions. These points simply illustrate that an anatomical portrait of the human form with the form standing vertically in some relation to the cosmos has an important effect on the sequence and movement of human response to the environment, yet that organization of response is somewhat different from the sequence determined by the light phase and time-dependent cycle of the seasons. In particular, the vertical description of the world derived from the number three, heaven, horizon, and earth is in contrast to the horizontal/directional order of the cardinal points—north, east, south, and west. As developed above, these two descriptions of the same cycles describe two different organizations of the movement of chi. These distinctions—between vertical and horizontal, space and time—play out in greater detail when the five elements and six configurations are compared.

The theoretical crunch arrives in the specifics of acupuncture meridian theory in their relations to natural cycles. Are the meridian functions related by time or space, functional sequence, geometrical planes, or anatomical relations? This problem reduces to the question, In the afternoon (3:00-5:00 during the late summer period), does the earth phase (five elements) or the metal phase (three yins and three yangs) take precedence? At this phase of the diurnal cycle, the sun is still above in heaven (upper body), visible in the sky, and generating heat. Seasonally, the days are long though declining in length, and the weather is hot, warm, and yang. In the anatomical, the above-to-below relationship of the period is yang. However, in east-to-west or horizon relations, the period is relatively yin or in decline but not very far into the process that dries plants and exposes them to killer frosts. These two perspectives appear to clash even though it is their combination that yields the characteristic facts and correlations that result in our descriptions. Much of the complexities of meridian theory and meridian sequencing derive from this tension in the declining phase. Appropriate to yin, there seem to be two ways of looking at the declining cycle. If taken as a whole, the declining block of *tai yin/yang ming* functions makes sense; but in characterizing the first phase, the upper-to-lower and phase sequence seem to be in tension.

The outcome of this tension is at least two profoundly different sets of indicators for the use of horary points and entry-exit points to meridians, which in their turn, as we have seen, will both refer to the cosmic order. If either the anatomical sequence or the phase sequence is emphasized at the expense of the other, the system breaks down. Understanding the three-dimensional variables in the system will help to resolve the specific contradictions, and the rest of this chapter will use three different spatial models to show how traditional thinkers used numbers and human anatomy together to create a fit.

Three different approaches to the relations between models of five and six may improve our understanding of how five evolves into six. The first approach in section 3 below has to do with a spatial model for the three-yin and three-yang blocks; the second in section 4 examines a clean vertical differentiation of heaven and earth implied by the stems and branches. And in section 5, a third expands on the vertical differentiation of function and structure to show that the circuit phases create a five-level model that allows for regularity and variation in the annual cycle. As a basis for this discussion, figure 17b above shows the five-element distribution and 16b the three-yin and the three-yang anatomical distribution, where the conceptual differences in the declining cycle are evident.

III. Three Orbs Conceived in Three Dimensions

One of the conceptual limitations against understanding relations between functions of qi and physical structures is the tendency in historical periods to draw cosmographical models in two dimensions and the difficulty of applying geometry to moving three-dimensional forms. Calculus has helped but also has obscured in a wealth of detail the importance of whole numbers to the shape of things. Figure 19 takes a fresh approach to the seasonal sphere by drawing the proposed three yins and three yangs as three-dimensional functional orbs.

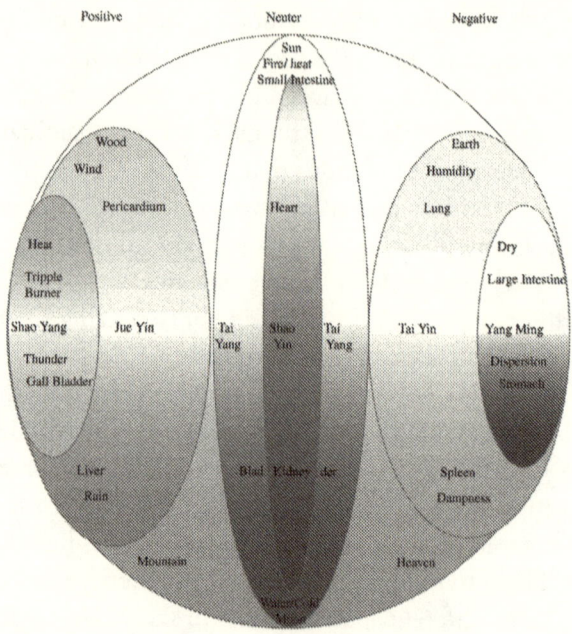

Figure 19. Proposed Three-Yin and Three-Yang Sequence Shown as Semi-independent Orbs with Their Own Internal-to-External, Yin-to-Yang Equilibrium

The colors and shapes of figure 19 attempt to capture the living interactive shape of the whole circuit. These orbs are seamlessly continuous and adaptive, which means that any facet may surface under certain cosmographical pressures and still maintain its equilibrium with other parts of the system. The drawing describes three blocks of meridians that represent the three vertical planes of figure 11 now distributed in space near their functional cardinal points; but they are functionally distributed in time as rising light, central equilibrium, and declining light blocks—left, center, and right. The drawing implies that in any phase such as the late afternoon phase, the whole system of *tai yin* and *yang ming* is implicated, not simply the earth or metal element portion. This, to some extent, places the question of metal (three yins and three yangs) versus earth (five elements) into a wider shared context.

Any moment in the cosmos has its own complete spherical equilibrium, so this drawing is a conceptual distribution of the factors in different phases. In phase, each of the three blocks in a way rises to the condition of full manifestation and in so doing encompasses the whole, larger sphere symbolized by the prime vertical. Naturally, this manifest condition suppresses the other blocks of function, which in their turn operate internally in the present as structural limitations on the manifestation of the particular phase or in time as residual effects from the past or potential outcomes in the future of conditions in the present. Each phase represents a particular kind of balance between yin and yang appropriate to the season or time of day; and in a very real way, the equilibrium, say between *shao yang* and gallbladder, at the spring equinox or dawn where light and warmth symbolically gain ascendancy over darkness and *jue yin* liver, which contains the residual coolness and moisture of winter in the duration of darkness keeps heat in check and brings cooling rains.

The way the whole cycle achieves homeostasis is one thing, and this is reflected in the annual growth cycle implicit in the meaning of the twelve branches. The tripartite divisions give us important insights into the way energy passes from one meridian or phase to the next and into the equilibrium to be found in waxing and waning phases, which in turn greatly inform the way the four branch-derived phases of each interact.

All ancient systems understood the oscillating pattern in nature and followed a positive/male phase with a negative/female phase to reflect this. This underlying process must somehow be reflected in any valid portrayal of the twelve-phase sequence because this rhythm closely represents the oscillating periods in nature, because each contains the rhythmic pulse of a complete lunation consisting in turn of a waxing or positive phase and a waning or negative phase which at the least provide the basic time standard for evaluating natural appearances and at most have a bearing on real conditions. The advantage of the six-phase portrait is that it combines in each division two lunations, a positive and a negative (back and front), which in turn are connected in a similar pairing (above to below); and therefore, each block (back to front, up to down) creates a whole in miniature that encompasses and explains the rhythm of a greater 120-day period. These three 120-day blocks each have unique characteristics.

The Rising Block

The rising light cycle may be viewed as an organization of time according to principles of five elements or of space according to the sequence of three yins and three yangs. Five elements begin the rising cycle with wood and then moves to fire. Wood has two branch/meridians in it—yin, foot *jue yin*, liver and its partner foot *shao yang*, gallbladder. The wood-foot meridians operate below the horizon or underground, organizing the energy of morning before dawn and early spring before sprouting. The fire branch/meridians that extend the *jue yin* and *shao yang* of the foot are their ministerial fire hand counterparts that represent the rising light and heat of spring above the waist/horizon, in the daytime, and late spring after the equinox. In the rising cycle, then, the phase sequence based on time and the phase sequence based on space are somewhat compatible because although the sequence of phase associations is different, they both assume the same one-to-one compatibility between the standing figure of anatomy and rising light of the cosmos.

The preferred beginning in both sequences is liver. The five-element sequence beginning with liver proceeds in the next phase to the yang partner gallbladder, then in the standard entry-exit sequence to three heater, and pericardium (fig. 20). Detailed explanation of entry-exit point relations occurs in chapter 12; here, it is sufficient to establish the coherence of the three yins and three yangs conceived as three seasonal blocks with yang and yin features.

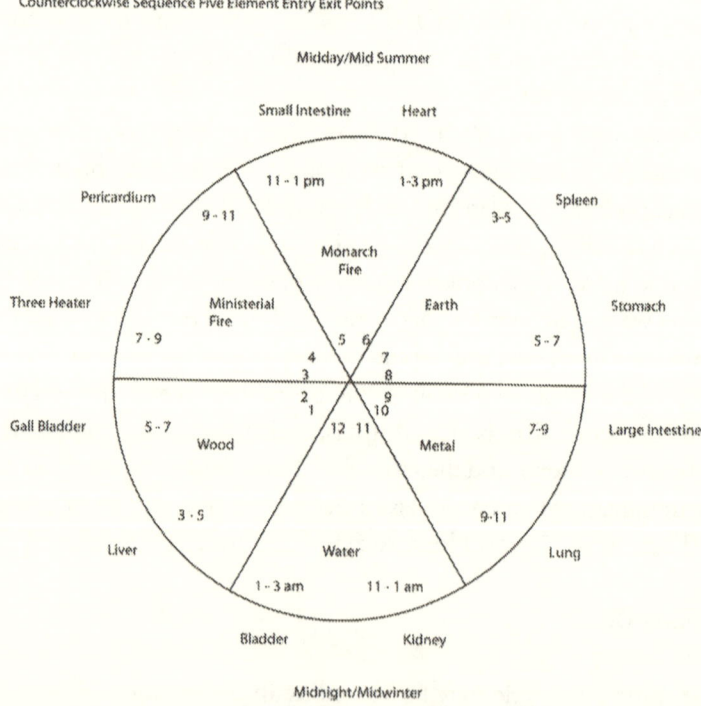

Figure 20. Standard Five-Element Sequence on Seasonal and Structural Harmony of Function and Maintaining the Standard Pattern of Meridian Entry and Exit points in Counterclockwise Motion

This sequence is chosen because it places the yang meridians at the active transition of dawn/spring equinox, and while it preserves the element sequence in a clockwise motion, it preserves the standard meridian entry-exit sequence in a counterclockwise motion. This sequence is compatible with the three-yin and three-yang anatomical sequence of entry and exit, which beginning with liver proceeds vertically upward to the partner meridian to liver, pericardium, then adjacent, three heater, and finally back down the gallbladder (fig. 21).

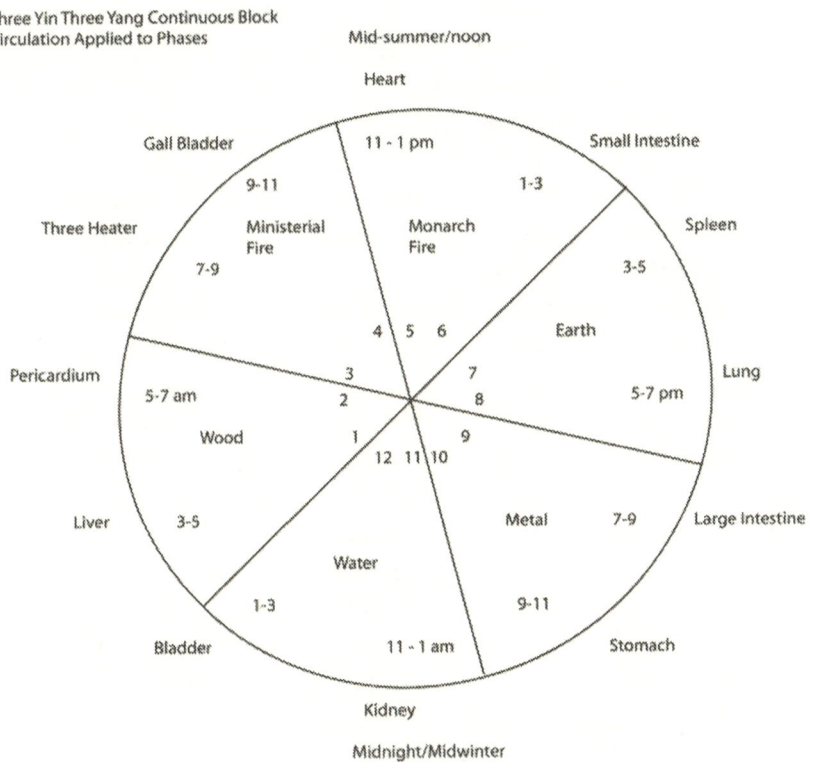

Figure 21. Implicit and Continuous Three-Yin and Three-Yang Entry-Exit Cycle and the Three-Yin and Three-Yang Entry-Exit Pattern Projected onto Temporal Phases

The continuous cycle of chi through each block following the entry-exit sequence of three-yin and three-yang meridians creates a counterclockwise flow of chi that compensates for the solar action in the macrocosm. Although in figure 19, gallbladder, a foot meridian, is above the horizon, which causes it to lose its structural integrity, that placement places both yin orbs liver and pericardium below the horizon and both yang orbs above. What we are looking at in figures 18 and 19 are two valid windows into the above-horizon to below-horizon relations, the former having to do with balance between functions of celestial time (five-element sequence) with fire near summer and terrestrial space (six configurations legs below arms above), the latter nineteen showing the effect of the figure lying facedown with arms spread instead of standing with the qualitative yin functions closer to winter and below yang functions above. The former is a vertical image and the latter a horizontal image of the same totality, and it is important to see the three-yin and three-yang portrait as combining the horizontal emphasis of four and five congruently with the vertical distinctions of heaven and earth in a graduated sequence that matches human anatomical topography. Figure 18 is the more useful drawing because it allows us to assume (b) as an ongoing internal circulation and focus on the simultaneous effects of the solar on the actual position's physical structures (three yins and three yangs, legs below, arms above) in their proper positions and temporal sequence of water, wood, and fire in the proper sequence.

The entry-exit sequences differs between the two systems, the five element with the standard sequence going backward as liver, gallbladder, three heater, pericardium, the three yins, and three yangs moving forward internally from liver to pericardium then the standard sequence, three heater, gallbladder. These two models for phase activation and entry-exit-point relations clash, and this calls into focus the question of what mechanical relations activate points and establish communications between different meridians and their functions when phase transitions occur. This will be examined in detail below in chapter 12, where the twelve branches are fully articulated. Here, it is sufficient to say that in an open circuit, it is possible for the energy to flow in a standard pattern of structural relations that conceptually can allow certain phases to be activated by the solar and lunar wave according to the physiological effects of light and the active and passive functions of the phases of the day.

The Declining Block

The portrait of declining light of the day or year again can be depicted in terms of time and five-element phases or in terms of space using the three-yin and three-yang relations. The outcome is somewhat different than in the rising cycle because, as described above, the above-to-below relations are reversed in the descending five-element movement from earth to metal. Earth meridians are leg meridians, and they appear first in the declining cycle in time periods when days are longer than nights and in the afternoon rather than after dusk when the sun has set. In contrast to standing, anatomical cosmic human metal is upper body, daylight, and afternoon; and earth is below the horizon in a way appropriate to leg meridians. This contradiction demands that cosmologists and physicians differentiate the mechanisms that explain these different processes.

One direct way to differentiate the rising from the declining block is to recognize in their different character with rising light demonstrating an upward movement as the sun rises in the sky toward midday and heaven and with declining light demonstrating a downward movement as the sun sets in the sky toward dusk, midnight, and earth. It may be that the earth emphasis of the declining cycle causes the horizontal perspective of cardinal points and five elements to be ascendant over the vertical perspective. Thus, with the body lying facedown with the head (sun) toward the west rather than the east with arms spread, the yin functions and meridians will be down, the yang up. The difference is in the orientation of the head, and from this view when projected on the whole cycle, the morning sun will strike the limbs first and the upper body last. Alternatively, the head could face north with its implicit downward motion. Shifting the emphasis on the declining cycle to the horizontal earth view rather than the vertical heaven view is a simple yet sophisticated way to display the space and time comprehensively in a model that is so hampered by theoretical compression from twelve to six functions. In this way, both the vertical and horizontal outlooks are efficiently articulated, each in the zone most appropriate to the functional imperatives.

Some parallel insights emerge when the declining light cycle is viewed through the lens of the vertical human body and as a mirror image of the rising cycle. The standard element-phase sequence is not immediately so structurally satisfactory as was the ascending cycle, owing to the yin and leg meridians being above the horizon. This arrangement can be rationalized satisfactorily. Heaven forms a sphere. If the rising cycle has the head facing up, the direction of the declining cycle will have the head facing down (fig. 22).

Figure 22 legitimizes the presence of spleen above the horizon and at the beginning of the descending sequence in any model. The element sequence in this model mirrors that of the rising cycle that begins with the yin foot meridian, *tai yin* spleen. The standard element sequence is preserved as the sequence of earth, metal, and water; and the standard entry-exit sequence operates but perfectly in reverse as was the case in the rising cycle: spleen, stomach, large intestine, lung. The three-yin and three-yang structures and continuous entry-exit equilibrium is maintained as a clockwise, forward sequence with spleen, lung, large intestine, stomach.

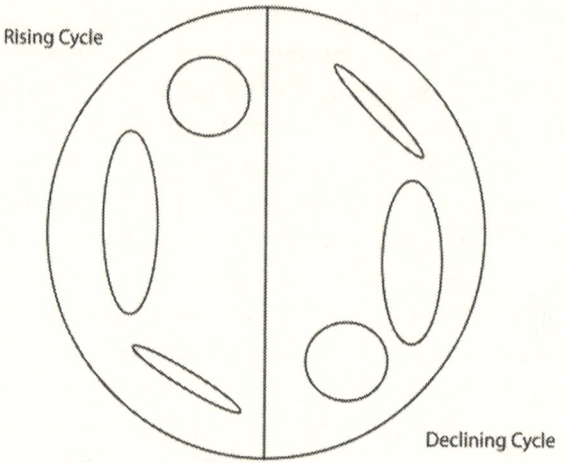

Figure 22. The Mirror Cosmos with Head Up for the Rising-Sun Sequence and Head Downward for the Declining-Sunlight Sequence

The spleen position in any model respects relations between the body length of three-yin and three-yang meridians even when these relations are superseded by the desire to unite the yin and yang partnerships within elements/phases. That is, spleen is followed by its arm partner, lung in the three-yin and three-yang sequence but is followed by its earth's yang leg counterpart in the standard five-phase model. The phase sequence described by the three-yin and three-yang sequence is shown in figure 16b.

Both the horizontal and vertical relations that develop with a starting point in *tai yin* spleen are credible in the same ways as those of the rising cycle they mirror. However, it is important to remember that the celestial stem mandate is the active mandate determined by the arrival and departure of the solar wave, and the three-yin branch structures are passive recipients. From this perspective, their structures indicate clear lines of continuous internal energy circulation and relations between anatomical position and elements, but the blocks and phases are more likely to be activated according to the will of heaven. If one is represented sequentially in a diagram, the other suffers and is suppressed in the representation, though both relationships function equally in nature and the body remains patent. Here, as in the case of the rising cycle, both the five-element sequence honors in a mirror image the vertical stem-based representation of the division by six while the three-yin and three-yang sequence bows to the plane of the horizon.

This is not the full story of the declining light cycle because the declining cycle is not only different in orientation but also different in kind. It is more yin than rising cycle with its vertical north-south orientation, and its directionality plunges to the horizon and earth rather than rising out of it. Therefore, the declining cycle is fundamentally tied to the horizontal plane and east-west direction in ways that the rising cycle isn't. Its functions gravitate to the middle of the body and deal with rest and digestion more than with action. The cycle progresses toward sleep, which again supports the implication of the horizontal, and this orientation is different in kind.

The basic three-yin and three-yang portrayal with the human standing vertically facing south as in figure 15 creates an element sequence of metal, earth, and water in the declining cycle. This structure makes physical sense because of the parallel positions of fire/heart and metal/lung above the waist as are their meridian corollaries. The heart occupies the center, and its meridian is medial when arms are down and lateral when arms are above the head with palms forward; the lung occupies the sides, and its meridian is lateral on the arm when arms are down and medial when the arms are above the head. Facing in any direction, the heart is medial to the lungs. The declining sequence in this model makes the entry-exit progression move forward in the standard pattern for the element orbs from lung to large intestine, stomach, and spleen. This provides a mirror reversal for the reverse order of the standard entry-exit sequence for orbs in the rising cycle that began with liver. However, the three-yin and three-yang sequence provides a forward motion that appropriately is symmetrically structural to left and right with the rising cycle: starting with liver and moving to its *jue yin* partner pericardium; then the cycle progresses in the normal entry-exit sequence of three heater, gallbladder; then restarts in a continuous cycle with liver.

The relatively yang partner meridian to metal *tai yin*, lung, is metal *yang ming*, large intestine. As the yang arm meridian in the upper body, it, nevertheless, refers to an organ located below the waist. This combines with two factors to suggest that at the very least for the declining cycle, the two sequences metal, earth, and water and earth, metal, and water should be considered simultaneously. Moreover, it is even more likely that the horizon-based earth, metal, and water should be considered the dominant pattern. The reverse order of the standard five-element sequence of entry-exit points for the rising cycle should continue in the declining cycle because this reverse order is consistent with the direction of the rotation of the earth, which in the diurnal cycle determines the sequence of the orbs.

The sequence of phases activated by the sun and the reverse terrestrial rotation indicated by the phases of the moon would be consistent with the entry-exit sequence if all orbs conformed with the reverse order when moving clockwise and conforming order moving counterclockwise around the cycle 2. The stomach meridian has the sea points on the shins that refer to the small intestine and large intestine. The metal point on the *tai yin* spleen meridian is called great white, the metal color with obvious reference to the metal aspect of the *tai yin*. These small indications show reciprocal metal and upper-body organ references on earth's leg meridians. These are there explicitly to accommodate the five-element sequence. These points allow treatment of upper body and metal conditions in the afternoon phase through the vehicle of partner meridian points, based on the downward orientation of the cosmic human. The position of the cosmic human in the declining cycle is with legs up and head down in accord with the horizon-based system of cardinal points and five elements. The vertical structure also is present but implicit and internalized.

The dominance of the horizon and earth in the declining cycle requires that given a choice, the system based on the horizon should dominate. The vertical and structural implications of the three yins and three yangs are present as indicated by the metal lung above. But this is an interior *zhang* organ influence, whereas the yang and obvious outer manifestation is below where the downward and outward eliminative function is natural and more fully manifest. This location is deliberately strange in a hand meridian that is located above the waist and refers to the afternoon and obviously carries cosmographical freight that has determined the association.

The Central Block

The vertical central block and its entry-exit mechanisms suggest an explanation of how the mandate of heaven continuously regulates the actions of the meridians in the rising and declining blocks. Moreover, these central meridians perched on the vertical axis have multiple close connections with the eight primary meridians that define quadrant and the spatial organization of the body. They seem to mediate between these eight oceans, which express basic relations between heaven and earth and the four directions and the particulars of meridians associated with bodily functioning.

From the standpoint of the five elements, the new model positions are hand *tai yang* monarch fire and hand *shao yin* monarch fire at noon and midsummer while foot *tai yang* water and foot *shao yin* water are placed symmetrically on the midline below at midnight. In contrast, the standard three-yin and three-yang positions and the positions shown by Porkert and those on the standard twelve-phase Chinese clock are asymmetrically placed in the late afternoon.

This differentiation confirms the absolutes of the vertical distinction between day and night as reflected in the body in the distances between the hands and feet when hands are above the head, and the prime meridian creates a central vertical axis in the middle of the body. The *shao yin-tai yang* meridians are continuous on the body on either side of this central axis, and when this principle is reapplied to the cosmos, the hand and foot functions manifest in the basic cardinal-point model of five elements and in the proposed twelve-phase derivation must be linked vertically in space but totally separated in time.

The meridian and entry-exit sequence in the central block for the standard five-element model would be in time for water to unfold as bladder/kidney and fire as heart/small intestine. The proposed three-yin and three-yang sequence in time would be the same reverse sequence shown in the rising and declining blocks, that is, kidney/bladder and small intestine/heart. The three-yin and three-yang internal circulation in the block of meridians would, as with the rising and declining blocks, begin with the *shao yin* foot meridian kidney and proceed to its structural partner *shao yin* heart, then in the standard sequence *tai yang* small intestine (fire, noon) and *tai yang* UB. Each phase would be activated by the solar wave, either at midday or midnight according to the proposed five-element orb sequence.

The concept that the structures of the body can be functionally connected, though at opposite ends of the body, by geometrical planes and that they are activated by the solar and lunar motions is a common sense of one in cosmology, but it may be unfamiliar to clinicians who are depending more on proximity of functions. This may have caused them to place kidney and bladder (water) functions next to their fire meridian partners in the late afternoon positions of the standard Chinese clock.

The role of the central block is unique. The celestial stem model will emphasize the sun and refer to *tai yang* fire as the primary reference point. The terrestrial branch models will inevitably start with water, earth, and moon as the primary reference point. These are two different starting points for evaluation of cycles and represent more absolute differentiations than dawn and dusk. Interpreting experience from the point of view of the heart is different from interpreting experience from the point of view of survival. A solar-dominated cycle may subordinate *shao yin* and *tai yang* water as a mere complementary partner to the sun and place it in the afternoon position, but a basic understanding of the four directions and the vertical importance of the opposition of heaven and earth as monarch fire and water in the division of six make this standard subordination philosophically untenable. *Tai yang* may begin at the height of noon or summer, but then being the great yang that encompasses all other functions, it necessarily descends all the way to the deepest level of yin and partners with it as *shao yin* water. Tradition meridian positions on the body recognized the integrity of this relationship, but the phase model that places the deepest yin, *shao yin*, that properly represents earth and midnight in the afternoon, loses integrity.

When viewed through the lens of the spatial three-yin and three-yang relations with the sequence starting as usual at the yin foot meridian appropriate to terrestrial branches, the sequence begins foot *shao yin* kidney, extends to its *shao yin* partner heart, and then descends through the small intestine to its partner bladder. In nature, this circuit symbolically encompasses the entire movement from midnight to midday and the descent back to midnight. As with the associated meridians, all other functions are encompassed by these. Perhaps, its directional imperatives were implied among those who knew, leaving others to flounder. In any case, the essential point is that the sequence, starting with the heart, is the same in the proposed and traditional clock. Starting with kidney is the ultimate traditional branch expression of the yin water aspect of the pairing dominating the whole cycle. Starting the proposed depiction with the heart at the midheaven or continuing with tradition to place water adjacent in the afternoon position forces the terrestrial cycle into conformity, however mangled, with the stem cycle progression of heaven. Forcing the system in that way, however, clearly strips the terrestrial cycle of its deep root in the earth, in darkness, in mystery, and in water.

Starting the cycle in water is a reversal of the solar order and by having kidney and heart associated with the rising cycle, means that they act as a break or counterforce to the rising cycle. Together, they embody yin and encompass the entire descending cycle and its drive to darkness and introspection. The same may be said of *tai yang* tracking with the descending cycle. This too implies a control of yin. Clinically in acupuncture, this is exactly the function of *tai yang*, to regulate the internal organs that all have corresponding points; exactly the function of *shao yin* is to regulate and restrain excess yang and rising heat in the body and to control even the ultimate extension of yang, the shen.

The dramatic feature of the central block, then, is its regulatory and counterbalancing role to the rising and descending blocks and their functional correlations. The central block has nothing to assert, has no directionality of its own; but it does seem to represent homeostasis in a fundamental way similar to that implied by the names of their implicit governor and conception vessels meridians on the centerline.

The structure and separate phase activation of the central block highlights that it is the solar wave and counteracting terrestrial cycle that activate points in phase. Otherwise, the circulation and counterbalancing within *tai yang* and *shao yin* must be continuous. This continuous internal circulation and connection happens through entry-exit points different from the standard entry-exit point sequences. If we start the sequence with kidney 1, the entry point on the foot meridian, it assumes that UB 67, the exit point, will exit continuously to kidney 1. Similarly, the rule that exit points must be anatomically near entry points suggests that kidney 27 is the appropriate exit point in the trajectory to its partner *shao yin* heart 1. From heart 1, the entry-exit sequence follows the traditional pattern—Heart, SI, UB—and subsequently to the continuous internal cycle of the three yins and three yangs.

The dramatic feature of using these points or of selecting points in phase would be the ability to create order and to counterbalance excesses of yin or yang, particularly in phase at the midday and midnight periods.

This cursory examination of the central block entry-exit points introduces the idea that some entry-exit points may be more critical than others owing to a confluence of internal and external connections. The idea that energy from the rising and declining cycles may regularly or continuously circulate through the central *tai yang-shao yin* combination is intriguing. The possibility justifies its two widely divergent and encompassing heaven-to-earth time periods in the circle, and it is interesting that the possibility emerges simply from its position between the rising and declining orbs and on the medial surfaces of the body and central, internal positions. The three-yin and three-yang system also introduced the vertical dimension to the central block, and this conception lies at the heart of the number six and its important distinction between ministerial fire and monarchial fire. The extension of this theory into the clock establishes the vertical midday-midnight axis as a single pivotal system that houses the concept of three dimensionality in a two-dimensional circle. In this case, the prime meridian (governor vessel and conception vessel on the body) is the medial line, registered as a line between time periods on either side of the midday-midnight axis.

Summary about Rising, Declining, and Central Blocks

Overall, the concept of three distinct blocks of vertically oriented meridians that works synergistically within themselves and fully projected as the total equilibrium within whole phases of cycles is a major contribution of the number six. The blocks that are out of phase are subordinate and implicit, which means that in phase the rising, central, and declining blocks are capable of maintaining the yin and yang equilibrium of the whole system. This makes the block partnerships of paramount importance for visualizing how to restore balance in patients. The various models for the entry-exit sequences for the rising and declining blocks will be studied in their turn in chapter 12, where individual meridians can be considered. The principal insights of the division of six are that it displays openly the vertical and structural consequences of the five-element phases. It defines the zones of the body in terms of the horizontal and vertical implications and forces the five elements into clear the symmetrical yin and yang relationships. Now you can identify a wood causative factor and know what plane of the body above and below and left and right that will treat it.

The blocks exhibit for the most part the compatibility with standard entry-exit point relationships once it is understood that the terrestrial cycle is fundamentally operating in a reverse direction and as a counteracting force to the celestial cycle. When the rising and descending cycles are taken together, the three-yin and three-yang portraits highlight some really significant alternative conceptions of entry-exit point sequences. Of greatest importance will be those points that link the three blocks to each other at the major transitions of the day and year.

From another perspective even though the rising cycle sequence begins with liver, to be continuous, the gallbladder 41 exit point at the end of the rising cycle must exit continuously to liver 1, which is the standard sequence. The only departure in the three-yin and three-yang anatomical model from the standard entry-exit cycle is the exit from liver 14 to entry at pericardium 1 for the sequence to work. This is thought provoking and more than justifiable on the basis of compatibility and proximity than are the traditional notion that liver 14 exits to lung 1, which is incompatible by element and more distant. The three-yin and three-yang one-meridian theory is behind it, and the two points are physically closer together.

The same considerations extend to the three-yin and three-yang portrait of the declining block sequence, which shows spleen 20, not 21, exiting not to heart 1 but internally in the block to lung 1, which meets the same standards of proximity and compatibility that we saw for liver to pericardium. As before, the last point in the foot-to-hand-to-foot cycle is stomach 42 exiting to spleen 1, which is the standard sequence.

Of similar importance is the recognition that acupuncture points may connect meaningfully with many other points but in an organized way according to regular cyclic impulses. With this in mind, the standard five-element, orb-driven, entry-exit sequence can be seen as just as valid and useful as the entry-exit sequence driven by the three-yin and three-yang anatomical structures.

The primary difference between the five—and sixfold yearly divisions was that the paired stems underlying fivefold division referred primarily to time and active energies, while the paired branches refer to structive or spatial organization whose basis is the four directions with up and down. The result of divisions by five and six is a slightly different configuration of element phases or periods. Fire is expanded in its seasonal presence and the other orbs contracted. However, the seasons are more symmetrical around the cardinal cross with its solstices and equinoxes and, therefore, more accurate as seasonal representations. However, the different time and space orientation reflect the primary dichotomy between the character of heaven and earth. Treated together, they represent a whole system, apart, of an anachronistic set of standards—either too spiritual or too mechanical. Like all odd even-number divisions of the circle, they carry an interpretation either of yang or yin respectively. Nevertheless, the movement from five to six describes not a shift in reality but a shift in point of view. The rising and setting of the planets, the moon, and stars were the great timekeepers of agricultural society, whereas the horizon and the directions were the great landmarks of space in a timeless continuum of experience. There can be no reasonable explanation that justifies a model of the sixfold terrestrial sequence that departs from that dictated by the heavenly cycle or the basic scheme of increasing and decreasing light.

IV. The Circuit Phases and Levels

The last section reduced to two fixed patterns of five elements derived from the celestial stems and six configurations derived from the terrestrial branches. These describe in outline the regular rotation in heaven, meteorological conditions, and transformations in nature associated with the growth cycle. Clearly, astrologers and physicians needed to go beyond the simple process of characterizing and predicting the regular seasonal movement in order to predict and explain variations that occur from year to year. One year may be colder, dryer, or rainier than another; and this begs explanation that is unavailable in the regular arrival and departure of certain fixed stars or lunations. This aim was identical to increasing interest from the third century to the early modern era of explaining variations in individual destiny for all those people born in a given year under the same stem and branch signature.

Some systems that attempted to account for variations from year to year created an implicit three-dimensional and real-world reference in their two-dimensional drawings, and this section will break these out into the open by relying on the basic vertical and dimensional differentiations of the celestial stems and terrestrial branches. Porkert displays early attempts to include variables into the standard cycle. His figure 14 shows stem/element and branch/configuration of the year in the uppermost midheaven phase of the standard three-yin or three-yang annual sequence (fig. 23). Astrologers dominated by the solar perspective viewed this as the controlling segment of the six yin/yang configurations. Figure 21 below imposes the annual branch signature from the sixty-year cycle in the upper segment. Its opposite terrestrial branch dominates the midnight-midwinter position directly below, leaving the standard sequence in place in the other phases in order to visually impose the unique impulse of the year and its likely seasonal disruptions that would ensue from its conflicts or harmonies with the regular seasonal pattern upon the regular cycle.[77]

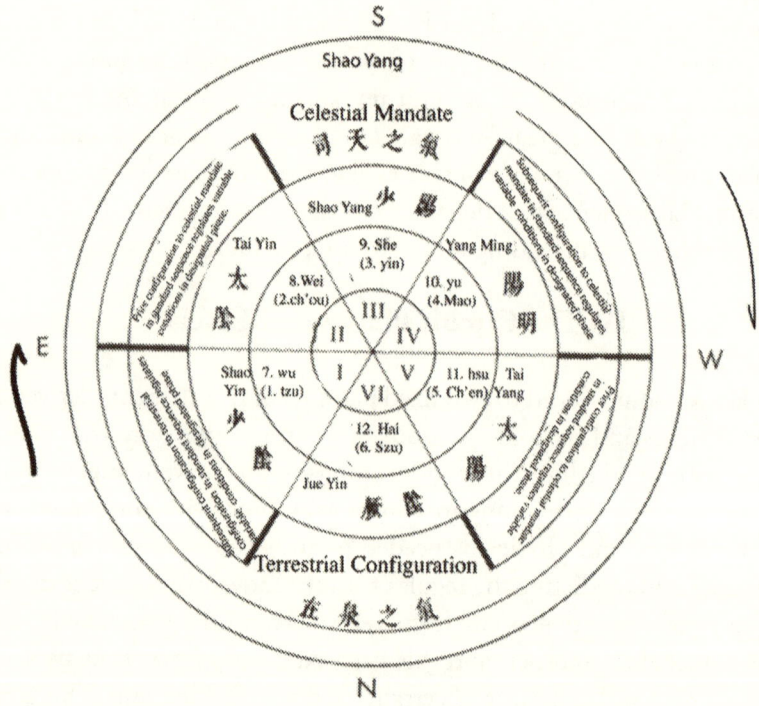

Figure 23. Figure of the six divisions (inner circle), the terrestrial branch combinations (second circle), the "fixed" cycle three-yin and three-yang assignments in I, II, IV, and V with the stem branch combined annual designations in III. (the midsummer, midday position) with the annual designation opposite in VI at the midwinter/midnight position in circle 3, and the discrimination of fixed cycle from the unique variable designations in circle IV

This type of intrusion can be confusing to historians who are not astrologers or practitioners of naturalistic medicine because it assumes without stating the regular movement of terrestrial branch and seasonal changes. This is an implicit two-tier system with the standard cycle on one level and the variant cycle imposed by the larger system on another. Again, the fixed cycle is implicit in such a drawing, not absent. The complex drawing that includes the stem and branch of the year in the heaven phase (midyear, midday) of variant explanation is better discriminated as part of a two-level system of influence that include in a drawing on one level the standard and variant motion (fig. 22). This applies to the celestial stems and to the terrestrial branches yielding a four-level system.

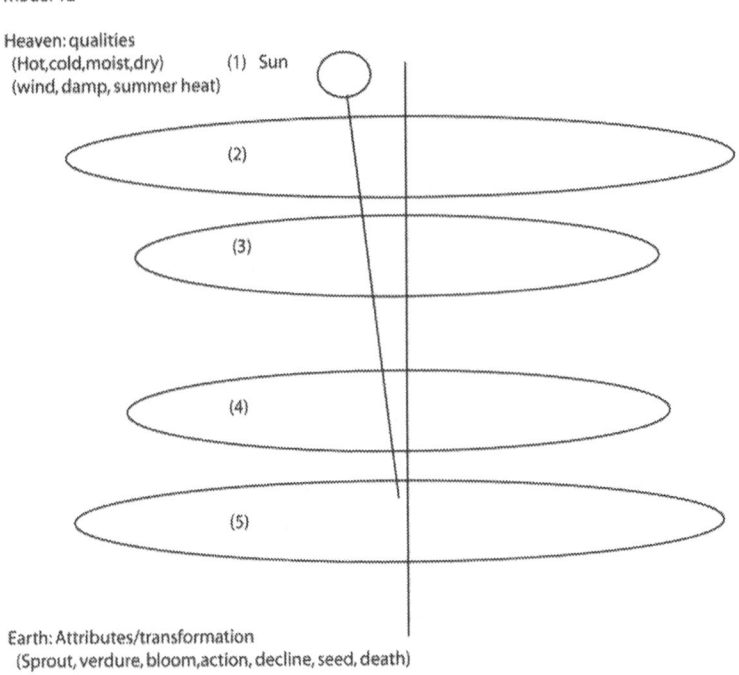

Model 12

Heaven: qualities
(Hot, cold, moist, dry)   (1) Sun
(wind, damp, summer heat)

(2)

(3)

(4)

(5)

Earth: Attributes/transformation
(Sprout, verdure, bloom, action, decline, seed, death)

Figure 24. The Four Levels, Two for Six Terrestrial Configurations and Two for the Celestial Stems, Each Expressing the Standard Cycle and the Variant Cycle

This representation of the terrestrial configurations can be clarified somewhat in a redrawn two-dimensional model if the steady rotation of the background celestial five-element model is included in the center, if the matching terrestrial configuration next to it, then by placing the variant impulse as an outer circle that can change its orientation though not its sequence from year to year. The elements and configurations of the variable wheel combine with the central two circles uniquely in each phase of the year figure. Figure 6 is a clearer set of relations than those shown in figure 21.

Figure 6 shows the six yin and yang configurations, but it shows them more directly in relation to the background cycle of celestial stems. This depiction is an attempt by traditional philosophers to show (1) the annual governing energy, (2) steady heavenly motions and cycles of the seasons that recur annually, and, finally, (3) the tiered secondary stem and branch cycles that account for variations. The details in these arrangements become very elaborate and as we see in figure 21, an attempt to show these three levels operating within the six-configuration model. Figure 21 shows in particular that in some models, the imperial configuration of the annual pair is imposed on stage three, south, heaven; and its opposite in the great cycle is imposed on the opposite side of the circle. The other standard phases fill in the remaining phases in their standard configurations.

The tiered three-dimensional figures 8 and 22 may be a better way to visualize the content of the imperial configuration because it shows each component separately. If we understand that the annual pair governs the axis and totality of the year and that this gives a specific bias that is most evident at the peak phases of the year near the summer solstice and winter solstice, some of the complicated elaborations become less troublesome. Figure 5 in chapter 12 shows the four levels broken out into detail in relation to the horizon. This five-tier structure with the admixture on the horizon in the middle, heaven above, and earth below, both fixed and variable provides another dynamic three-dimensional view of the relations between divisions of five and six or stems and branches. The models and approaches shown in these sections—3, 4, and 5,—all points to the need to see five elements in an intimate multidimensional partnership with the six configurations.

Section 5 sets the stage for later more detailed discussion in chapter 12 of relations between blocks, timing of treatment according to phase activation, relations between the clockwise motion of the sun, the five elements, and the variable terrestrial cycle for the year as opposed to the implicit counterclockwise motion of the fixed terrestrial pattern. In this section, it is sufficient to recognize that treatments are needed to be adapted to the unique environmental conditions of a given year with ready recollection of the basic reliable movement of the elements during the year.

The six configurations allow us to translate seasonal change into specific meridian treatments on appropriate levels of the body. This can only be done if it is understood that the blocks of meridians represent the total equilibrium of yin and yang when they are in phase. Their roles contract correspondingly when out of phase. This means that in cosmographical treatments, timing is everything that it will allow the manipulation of whole systems of relationships during certain periods through certain congruent vehicles but not otherwise.

The counterclockwise arrangement of standard entry-exit acupuncture points hinted that the three-yin and three-yang terrestrial cycle has an incipient counterbalancing role to the solar cycle at the same time that it conforms with it. When the horizontal pressures in the structure flips the rising and declining blocks, this to seems to have the aim of creating congruence between the sun's arrival at some point of the shen cycle and cosmic human that also activates the terrestrial counterpart in sequence. The standard clock fails to conform to reality because it reverses the block sequences to force the counter cycle into conformity with the appearance of a dominant clockwise solar motion. The depiction unfortunately is treated as a rigid law operating on a two-dimensional plane.

The concept of a clockwise celestial motion that conforms to the wu hsing creative cycle that simultaneously works with a counterbalancing counterclockwise terrestrial cycle is simple. The same cyclic portrait of the cosmic human appears to rotates in two opposite directions so that all functions continuously counterbalance each other in all aspects: upper to lower, left to right, and inside to outside. In reality, it is simply a product of the earth's rotation that makes the sun appear to move in one direction that at the same time it logs a steady progression along the ecliptic in another. The sun moving east to west casts a shadow on a sundial that moves west to east. Nevertheless, beyond the fundamental recognition of balance, this conception is three dimensional, and it becomes difficult to represent logically or visually without breaking it down into essential components. This, it would seem, is the basis for most misunderstandings of Chinese cosmology and its root order imposed on acupuncture physiology.

The proposed model is the result of careful presentation above of different projected models for the rotation of the twelve terrestrial cycles in comparison to the root's seasonal cycle based on a three-dimensional conception of clockwise celestial and counterclockwise terrestrial apparent motion. These concepts are enshrined in the differentiation of stem and branch numbers, and they play out directly in the differentiation of wu hsing and three yins and three yangs. The important variations between model-examined gaps between traditional models that emphasized are (1) phase evolution (time) versus anatomical (space) sequencing, (2) the difference between the functional imperatives of rising yang versus increasing yin, (3) the essential character of the horizon in the east as a yang rather than yin point of transition and west as a yin rather than yang transition, (4) the functional role of the vertical axis as a passive hinge and an active phase, and(5) the discrimination of three-yin and three-yang orbs into three blocks of two, one rising, one transitional/central, and one descending with implications for the division of the central block into two connected but temporally separated functions.

Analysis of these structural alternatives led to some basic questions and insights into how acupuncture points and meridians function. It was proposed that (1) acupuncture meridians and points are activated in phase sequences by the sun in the daily and seasonal cycle, and the moon in the lunation cycle; (2) acupuncture meridians and points are activated by structures according to the dominant phase of cosmic organization at any point in time that is the east and west horizon on the one hand and the midheaven and nadir axis on the other, which in their turn are conditioned by planets that occupy those positions during a particular phase; (3) the meridians and points are activated individually in phase but also with their immediate three-yin and three-yang partners as blocks of meridians during the greater rising (*shao yang, jue yin*), central (*tai yang/shao yin*), and declining periods (*tai yin, yang ming*) of any cycle; (4) the solar and lunar waves activate meridians independent of the entry-exit patterns of meridian interconnection, and the wave of activation is accessible through any point on the meridian but particularly the horary and source points; (5) exit points of meridians can exit to more than one entry point and that exit relations have both phase and proximity considerations that determine relations between meridians; (6) the entry and exit points to blocks of meridians have overarching significance for the periods they represent; and (7) blocks of meridians have a continuous internal circulation that is active in phase and quiescent when out of phase. Blocks of meridians may be accessed through different entry points at different phases of the greater cycles. These insights will be developed below. They have significance for astrology because the relations in the body can be projected as a means of clarifying apparent conflicts between different cosmic models and significance for medicine by offering a more powerful and adaptive theory and practice that resolves contradictions inherent in present theory.

V. Additional Rationale for the Proposed Six Configurations Model

One solid physiological foundation to the propose three-yin and three-yang sequences that aligns with the five elements derive from the traditional Chinese medicine (TCM) study of pathogenic processes as they progressively invade the body in regular ways, often in the sequence *tai yang* (small intestine, bladder, head, neck, upper back), *shao yang* (three heater, gallbladder, head neck shoulders, diaphragm, to *yang ming* (large intestine, nose, eyes, throat, chest), and then stomach before they penetrate to the interior or yin of the body to become chronic disease. The relations between yang and yin in the defensive perimeter of the body reinforce the idea of yang above and yin below, *tai yang* as perimeter, *jue yin* is medial, and *yang ming* as central that we have seen as a pattern above. The typical sequence of pathological insult through the yang meridians show the same essential relationship—exterior to interior, yang toward yin—that is apparent in the macrocosmic relations where the increase and decrease of heat, wind, damp, and so forth also follow regular patterns in relation to each other. In humans, typical live, active daytime patterns give way to passive and digestive nighttime patterns.

Moreover, the meridians show some discernable excitation in their horary period. Symptoms that are associated with meridian functions frequently amplify or diminish during the associated two hourly diurnal periods depending on the relative repletion or depletion of the system. Blockages at the entry and exit points of meridians are important to clear, and these blockages are evident on pulses and by increased appearance of symptoms at the points of temporal transition. People worldwide eat their primary meal in the evening, a period associated with relaxation and therefore digestion, whereas people worldwide rise in the morning, eat a secondary meal, and plunge off into activity, conquest, and fight-and-flight behavior. These natural human responses follow the pattern of the diurnal cycle. This pattern is the same as the larger seasonal pattern where people work to grow crops in the summer and consume them in the winter. This model satisfies the standards established by William Lowe for analysis of the validity of relations between the five-element correspondences and the derivative six configurations as functional categories.[78] He asserted that owing to the macrocosmic to microcosmic one-to-one relationship that the balance between functional qualities of arm, upper body (heaven) and leg, and lower body (earth) meridians must be balanced according to the principles of five elements. That is, fire element and heat in upper body meridians must be balanced by the macrocosmic qualities associated with the cool and damp characteristics of their leg meridians. The qualities of the proposed six configurations balance each other as follows:

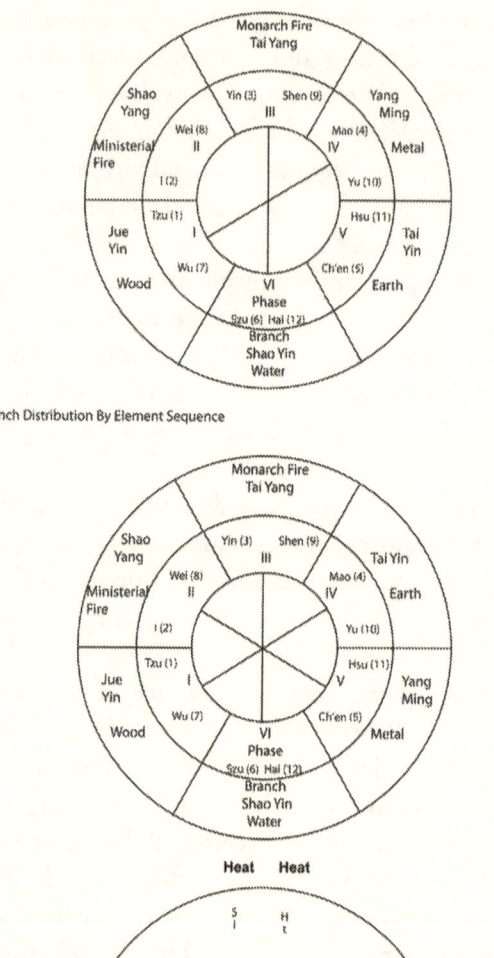

Figure 25. Relations between Five Phases Qualities Associated with the Six Configurations, Counterpoint to William Low

When aligned with the proposed macrocosmic pattern, Lowe's diagram takes on new life and clarified meaning. The three-yin and three-yang vertical meridian pairs show simultaneously as bodily and macrocosmic conditions rather than simply as a set of balanced, abstract qualities. The result is a better understanding of the mechanism of balance in the body and between the three yins and three yangs.

First, each yin is paired directly with a yang counterpart—one left, one center, one right—in the same relative positions as the bilateral meridians on the body and in the same pattern as the rising, central, and declining phases of the day/year. Clearly, the yang-to-yin relation between these meridian pairs creates a fundamental balance between one yang quality associated with the yang meridian and one yin quality associated with the yin meridian as in table 3.

Table 3. Proposed Alignment between Three yin and yang Orbs their Hand and Foot divisions and meteorolgical Associations

Hand/Foot	Lesser yang	Jue Yin	Tai Yang	Shao Yin	Yang Ming	tai yin
Hand	Moist Heat	Thunder	Hot	Hot	Dry	Dry
Foot	Wind	Rain	Cold	Cold	Damp	Damp

These single qualities reside primarily above for the yang quality and primarily below in for the yin quality, but the influence permeates the whole length of the body and macrocosm. Thus, in the left-hand pair, the fire/heat dominates the yang, while the controlling moisture of spring and rain, associated with the wood, regulates the yin orbs. At the same time, upper-body meridians and orbs display relatively more heat and dryness and relatively less moisture and cold than lower-body meridians. The combination of vertical and seasonal pairings yields a three-dimensional characteristic to the qualities that is not present so obviously in the four-directional model and its five-element counterpart.

Shao yang contains the three-heater meridian: an arm meridian associated with the fire element and the gallbladder meridian associated with wood element. Wind is commonly associated with the wood element and secondarily as a form of pathology. However, wind also describes movement and change, and it is one of Chi Po's six celestial configurations. It has a natural affiliation with the gallbladder and its yang (in the yin to yang partnership of liver to gallbladder) associated functions. Thus, the qualities balancing each other between the three heater and gallbladder are Lowe's heat and wind respectively. Liver also has wind associations, but in table 3, the emphasis is on moisture in a relatively more active frame of reference than that of the winter-water element. In Chinese medicine, it is the liver that stores and rejuvenates the blood; and if blood is patent, wind is held in check. The kind of moisture referred to here is that of the sap in vegetation with its analogies to blood in humans that life-giving moist nourishment that strengthens the plant or human. Thus, the qualitative balances are maintained, both yang to yang and yang to yin, in a way that is congruent with the principle signature of the season of spring (wind, rain, growth [movement and change], and nourishment) and the real conditions observed by TCM practitioners.

The second major contribution of the new diagram is its ready handling of the pairing of qualities within a single yang or yin orb. In this case, it indicates *shao yang* to be a pair whose dominant character is increasing yang both of the rising day and rising year, which begins as life and movement and culminates in heat. So, logically, heat is the dominant quality of the pair as in figures 5-3 and 5-8 is a secondary quality that to some extent contributes to and counterbalances heat depending on the degree of its intensity. Spring and fall both generate wind as the changing temperature of the atmosphere contests with the residual heat or cold in the earth. Wind is a factor in the autumn and contributes to the drying, dispersing, and descending of chi; but it operates as a secondary counterforce to the predominating yin.

The yin-paired meridian shares the same polarization of fire and wood in the pericardium and liver orbs as its yang counterpart. Lowe applies the same polarized qualities to both pairs, but the new drawing makes it clear that the entire yin lower-side and leg-meridian complex has a center of gravity in yin and water. The yin deals with storage, the yang with circulation—generally of fluids. Thus, *jue yin* generally, and the liver in particular, need to be appreciated for the braking action that it exerts on the process of rising day, year, and bodily heat. It contributes to the springtime mobilization of energy but gives it executive power because it adds a directional and controlling component to the otherwise rapid ascent and dispersion of fire. That is why I attribute the concepts of moist heat, of springtime thunder, and nourishing rains with *jue yin*. The fire element associated with liver's yin partner, pericardium, still has heat as its prime quality; but this heat is moderate relative to the surging heat of the yang partner three heater. The outcome of this second kind of insight is table 4. Each meridian can be seen in its total set of relations, which allows for a more refined qualitative judgment of its relative balance to its counterpart within an orb, its yin or yang partner, and partner orb, and the whole system of rising or falling yang and yin.

Table 4. Element and Qualities of the Three Yins and Three Yangs

Configuration	Arm/leg	Meridian	Element	Quality	Energy
Shao yang	arm	Three Heater	Fire	Heat-strong	Rising
Shao yang	leg	gallbladder	Wood	Wind	Rising
Jue yin	arm	Pericardium	Fire	Heat-Mod.	Gentle rising
Jue yin	leg	Liver	Wood	Thunder	Gentle rising
Tai yang	arm	Small Intestine	Fire	Heat—	Strong Ascending
Tai yang	leg	Bladder	Water	cold	Descending
Shao yin	arm	Heart	Fire	Heat	Mod. Rising
Shao yin	leg	Kidney	Water	cold Mod.	Descending
Yang ming	arm	Colon	Metal	Dry	Descending gentle
Yang ming	leg	Stomach	Earth	Humid	Descending gentle
tai yin	arm	Lung	Metal	humid	Descending
tai yin	leg	spleen	Earth	Damp	Descending

Placement in the diagram explains by position in the cycle the common clinical knowledge about the ascending and descending functions of the corresponding meridian/organ orbs.

Of particular interest is the gentle or reluctant rising energy of the *jue yin* in counterpoint to the gently descending energy of the *yang ming*. Yin (water) naturally descends; yang (fire) naturally ascends, yet in the spring, when yin either is temporarily overwhelmed by the power of rising yang and heat or when it simply adapts and moderates its influence, it counteracts the increase of solar radiation with inertia, drawing on the reserves from the preceding winter of moisture and cold in the earth. Similarly, that can be called sunset yang because of its western, evening, and autumn associations in this model; or *yang ming*, expresses the yang energy in a circumstance where yin is gaining the upper hand, literally, in the character, framing the yang like a window viewing it from an interior yin space. In this case, the descending function of the yang is a kind of spent yang just as dryness is the end product of heat; and as yang descends forced along by an increasingly dominant yin, it first reveals its greatness at the point of greatest contrast and tension with that yin and after the crisis (of sunset and autumn) then sustains a form of interior or introspective activity even when external capacities either are invisible or limited. An internal human activity is thought, and it is this internalization of yang that lends to the spleen and stomach meridians a traditional association with the spirit of contemplation or yi.

When placed in the normal element sequence of the five phases from which they emerge, the six-paired terrestrial branches yield a portrait in which all the ko cycle relations of the five-phase model yield oppositions in the six-phase model. The ko cycle movement is one way as follows:

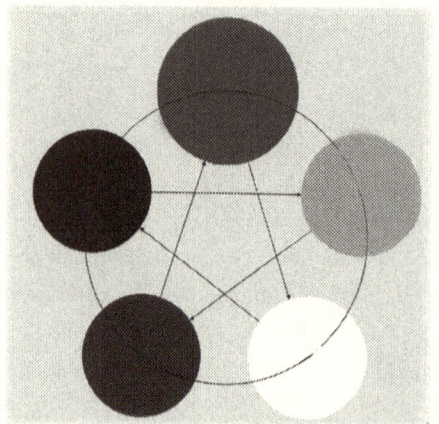 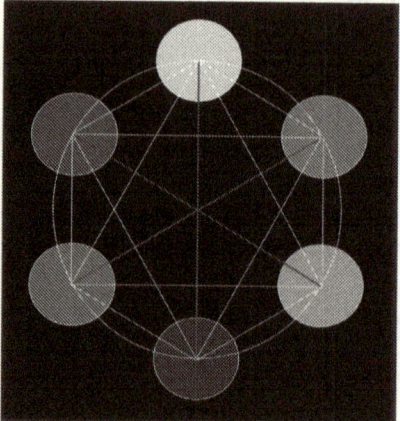

Figure 24. Ko cycle Relations in Five- element Phases and Six Configuration Oppositions

Table 5. Control (Ko) Sequence

Water	controls	fire
Wood	controls	earth
Fire	controls	metal
Earth	controls	water
Metal	controls	wood

When combined with the counteracting cycle, the ko cycle relationship, like all relationships in nature, must be considered a two-way action. Thus, the five-phase, two-way control relations from the ko cycle yield the following ten relationships in five pairs.

Table 6. Two-Way Relationships Inherent in Ko Cycle

Water	—	Mon. Fire
Wood	—	Earth
Min.Fire	—	metal
Earth	—	Water
Metal	—	Wood

Either of the two models in figure 24 produce the two versions of the six configurations dominant element in each configuration with the second element implicit.

The oppositions yield six dominant relationships in three two-way pairs, four dominant to implicit, and two implicit to implicit relations.

Table 7. Oppositions in Natural Order of the Three Yins and Three Yangs

Dominant		
Water	—	Monarch Fire
Wood	—	Earth
Minister Fire	—	Metal
Dominant to implicit		
Ministerial Fire	—	(earth)
Metal	—	(wood)
Wood	—	(metal)
Earth	—	(fire)
Implicit to Implicit		
(wood)	—	(earth)
(fire)	—	(metal)

These six dominant relationships all are ko cycle combinations. Metal and fire now have a dominant, dynamic, two-way oppositional structure similar in significance to water and fire, wood, and earth. The dominant relations express the essential triune balances between ascending and descending energies on either side of dawn and dusk and the equinoxes on the one hand and the absolute balance of above and below found at noon and midnight and the solstices. This dominant pattern in turn will become the basis for reevaluation of the Chinese clock.

The dominant-to-implicit relations reduce to metal to wood, which also is a ko cycle relation and ministerial fire to earth, which is a departure. Fire is the mother of earth in the five-element model, not in a ko cycle relation. The implicit-to-implicit relations of this opposition simply mirror oppositions available in the dominant pairings. This additional opposition was created by the special extension of fire in the seasonal division from one fifth in the five-phase model to two sixths in the six configurations. The upper-to-lower-body relations within an anatomical portrait of the cycle, however, underlines the importance of this opposition. While metal dominates in phase, earth nevertheless is present in partnership with it and fulfills the needed counterbalancing action on the terrestrial level even though it is not the celestial standard in phase. This augments the exterior to interior relation between ministerial fires as the dominant element in phase, to earth, which is the subordinate element in phase, completes the balance, and symmetry of the opposition.

The relation between spleen and pericardium in matters having to do with regulating digestion and descent of nourishment now can be seen not simply as a mother-son relation but as a counterbalancing expression of late spring to late fall and between the interior controls of pericardium (mild descent) on late spring upsurge and the interior control of spleen (mild ascent) on the late-fall decline to dryness and cold. The relationship of spleen to pericardium shows a synergistic balance between internal regulating functions that assist each other to counteract the actions of their dominant in-phase partners. Indeed, the directionality of spleen is similar to the sum actions of pericardium with three heater, a gently ascending chi, just as pericardium may be regarded as acting in a way similar to the sum of stomach and spleen, a mildly descending harmonizing assimilative chi.

The further differentiation of fire between qualities of monarchial and ministerial fire allows this to happen without departing from conformity with nature. The seasonal and anatomical opposition between fire and earth registers an important way that stored internalized heat and cold (fluid/chi balances) interaction with each other. This reflects the mechanisms that allow humans to gradually adapt to their internal thermostat to greater heat and greater cold; this process includes digestion, storage of fats and rates of use, and storage of sugars. This new ministerial fire may be seen as an extension of the wood element and the pair as a subsidiary set of forces that regulate the ascending and descending energies evident but not explicit in the five-element scheme. If the implicit (in parenthesis) weaker partners are considered as most ko cycle relations, they are expressed as oppositions in the proposed six-phase picture.

The laws of the five elements are greatly clarified by their depiction in a six configurations format because it brings them under the more general umbrella of yin and yang and the dualistic philosophy that governed almost all traditional philosophies whether east or west. Aristotle, like Lao Tzu or Confucius, was convinced that all motion are derived from the dynamic opposition of qualities in relationship. It may also be that in the six-phase system with its focus on the vertical dimension that an important elaboration of standard dualisms develops beyond the more static cardinal points of the four season models that had earth and horizon as its focus and center.

The first gain is the possibility of including the upright human figure and the vertically oriented functions of flows of chi and fluid into the cosmic scheme. The six configurations clarify what always was implicit in the water-to-fire opposition associated with north and south, whose limited meanings extended into a linear, nonevolutionary, five-phase models.

VI. Medical Implications of the Six Configurations

This section can only summarize what in the final analysis must be considered as a complete system of treatment based on the six configurations rather than the five elements or eight principles. I will treat some of the highlights in sequence:

1. The principle strengths of this window on the body results from the ability to view the body structurally as well as functionally. We can break past the element points on meridians to an understanding of the relations between whole planes of the body and the five elements. The topography of the body and the affected zones literally describe the condition and these positions link directly to the elements and the relative balance of hot and cold and moisture and dryness in them and their meridian corollaries. a. Now to treat a hot or dry condition in a particular organ and element, we can draw on the lower body partner to compensate where before in a five-element configuration, we would perhaps select the relatively yin points on the primary affected meridian. b. If it is springtime or morning, the whole trajectory of *shao yang/jue yin* can be harnessed using leg meridians or hand meridians according to phase and using local points that match the level of the sun in these larger cycles: in summer treating above or compensating below on the legs in complementary trajectories—*jue yin/shao yang*, late spring; *shao yang* and *tai yang*, central noon or midnight; *tai yin* and *yang ming*, declining waning phase of the moon; and all such combinations.

2. Six uniquely differentiates the middle from heaven and earth in a way that creates a new fundamental polarization between wood and earth, which is markedly different and more localized than the more basic differentiation of wood from metal and moisture from dryness found in the division of four. Wood remains associated with rising activating energy, storms and

moisture, earth with calming, equilibrium, clear differentiations and dryness; but the differentiation is more localized and intertwined. Mist rising from the earth on a clear night or morning and clouds dropping rain on an overcast day are gradations of earth and wood respectively of relative dryness and moisture, not of heat-induced desert dryness and water. Dampness now can be differentiated from actual rivers and streams in the lower *jiao*, soggy earth, mist and rain in the middle, and clouds above. The verticality of the system places the dampness in the middle first as a more limited equilibrium and second as a process that can be treated in the springtime one way and in the autumn another or a process that can be treated by drawing on the drying capacity of upper-body meridians or the moistening capacity of lower-body meridians that are associated with the particular problem in the middle. The opposition still reminds the practitioner that relative dampness will tend to result from liver metabolic imbalances and dryness from spleen and stomach imbalances. Treating the gallbladder may do more for dampness and fat/sugar metabolism than treating the spleen.

3. Ko cycle relations are elaborated into oppositions: a. The key advantage of oppositions is the ability to translate them into pattern treatments that cause balance. Operating as three yins and three yangs, the oppositions are structural oppositions, both horizontally and vertically on the body as well as being oppositions of principle or element in the seasonal cycle. Thus, element conflicts can be resolved by strengthening one side or both of the element conflict and the body at the same time. If monarch fire is weak, it can be supported by drawing on its partner yin or yang arm meridian drawing from its opposite horizontally—the most lateral *tai yin/yang ming* meridians or vertically its yin yang partners or both *tai yin* and *yang ming* on the leg as before but also now on the leg outside yang or inside yin as appropriate. b. The ability to use ministerial fire to balance metal problems and the reverse is a new, profound treatment insight that harks back to yet clarify the original wood-metal opposition of the cardinal points. Emotions from ministerial fire pericardium can stagnate fluids and create damp accumulations in the lungs while the dryness and heat of three-heater imbalances may cause disruption chi exhaustion and dryness. In reverse, the lungs whose spiritual perspective is generally its weakest or most incompatible with the late spring with its unbounded physical growth and accelerating heat, can over control and create too many dry clear days, can bring too self-conscious meaning or restrictions, can be too cool, too mature, too cynical for the spontaneous

expansion of the *jue yin/shao yang* hand meridians. Supporting the fire may break this inhibition. The reverse is true—that all the life and activity can exhaust the metal, leading to illness and collapse in the fall. One-to-one structural connection between elements and body allow translation into points. This means that an earth causative factor can be differentiated into yang stomach or yin spleen for greater specificity:

1) If the problem is psychological and mental, it can be treated on the upper torso, head or hands and feet; if adaptive, the middle torso, forearms and calves; if having to do with instincts, ambition, and reproduction, and the lower torso.

2) If the person is thin and active, direct support can be delivered on the well-worn and overused stomach meridian, or the yin yang partner spleen can be chosen to counterbalance. If heavy and lethargic, the yang partner can be selected, or to activate further, the yang hand partner can give leverage; or to create a gentler movement toward a healthy equilibrium, the gentle hand partner yin meridian can be selected with local points.

3) If the treatment occurs in the morning, the time and space correlations can be used as corresponding wood and fire points on the spleen meridian, or the partner lung meridian can be used; or the earth points on the liver, gallbladder, pericardium, and three heater can be selected in phase. If the season is the spring season, the same principle applies. If spleen is particularly weak and it is springtime when yang is accelerating, this energy can be deflected to support *tai yin* by treating in the afternoon, by using wood points on the *tai yin* meridian; or if pulses indicate excess, the wood upsurge could be controlled through the wood point on the *tai yin* lung/metal partner meridian, and strength could be drawn from the reservoirs of the water meridians, which represent the earth in its deeper differentiation of heaven and earth to reinforce treating the source and earth points with points on the spleen that will help it to conserve its resources and control wood and fire, the water and metal points, the earth points on the earth meridians.

The point is that any treatment target can use seasonal or parallel or opposing points by function or structure on any appropriate point along the whole trajectory of the body, particularly using those larger trajectories of the rising, central, and declining orbs active during a particular period.

4. Differentiation of the three blocks, which allows for treatment of whole body through block windows in season and time of day but identifies larger circuits and key three-yin and three-yang entry-to-exit relations. Of particular importance are the entry-exit patterns revealed by the block of three-yin and three-yang meridians, which will be explained in chapter 12. These allow for the opening of entire circuits, which can be applied to any local problem in a related topographical zone or body plane: medial, central, or lateral.

5. Principal of three is elaborated meaningfully in relation to the arms and legs as the meridians—central, medial, and lateral—reflecting the three zones of heaven and earth, anticipating the division by eight. The principle of three allows a determination of level to be made by asking, "Is it a problem above, middle, or below?" This can result immediately in the leg or arm, partner decision being refined with relevance to hand, forearm, or upper arm depending on the physical target. Every element now has a housing on a particular level. Heart, fire, is on the upper body; and it can be counterbalanced, therefore, on the hands and feet points for congruent treatment, foot points for equilibrium, or upper legs and arms and with more emphasis on the leg meridian points for a counterbalancing treatment.

6. Pulses. The division of six is the basis for the pulses of two depths in three positions bilaterally. The chi and blood levels represent the specific yang and yin organ partners of the three yangs and three yins. This reflects in reverse the basic back-to-front relationship that distinguishes the three yangs from the three yins. Similarly, the left pulse reflects the organs on the right side and the right those more the left. The left pulse contains the extreme polarity of monarch fire to water and describes the medial and vertical movement of the body, while the right side pulses reflect the horizontal emphasis of *tai yin* with its bilateral lung and lateral position of meridians. These reversals and relations between structures and functions are congruent—the interplay of time and space, structure, and function—that we have seen theoretically above. The pulses otherwise reflect the purely structural Fu Hsi organization and a one-to-one relationship between vertical position and organ. The colon is proximal to the stomach, and the gallbladder is proximal to the liver as you would expect anatomically. Once again, from the perspective of the six divisions, the vertical heaven-to-earth relationships are solid and clear; but the east, west, right, and left functions trade positions, and are not fully differentiated bilaterally, owing to the focus on basic zone structures and functionally driven differentiations.

The three yins and three yangs are important structures to consider in all acupuncture treatment. Fundamentally, these reflect in great detail the spatial symmetries on the body that can be used to augment acupuncture treatment. Partner meridians follow the same trajectory and plane of the body both above and below and on arms and legs. This means that they will work synergistically. When treating above, always consider below treating the same trajectory or its yin/yang partner. Also, consider the oppositional relations of metal or earth to wood or fire to create equilibrium. Consider which trajectory yin or yang or rising cycle versus declining cycle or the middle equilibrium is involved. Create balance in that orb, rising, declining, or central; and you have a major diagnostic key and method of resolution.

A very important key to using the three yins and three yangs is to remember the key transitional points, the entry-exit points to these cycles, particularly those between whole blocks, notably the importance of heart 1, kidney 1, and the three exit points from yang, namely, gallbladder 44, stomach 45, and UB 67, particularly UB 67. Other transitional points include the exit points of the three yin leg meridians, SP 20, LV 14, and KI 22/27 with the exit of the three hand yin meridians (HT 9, PC 8, LU 7) to the principal hand yang point si 1. These points when properly used can create equilibrium between whole systems—rising, central, and declining—and have wide effects on symptoms. So next time you use KI 1, think of augmenting it with UB 67. Next time you think of PC 8 or 9, think of using SI 1. There is no better way to increase the power of a standard treatment than by using its hand or foot, yin or yang partner, which inevitably creates balance in a whole system.

Chapter 7 七

The Number Seven, Lunations, and Time

Seven offers a special insight into whole number divisions of the natural cycles. More complicated than the number five, it seems obscure and mysterious and, at first brush, to have little possible clinical significance. This deficiency is owing to the close connection of the number seven to cosmology and to the use of lunations in clinical practice, which has diminished in the last three centuries. Chi Po stated above, "The manifestations of change consist in the perpendicular phenomena of the heavens and the physical shape of the earth (horizon), so that the seven planets (sun, moon and five visible planets) are moving according to the longitude and latitude of the cosmos and the five-elements are manifesting themselves on earth."[79] This intriguing quote suggests that the five elements themselves are the consequence of the movement in heaven, not the cause. This means that the bodies in heaven are the real reference of the stems, not the five-element transformations. For Chi Po, the basic significance of seven is in its association with the five planets and the sun and moon.

As a yang number associated with time, it falls between six and eight, which are the two spatial numbers that develop the structural implications on the human body of the vertical dimension. The vertical dimension is comprised of heaven and earth. If we ask the question, How would the principles of the number six play out in time? The immediate answer is that it would play out in very specific day-to-day terms having to do with the number seven. Chi Po described the six specific meteorological changes of the three yins and three yangs (wind, cold, summer heat, dampness, dryness, fire) and the six terrestrial transformations (metal, wood, water, earth, monarch fire, minister fire). Lingering in acupuncture theory are the seven externals (wind, cold, rising or flaring heat [fire], summer (heat, humid, dryness, moist), and seven internal causes of disease (restlessness/passion, fear, anger, excess joy/sadness, sympathy/worry, grief, attachment or anger, joy, anxiety, thought, sorrow, fear, fright[80]). These bear a close resemblance to one another and to the six internal and external causes identified with the three yins and three yangs. As the seven internal and external causes have lingered in the curriculum longer than the elements and meteorological qualities of the three yins and three yangs, this suggests that the principle of seven had more day-to-day usage in traditional clinical practice.

The only explanation for this is that the number seven is tied universally to the seven-day week (fig. 1). Each celestial body had governance of one day in the cycle, and all the associated meanings both meteorological and personal of the planets are retained in the names for the days. Moreover, weather patterns do occur frequently on a seven-day cycle that is associated with the moon. For periods of time in a given three-yin and three-yang phase, it can rain every Thursday or be cloudy every Saturday. Agricultural societies were organized by the phases of the lunations, and small differentiations about labor between different days in different lunations could make a big difference in crop yields and were, therefore, carefully tabulated. The seven-day week is associated with the larger four-week lunation, and the general meaning of lunar phases were imposed on the repetitive pattern of planetary associations. In every tradition, six provided the basic structure for applying element theory, but the added "day of rest" allowed for the perfect application of the specific meanings to the annual cycle.

	☉		☽		♂		☿		♃		♀		♄		
Night hours	Sunday		Monday		Tuesday		Wednesday		Thursday		Friday		Saturday		
Day hours	Day	Night	Day	Night	Day	Night	Day	Night	Day	Night	Day	Night	Day	Night	
First	☉	♃	☽	♀	♂	♄	☿	☉	♃	☽	♀	♂	♄	☿	1
Second	♀	♂	♄	☿	☉	♃	☽	♀	♂	♄	☿	☉	♃	☽	2
Third	☿	☉	♃	☽	♀	♂	♄	☿	☉	♃	☽	♀	♂	♄	3
Fourth	☽	♀	♂	♄	☿	☉	♃	☽	♀	♂	♄	☿	☉	♃	4
Fifth	♄	☿	☉	♃	☽	♀	♂	♄	♀	☉	♃	☽	♀	♂	5
Sixth	♃	☽	♀	♂	♄	☿	☉	♃	☽	♀	♂	♄	☿	☉	6
Seventh	♂	♄	☿	☉	♃	☽	♀	♂	♄	☿	☉	♃	☽	♀	7
Eighth	☉	♃	☽	☿	♂	♄	☿	☉	♃	☽	♀	♂	♄	☿	8
Ninth	♀	♂	♄	☿	☉	♃	☽	♀	♂	♄	☿	☉	♃	☽	9
Tenth	☿	☉	♃	☽	♀	♂	♄	☿	☉	♃	☽	♀	♂	♄	10
Eleventh	☽	♀	♂	♄	☿	☉	♃	☽	♀	♂	♄	☿	☉	♃	11
Twelfth	♄	☿	☉	♃	☽	♀	♂	♄	☿	☉	♃	☽	♀	♂	12

Figure 1. The Western Planetary Rulerships of the Seven Days of the Week in Twelve Two Hourly Periods.[81]

The number seven expanded on the cyclic divisions and meaning of the five planets and their corresponding earthly elements by adding the sun and moon. This admission accomplishes for the celestial bodies what already has been achieved for the terrestrial branches in the number six, the inclusion of the vertical and overarching *tai yang* and *shao yin*, monarch fire (sun, yang) and extreme yin (moon, water), that established the vertical in its yang and yin aspect. The sun and moon in the number seven act as equal planetary bodies, but they carry with them the freight of being the ultimate symbols of yang and yin. Thus, the number seven somehow recalls this but imprisons these greater meanings in a more limited day-to-day process and sequence that result in the generative cycle of the seasons.

The number seven plays out cosmographically not only as four weeks each with seven days, but correspondingly as the four seasons, each of which contains seven divisions of the heavens. The cumulative annual total of these divisions or constellations is twenty-eight. The Chinese zodiac has twenty-eight designated constellations that range above the ecliptic among the circumpolar stars that are visible throughout the year in the Northern Hemisphere. Chinese astrology was less concerned with the heliacal appearance and disappearance of fixed stars along the ecliptic and in their system positions of the sun, moon, and planets could be measured at any time of the year.[82] The moon, which traverses approximately twelve degrees along the ecliptic each day, occupies the equivalent of one constellation each day over its approximate twenty-eight-day cycle.

Chinese constellation

Three Enclosures (三垣): Purple Forbidden Enclosure (紫微垣) | Supreme Palace Enclosure (太微垣) | Heavenly Market Enclosure (天市垣)

Twenty-eight Mansions (二十八宿):

Azure Dragon (East) (青龍): Horn (角) | Neck (亢) | Root (氐) | Room (房) | Heart (心) | Tail (尾) | Winnowing-basket (箕)

Murky Warrior (North) (玄武): Dipper (斗) | Ox (牛) | Girl (女) | Emptiness (虚) | Rooftop (危) | Encampment (室) | Wall (壁)

White Tiger (West) (白虎): Legs (奎) | Bond (婁) | Stomach (胃) | Hairy head (昴) | Net (畢) | Turtle beak (觜) | Three stars (參)

Vermilion Bird (South) (朱雀): Well (井) | Ghosts (鬼) | Willow (柳) | Star (星) | Extended net (張) | Wings (翼) | Chariot (軫)

Figure 2. The Twenty-Eight Chinese Constellations[83]

Thus, the divisions of the Chinese heavens portray the heavens in terms of a day-by-day circuit of the moon in its monthly path through the zodiac. This cycle then generates the seven-day week and a calendar that is tied to the waxing and waning phases of the moon. Clearly, in Chinese culture and medicine after determining the five-element bias in a patient, the position of the moon in a lunar mansion and the phase of the moon in its solilunar cycle became important for diagnosis and treatment.

The moon was important both in European megalithic and early Chinese culture. The position of the moon in relation to the dominant portions of the sky, the timing of its risings, and its conjunctions and geometrical relations to planets defined symbolically to traditional thinkers, the operational conditions of any period.[84] Professor Thom mapped the territory of how lunations were observed and measured in early stone observatories, and John E. Wood in *Sun, Moon, and Standing Stones* summarizes effectively some of the key cycles.[85]

Wood shows that the orbit of the moon rotates in the same direction as the sun but that the plane of the orbit is 5.14 degrees from that of the ecliptic (fig. 1). This means that at different points in the year and over a period of years, the angle north or south of the celestial equator can vary as much as twenty-nine degrees north or south during the peak of the differential and nineteen degrees north or south at the period of least variation (fig 2). He mentions this movement as part of the pattern of the Metonic cycle, which describes how every nineteenth year the phases of the moon repeat on the same day, though not at the same time, because nineteen tropical years (years of solar returns to the equinox points) equals 6,939.6 days while 235 lunar months equals 6,939.69 days. These changes in elevation of the moon were very visible in the Northern Hemisphere and made a great difference in early civilizations dependent on night hunting. The elevation of the moon at different times of year allowed for certain practical kinds of planning.

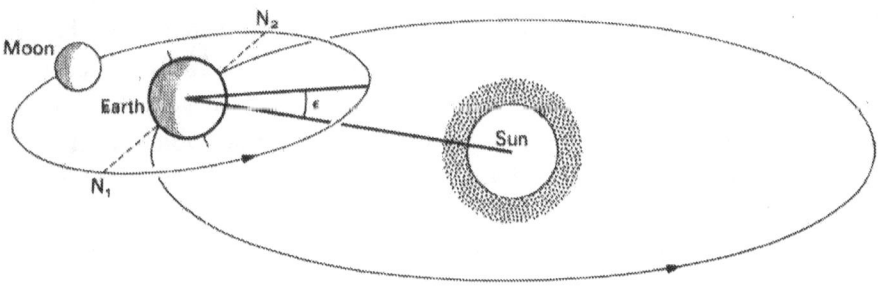

Figure 3. The Orbit of the Moon from Woods[86]

This shows the inclination of the moon's orbit to the plane of the ecliptic and its orbital direction similar to the sun.

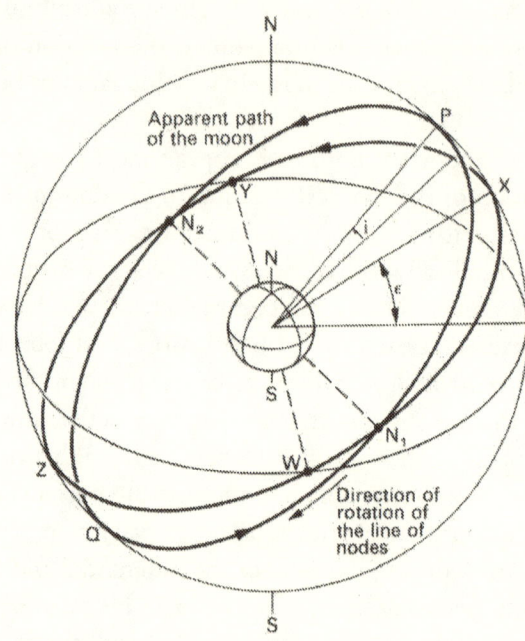

Figure 4. The Range of the Moon's Orbital Plane in Relation to the Ecliptic Showing Lunar Nodes and the Basis for the Metonic Cycle from Woods

The second cycle recorded in megalithic monuments was the eclipse cycle. Lunar eclipses are fairly frequent, and they occur regularly when the full moon is near the moon's nodes. The moon's nodes are the points where the plane of the moon's orbit with its 5.1 degrees of tilt from the ecliptic intersects the plane of the ecliptic. The Babylonians discovered the eclipse year, which is called the Saros Cycle, and the duration of this cycle where the eclipse recurs on the same day of the year is eighteen years eleven and one-third days. This cosmic pattern causes lunar eclipses to recur in a pattern twenty days earlier each year owing to the constant progression of the nodes along the ecliptic, which shortens the annual solar, return to the node by twenty days. Wood says, "The time it takes for the sun to traverse from one ascending node to the next is 346.78 days. This period is called the eclipse year."[87] Nineteen eclipse years equals 6,585.78 days while 223 lunar months equals 6,585.32 days.

These facts about the moon show a certain level of geometrical understanding in early civilizations, and particularly the stone circles that measured these cycles were constructed using a knowledge of triangles and proportionality that rely on the principles of the number three. Woods supports Hoyle's argument about how to use Stonehenge to record these cycles as follows:

In Hoyle's suggested use of the Aubrey Holes, the circle of the holes represents the ecliptic. He says markers can be placed beside the holes to indicate the positions of the sun, the moon, and the ascending and descending nodes (of the moon). A typical arrangement is shown in figure 4.11 (fig. 5), where S and M are the sun and moon markers, and N1 and N2 the nodal markers. S and M are directly opposite each other, because the figure shows how the markers would have been placed at the full moon. The sun marker is moved anticlockwise by two holes every thirteen days, making one revolution in 364 days, very nearly a year. The moon marker is moved anticlockwise two holes each day, making one revolution in 28 days. The nodal markers always keep on a diameter and they are moved clockwise one hole every four months, thus making a revolution in 18.23 years. These are reasonably good approximations to the three important periods. If the markers are initially set correctly, then every time S and M are opposite, and the nodal markers are within one hole, there will be an eclipse of the moon.[88]

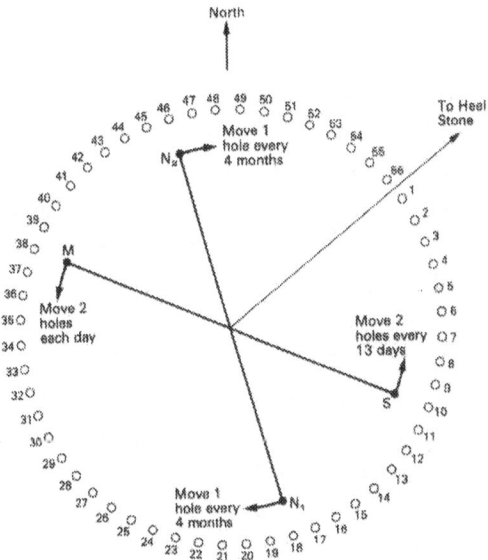

Figure 5 (4.11). Hoyle's Method of Using the Aubrey Holes as an Eclipse Predictor

The record of the progress of the sun and moon using the Aubrey holes at Stonehenge both proceeded in the same counterclockwise direction but at different rates. Of particular interest, here is the way that these cycles are measured with the combination of clockwise and anticlockwise cycles. The clockwise solar motion in heaven is recorded counterclockwise on earth. The moon's nodes move clockwise on earth while the apparent motion of the waxing moon is counterclockwise. These observations about the solilunar cycles may account for the mentions of two motions here as the clockwise sheng cycle and there as the counteracting cycle, here as the solar progression of the year according to stems and seasons, there as the terrestrial progression that begins on the counterclockwise side of the midwinter pole. In the Chinese horoscope, like its Western counterpart, the twelve spatial house divisions and their meanings run counterclockwise so that when the wheel is rotated clockwise, each would come up in turn.

I. The Number Seven

Chinese numbers too suggest that thirty years was considered a generation and that thirty years was that active portion of a life preceded by youth and succeeded by old age. The sixty-year period roughly approximated two complete Saturn cycles (approximately twenty-eight years each, one complete progression of lunations of twenty-eight years), which in turn reflected the lunations and matched the twenty-eight lunar mansions of the Chinese constellations referring to the same. Sixty years round out and project the duration of three lunar periods of nineteen years each; and this would conform with the heaven, horizon, and earth cosmology of three that permeates Chinese thought.

The number seven reintroduces time by making a transition from seasonal time to monthly and daily time considerations. This results in a greater sense of detail, which in turn gives a precise picture of what is happening during each of the twelve lunar months that make up the year. Instead of seeing four seasons of three months each, we look at each month of twenty-eight days and see it as a four-phase lunar cycle. Four weeks is a complete light cycle, a miniature of the larger seasonal cycle, and the weekly cycles itself interacts with the seasonal cycle. This means in practice that a dark of the moon in winter will be somewhat different in its effects than a dark of the moon in the summer. It will be colder and more life threatening at a minimum. The treatment implications for this are important. We want to tonify or sedate in relation to the moon; we want to avoid bleeding at certain times—particularly at the dark of the moon, half-moon, and full moon. At the most basic level, the number seven shows us how and when to employ the three-yin and three-yang body meridians and, in which segments, when to emphasize above or below the horizon waist in conformity with the lunar cycle that is augmenting below or above in a way similar to the seasonal cycle.

The number seven, [七] ch'i (1), Wieger lists as a derivative of radical 1, the single horizontal line. He also includes as derivatives: wan (4) meaning ten thousand things; *ting*(1) the fourth celestial stem; *san* (1) the number 3; the numbers 20, 30, and the character *shih* (1), which means "an age, a generation of thirty years, the world," and *ping*, the third celestial stem.[89]

"The vertical stroke with a tail has relations to radical 2 the vertical line, to weave, or to connect heaven and earth that is closely aligned with the symbol for man and tzu the first terrestrial branch symbol. The downward stroke tails off to the right like radical five, which implies germination or movement particularly in the soil or in the body as (swallowing) owing to influx of chi from heaven. Similar strokes to that of I can be seen the second celestial stem and the number nine Chiu."

The number seven then implies a movement of energy from heaven down through the horizon into the earth, which creates a generating effect. The single line can also simply mean the whole or the one that has a movement with a generating effect or transforming effect in time owing to the tail. The concept of movement and life is appropriate to the yang odd number whose root is the one.

The number seven is the sum of two and five and, therefore, implies the idea of yin and yang with the notion of the five elements. Applied one way, the yin and yang applied to the five elements yields ten but taken at face value as the contribution of the sun and moon to the five-element cycle, this combination profoundly expands the generative and sexual implications of the five elements. This notion is verified by the cosmographical structures that result from applying the number seven directly to the circle of the day, month, or year.

II. Cosmographic Structure

On the simplest level, the number seven divides any cycle into more detailed divisions of approximately fifty-one degrees, and these degrees will render the cycle in greater detail in each segment. Being a yang number, we can assume that the number seven will describe a dynamic process, and that process will have to do with the solilunar cycles. Generally, cosmologists do not use the fifty-one-degree divisions as seasonal markers or even as divisions of the day, but it has astrological significance as an expression of grace or intuition that resolves complications of the 45 degree relation between two bodies and anticipates the 60 degree phases. It is enough to recognize that even here, seven is an accelerated extension of powers that depend on the basic organizational stability of the number six, not a theoretical departure.

The division of seven produces seven equidistant points on the circle. If a horizon line is drawn provisionally as a dotted line and in keeping with the five-element precedent of linking yang to heaven and celestial stems and, therefore, of orienting the solar fire phase above on the prime meridian, the result is figure 6.

Seven: Full Moon late Heaven above

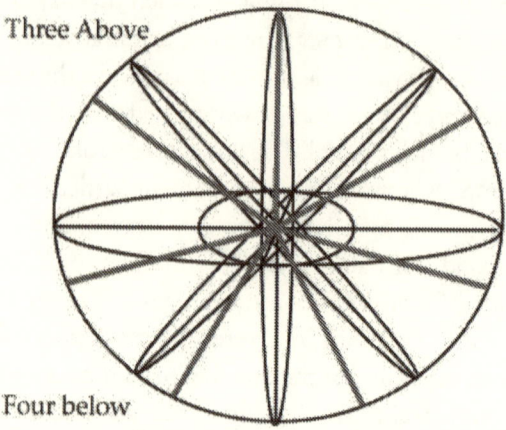

Three Above

Four below

Figure 6. The Distribution of Seven with Three Phases above and Four Below

The interesting thing about this image is that the two horizontal divisions are close to the horizon but not identical with it. The emphasis of the image is on heaven and earth, not the horizon, in keeping with the symbolism of six, but the directionality is unidirectional. Three above refers to heaven, and four below refers to four and earth; the single line identified with directionality and center is above on the prime meridian. If a human figure is superimposed on figure 6, the seven divisions seem to describe a head, four limbs, and the two lines below the horizon that roughly, but not exactly, replicate the horizon (fig. 7).

Seven: Full Moon late Heaven above

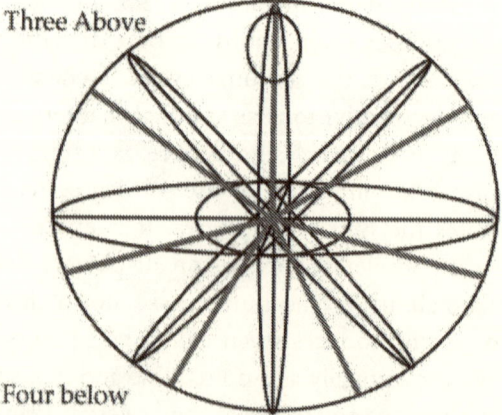

Three Above

Four below

Figure 7. The Stick Figure Showing the Anatomical Implications of Figure 6

This figure basically verifies the notion of the five-element sequence symbolized by the human position in relation to the horizon, which divides the cosmos into heaven and earth (2), symbolized by the two spokes near the horizon. The numerical value in seven of five plus two is symbolically confirmed to some extent in the drawing.

The idea of heaven and earth in six is a terrestrial configuration, not a celestial one. It included for the first time, after four and five with their examination of the horizon, both up and down, heaven and earth. Figures 6 and 7 only emphasize the sun and heaven and, therefore, seem incomplete as an extension of the number six. Also, because the two includes symbolically the sun and moon with the five planets and by inference their elements, it is clear that the number seven is complete only when it is drawn in its vertical reference with the head downward as in figure 8, which complements figures 6 and 7.

Seven: Full Moon Early: Female Earth Above

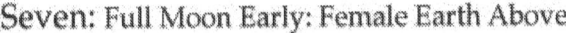
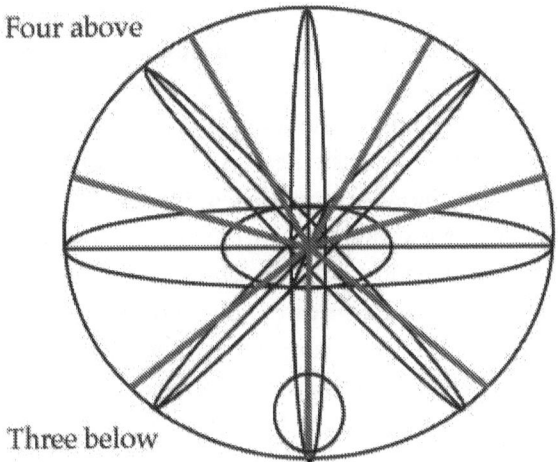

Four above

Three below

Figure 8. Stick Figure of the Seven Divisions with Head Downward in Conformity with the Lunar and Terrestrial Roots

Figure 8 also can be drawn with the image of a human being retaining the vertical orientation of figure 7, but the implications of this diagram are that of a male figure standing upright.

Seven: Full Moon Early: Female Earth Above

Four above

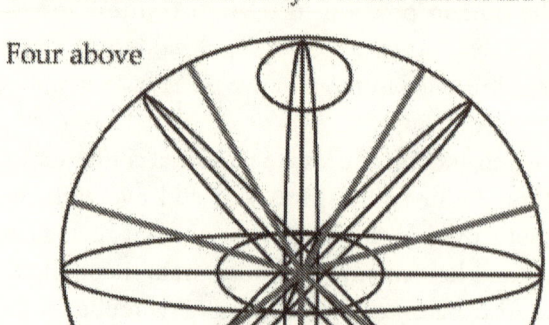

Three below

Figure 9. Male Stick Figure with Yang Principal under the Earth

Two very obvious philosophical and medical ideas result from these dual images. First, in a very specific way, the images can represent male (sun) and female (moon) generative principles. According to the seven divisions, gender is determined by the lunations, which have variations in their displays. The declination of the moon can fall far above or below the ecliptic at different phases of its nineteen-year cycle. The full moon sometimes is delayed below the horizon and sometimes emerges early above the horizon depending on one's location on the earth and the inexact conformity of the full moon with the earth's rotation (dawn and sunset). The late arrival of the moon is similar to the figure with the horizontal divisions below the imaginary plane of the horizon (three phases above and four phases below, female); the early arrival of the moon above the horizon is similar to the figure with the horizontal divisions above the horizon (four phases above and three phases below, male). Clearly, in general, with the sun farther north, the sun will set later; and the moon is more likely to appear before sunset, and this will result in the yang moon more frequently. The number seven and its divisions seem perfectly to describe the asymmetry of the lunations, which now can describe very specifically the sexual aspect of the generative cycle with its regular oscillations between yin and yang that depend on the motions of the five planets in the context of the seven-day week and the quarters of the complete lunations.

The second, the reversals of yang and yin with positions, have obvious sexual references, which explain some of the mystery associated traditionally with the number seven (fig. 10). In seven, the original differentiation of two, which medically translates on its simplest level into male and female patients, is extended into the generative functions and their relation to the lunations. Now the complete individual (five elements), with an upright mobile body that conforms to the natural order (6), can reunite in an image of wholeness to generate life. The bodies are independent and separate but, nevertheless, can in certain circumstances return to one and, in so doing, become capable of creation.

This is the mystery of seven and the source of many of its taboos.

The male and female models when combined, produce a complicated fourteen-point figure in which the basic models merge symbolically as one, but also, the two perspectives of heaven and earth applied to each are reproduced as twenty-eight divisions. This means, for instance, that the male figure merges with the female figure with heads facing in the same direction and heads facing in opposite directions and the reverse; male merges with the female as male and as his own intrinsic female, therefore producing the four basic circuit conditions of the day, month, or year—yang within yang, yin within yang, yin within yin, and yang within yin. The comprehensive generative meaning intrinsic to the structure of the number seven is played out fully in relation to the cardinal points, and the phases are played out in seven parts in each of four divisions of the monthly cycle, which then is projected on the annual cycle through the twenty-eight constellations. The numbers mesh so beautifully as to make the numerologist wonder whether numbers or the order of the world came first.

Seven: Union of Male and Female Phases: Creation

Three Above and
Four Below

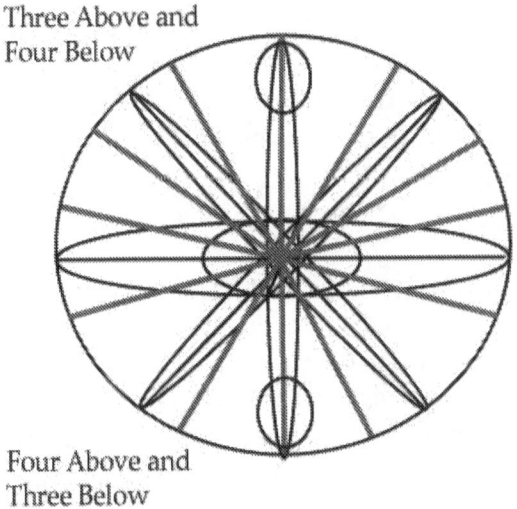

Four Above and
Three Below

Figure 10. The Two Facets of Seven in Union

III. Evolutionary Patterns in the Lunation Cycle

Since the number seven has a close association with the cycles of the moon and its material effects, some basic astrological conceptions of these effects are worth reviewing. The lunation cycle describes the sun-moon dynamic that is responsible for generating and maintaining the cycles of life on earth. Life in nature has two levels of function: subtle and gross or invisible and visible. The day-ruling sun regulates the visible and rational; the night-ruling moon organizes the invisible and instinctive (yang and yin). In harmony with the writings of Carl Jung, the character traits corresponding to day are relatively more extroverted; those corresponding to night are introverted. They both quicken intelligence but using different functions directed toward different aims. The sun and moon are active at every phase of the lunation cycle, though at night, the sun's presence is only implicit in the lunar reflection; and during the day, the vestigial moon and continual tidal motion register her presence.

Astrological "influence" ultimately always is local. Local climate and culture over time describe the scope of typical human responses to cyclic processes. The moon symbolizes tribal and cultural memory, and it always has told hunter-gatherer and agricultural societies what to do next given the time of year and phase. Cultural memory of this sort defines the moral and survival patterns in mammalian roles of mother, daughter, leader, warrior, and teacher. The rhythms of the moon give pace to circumstance and remind people of the phases of events, helping them to set goals and plan. People can plan to accomplish things by the next full moon. It is during its cycles that the details of life unfolds in particular ways. Moonlight generates the atmospherics in circumstances that we remember, and the dim waning moon on a winter night or a triumphant full moon path on the summer waves frequently set the tone for our small events.

Traditionally, the moon has been analyzed from two perspectives:

1. Structure, the perspective of twelve annual lunations and their general seasonal features and meanings. This has resulted in the zodiac sign and Chinese lunar constellations that lend interpretive meaning to positions either in the birth chart or any meaning given to a present circumstance. In every society, the moon was considered the ruler of the night, the sun of day, and this had direct significance in interpreting natal charts, where the moon was considered more important than the sun in a night chart.

2. Process, the perspective of phase relations between the sun and moon. The unfolding lunar phases in each cycle are consistent in each of the twelve annual lunations. They can be viewed as four phases that are easily discernable owing to the different shapes of the visible moon. It also can be viewed as six, eight, or twelve phases just like the annual cycle and with similar interpretive meanings. In astrology, for the most part the interpretation, of all other phase relations between planets, the phases of the lunations are the foundation for the interpretation of all other phase relationships between celestial bodies.

The Basic Structure of Lunations

Each year, the twelve monthly cycles of the moon define periods of natural growth and decline. The oscillations of lunar light and darkness regulate reproductive cycles and alternately accelerate and restrain the unending circle of seasonal transformation. By so doing, they account for much of the variety in the ecosystem. They also outline important epochs of the human psyche.

The sun and moon epitomize opposing characteristics within the spectrum of perceptible qualities. The moon's light is variable, retiring, reflective, cool, and intermittently visible at night, silvery in color; its evolutions quantify time and influence personal feelings, impulses, and moods (yin). The moon allows hunting of nocturnal animals and governed in traditional hunting societies' periods of nocturnal activity and rest/reproduction. The sun predictably is visible during the day; its light is constant, expansive, productive of heat and dryness, and its color is that of flame. The sun's unwavering brilliance sustains all the manifold external natural and social activities that occur during the day. It activates the will and intellect in humans and prompts activity. Yet the interaction of the sun and moon is harmonious. Their motions in light and darkness are regular and useful. The moon symbolizes and enforces cooperation as it reflects the sun's light and elaborates the solar purpose by transmitting different meteorological moods, nocturnal emotions, and tidal variations.

The moon has always been closely affiliated with the growth cycles of nature. Chinese systems record the lunar journey using animal references, and one approach to the lunations divided the lunar year into twenty-four divisions in a way that emphasizes the subtle differences between waxing and waning phases of each of the twelve lunation cycle (chapter 11, figure 5 from Porkert's table 4).[90]

The lunar divisions have clear correlations to phases in the life of plants from dormancy to full growth and eventual decline. Waxing phases are given as yang and male significations as waning yin. These oscillations between waxing and waning phases provide an added series of pulses and pauses to the otherwise steady ebb and flow of the solar year. These motions conveniently correlate with the evident fluctuations in temperature, weather, and plant growth of the general seasonal motion. Parallels may not be exact due to the number of variables, but the rhythm is unmistakable. Empirically, we experience surges of increasing warmth in the spring and pulses of a few days of significantly increasing cold in the fall as weather caused by global temperature changes pass through. The common experience of local Chinese culture is enshrined in the twenty-four divisions of the year. Nature's transformations mirror the transformations of the moon's shape in its progression through each lunation.

The circle of the lunation can be divided vertically and horizontally (fig. 11). The monthly cycle has twenty-eight divisions or phases that give seven phases per quadrant or around each cardinal point depending on their distribution. Figure 11 imposes a solar standard of twelve phases on the lunar motion and also implicitly attributes seasonal and elemental qualities to the phases. Nevertheless, it is important to understand two things:

1. The waxing phase opens in a reverse direction to the sheng cycle, and although moon traverses the sky in an east-west direction owing to the rotation of the earth, the angle between the sun and the moon increases on the waxing phase in the reverse direction, owing to the orbital motion of the moon. This reverse motion is responsible for all the notions of action and counteraction yang in yin in the phase relationships of astrology and medicine, and intrinsically, it is the reason the terrestrial branch sequence begins with *tzu*, and the element blocks were reversed in their positions on the Chinese clock.

2. The distribution of the phases in figure 11 is somewhat misleading because each phase completely dominates the sky and relates to all the cardinal points, just as spring and autumn are comprehensive expressions of yang and yin and occupy all the quadrants. The relation of phases to the cardinal points then is symbolic. Figure 11 shows the increasing light on the western horizon to underline the differentiation between solar and lunar cycles, but I still associate the phase of increasing light with the wood element and mentally reverse the body or reverse the perspective of the viewer to accomplish a waxing phase and east set of associative meanings.

The horizontal division suggests that the sun dominates all phases with less than ninety degrees. All phases with greater than ninety degrees are dominated by the moon. When the moon is invisible or barely visible, it is near the sun and under its influence. When the lunar sphere is more than half-light filled and it is visible in the night sky for more than half the night, its imprint is greater than the sun. These relations have interpretive significance.

In a night chart with the moon less than ninety degrees, the sun is nearly as influential as the moon. In a day chart with a moon greater than ninety degrees from the sun, the moon is nearly as influential as the sun.

Amplifying these relations in the natal chart, it can be seen that in a day chart with a moon less than ninety degrees from the sun, the sun will dominate the moon excessively. The day characteristics will be strengthened at the expense of the lunar smoothness and sensitivity, and the individual may display exaggerated forms of assertiveness, denial of interdependency, receptivity, sensitivity, or a subordination of the subtle feelings. The individual may be careless of form, prodigal in expenditure, and insensitive to nuance and personality issues. External forms and appearances may dominate the mind, and lacking a center, the individual may become paranoid or seek security by manipulating external circumstances. A contrasting night placement of the moon under ninety degrees can lead to a supercharged emphasis on emotions—to rebelliousness, restlessness, and sleeplessness—as the moon attempts to wrest control of the individual psyche from the sun.

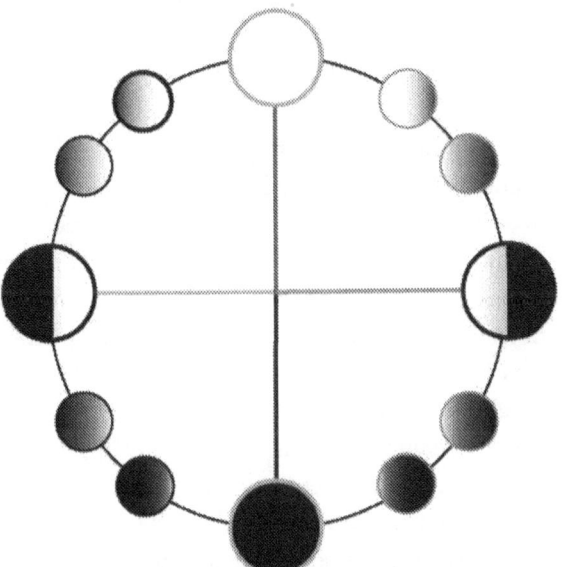

Twelve Lunar Phases
Reverse Motion
Each Phase Completes Circuit of Heaven

Figure 11. The Four Quadrants of the Lunar Cycle and the Twelve Phases

If the moon's aspects are greater than ninety degrees, the moon will dominate a night birth, particularly if it is in a more elevated position than the sun. Role responsibility, rectitude, awareness, subtle feelings, personality issues, dreams, fantasy, and the imagination may dominate the person. Consciously choosing to disregard ceremony, hierarchy, and objective standards about which they are fully aware, they pursue different goals according to their private vision. The prominent moon in a night chart gives great capability, charm, and intelligence. The individual gives close attention to delivery, to power, to the projection of an effective self-image as well as to the achievement of personally meaningful tasks. They must belong to the tribe, the company, or the community of interests. They understand responsibility, and instinctively, they adhere to the codes of nature and the collective agreements of those around them.

A full moon birth during the day heightens an individual's awareness of the need for balance between public and private worlds. Generally, people with this configuration pursue a normal course of social development, but periodically, they throw over these standards and do something unexpected in an attempt to achieve private fulfillment. When the moon is over ninety degrees, people will be relatively more objective than when the moon is less than ninety degrees. This is because the prominent lunar personality is complemented with heightened awareness of the surrounding world and a sense of mature individuality. The likelihood that the individual will see himself or herself as an actor in an expanded drama is great, and the individual is unlikely to think that the world revolves around himself or herself.

The vertical division of the lunation emphasizes the direction of movement, either waxing or waning. This directionality defines the aim of any aspect between the sun and moon, either toward greater independence and lunar fulfillment or toward reality, physical limitation, and psychic expansion—movement away or toward the sun. The correlations of the sun are those of insemination, protection, and organization. In astrology, the sun rules explainable phenomena, causative actions, principles of dominance, knowledge of external things, will, linearity and directionality, and assertiveness. The sun rules all external seasonal rituals, mating rituals, visible generative organs, and socially approved reproduction.

In contrast, the natural attributes of the moon are those psychic resonances in animals and humans that enable them to respond to intangibles—to anticipate floods, fires, and other forms of personal danger; to sense changes in the weather; and to understand the meaning behind rituals or the behavior of others. Where the sun is directional and linear, the moon is integrative and curved. The sun is energy and initiation; the moon is structure, memory, and habit (yang and yin).

Love at first sight, mate and leader selection, and luck have to do with the silent lunar knowing—the capacity to assemble from the inchoate forces of nature, a coordinated moral order.

IV. Intuitive Insights into Structure and Meaning

An intuitive early twentieth-century attempt to explain the meaning of lunar phases was William Butler Yeats's record in the book *A Vision* about his wife's automatic writing.[91] Though his phase descriptions are somewhat obscure, they capably describe some of the precognitive impulses and selections that regulate human aesthetic and feeling responses to the world. Such responses mirror psychologically those similar transformations evident in nature. Yeats records twenty-four divisions of each lunation, which I have altered in my own simplified code based on years of astrological observation into twelve phases in table 5.

Table 5. Summary of the Major Phases of the Lunation

Dark of the Moon

Phase I

Dark of the moon is about dismemberment and the dissolution of time. Living in the moment for the moment. The process includes complete subjectivity linked to complete objectivity and immersion in unknown interior motivations. It is the fool. It is destructive of form and social patterns.

Phase II

Instinctive activity. Living in the now of learned behavior the phase focuses on self-assertion and the subjective intellect with a lack of perspective. Natural spontaneity, self-will, full of saws and proverbs, living in the senses, but able to adjust to the environment. Lacking social or collective purpose, the person has a potential for insincerity: a wanderer, a worker full of life's energy.

Phase III

Experience is all-absorbing, subordinating observed fact, and even truth. The hunger for experience occurs in set paths established in the previous cycles and the cultural heritage. Originality is limited for action, and sensations are everything. Character originates in external experience but emerges from within the seed of an increased dynamism in expression. Selfhood emerging.

Half-moon

Phase IV

The half-moon square of ninety degrees. Transition from an identity based on norms, sensation, and experience to personality symbolizes the triumph of the individual life force over inertia and convention. Personal purpose breaks from the soil of the unconscious. Learned subjective behavior and instinctive motivation gives way to active projection of self into the outer world on the basis of inner drives. A crisis in growth and individuality results in insecure, unstable personalities but personalities with direction and aspirations. Personalities are complicated by inner tensions and a sense of weight or obligation that holds them back from full, responsible self-expression.

Phase V

Intense emotional power. Awakens a sense of beauty. Personal ambition. Self-centered drama, always feels he has the lead part. See the divine in whatever is the most creative act. This is a position of emerging joy, of aggression, and of defense—a period of flourishing expansiveness and possibility.

The Full Moon

Phase VI

The full moon. Self-sufficiency. Symbolizes a high degree of integration. The person is prone to self-conscious poses, plays a role in society without losing the self. Unity of being becomes possible. Hitherto, we have been part of something else but now discover everything within our own nature. Personality seeks personality. Every emotion begins to be related. It is as though we touch a musical string that sets other strings vibrating. This is the maximum outer extension of joy, completion, and individual possibility and freedom. The phase is one of joy and sadness mixed because the complex components of happiness seem ready to fly apart, and the tug of the unconscious and unmanifest self is felt in the midst of external fulfillment and capability. This often is the position of the poet who sees the irony of a limited self and self-awareness in an unlimited field.

Phase VII

Subjective power, intellectual and emotional projections. This is an image of softness with great perceptive power. "The surest guide is one's ideals, the best trait cooperation." The supreme light phase of the moon which brings the night into highest harmony with the day.

The person believes in the power of individual consciousness to dominate nature and institutional structures. There is reflective intellectual ability combined with capacity to act. Spirit and personality can work together to create a beacon for individual achievement, warmth, social purpose, and understanding.

Phase VIII

Diminishing subjective power. Reflection. Emergence of the contemplative life. Compassionate understanding and implacable regret combined to fuel a quest for meaning. Relationships are central not as vehicles for self-expression, self reflection, or conquest, but out of genuine interest and hope that the other may shine light on capacities and meaning as yet undeveloped. This is the position of the thinker, the counselor, or the strategist who brings groups together to forge social contracts. The cognitive powers are strong, allowing for a link between the head and the heart.

Phase IX

Will Power. Struggles with external fact as the personal and social order no longer can contain the variety of perception and consciousness. The emergence of organized forms in order to retain values of the cycle and season. The intellect provides cohesiveness and communication to compensate for what has been lost in unity of being. Light is declining toward darkness, and meaning, value, and economy of purpose are paramount.

Half-moon

Phase X

Waning half-moon, square aspect. The rational mind becomes fragmented, no longer retaining wholeness. Intellect fails, and surrender to external reality alone gives integration. Intellect is freed from all that was founded on personal desire and personality. Now the individual becomes part of the machinery of the cosmos or society, keeping only the spiritual quest separate in a profound dualistic split between substance and essence. This can be liberating, as the social order and work take care of themselves according to the limits of custom and circumstance while the spirit roams free. There no longer is the hope for unity, integration, and a physical/spiritual oneness to be found in the magic of the present. This view allows for mundanity and the spirit to coexist and go their separate ways. However, there is a deep dualistic conflict, a fear of surrender to fact, a sense of inadequacy and diminishment in accepting limited physical goals. Bits and pieces of ritual and magic are preserved to remind the self of what was once a present possibility but now only is a memory.

Phase XI

Knowledge of life sought in relation to the source of life. Factuality and interior communion. Dealing with the facts of circumstantial life offers the possibility of truth. Scientific truth is the shadow of such an alignment where only reality is considered a worthy subject. This phase cuts through the mists of social convention and aspiration and rediscovers purposes that lie deep within the unmanifested self-purposes that align with Dharma, the divine plan, and the expansion of an undifferentiated universe beyond time and space yet existent in a way discoverable only through undeviating realism. "Actions are seen for all their terrible self-interest and weakness with unflinching judgment. Man does not perceive the truth; God perceives the truth in man."

Phase XII

The individual's thoughts become an aimless reverie.

All of life is either sacred or profane, and the self cannot be distinguished from the movement of fate and the facts and fragments of inspiration or imagery as they present themselves. Surrender of the self, acceptance of destiny. The moon submits to the authority of the solar position and season, returning to unburden the content of the previous cycle, all its action and reflection, so that it may move forward in the new context that has emerged from their total experience in its relation to all previous cycles.

The contribution of Yeats's material was to shift inward typical astrological analyses of the moon and to focus diagrammatically on precise evolutionary phases of movement (fig. 7). These phase descriptions are exceedingly useful. Below, they will be complemented by a fuller explanation of their mechanics.

Dane Rhudyar in his *The Lunation Cycle* insisted on a dynamic approach to interpreting lunar phases. He views the phases as stages of relationship between the moon and sun.

The lunation is a cycle of transformations. The moon not only rapidly changes its place in the sky, but it also changes its shape to the extent that during a part of the cycle, it vanishes entirely from sight.[92]

The phases of the moon tells us nothing about the moon herself or the position of the moon in the sky. They refer only to the state of relationship between the sun and the moon.[93]

Rhudyar broke lunar interpretation from the confines of static zodiac and aspect meanings. Using his poetic prose, he characterized generally the essential themes of lunations and indeed most cyclic relations between celestial bodies:

> The solar power can, however, only be used and assimilated by the chaotic earth materials if it is released gradually during a process of organic unfolding and concept-revealing illumination, of which the waxing of the moon is the symbol. The waning period of the moon represents the dissemination of what has reached lunar integration at the full moon. The first quarter represents a crisis in action; the last quarter, a crisis in consciousness.[94]

V. The Moon Phases as Process

A traditional agricultural and astrological response to the increase of lunar size and light during the waxing phase was to assert that activity increased; people became more purposefully assertive, more independent, and more individualistic. Waning phases were thought to be more focused on completion, decrease, inward retraction, and a movement toward rest, fragmentation, and dismemberment of the personality. This model imitates the seasonal transformations, and it reflects the process aspect of the rising and declining seasons as opposed to the structural differentiation of heaven and earth. It assumes to its credit that the moon and personality are linked by continual rhythmic adaptation and transformation.

Individuals, like the moon, change their appearance and behavior in different circumstances. The personality is that ever-changing, ever-transforming matrix of learned beliefs, attitudes, thoughts, and habits through which we interact with the world. The words and symbols that distinguish facets of personality generally are metaphorically derived from simpler experiential roots.

The moon, personality, and nature are linked through the associative mind. Individuals process information differently, and there are at least two major patterns of assimilation—either through action or contemplation, gross senses or subtle senses. The introverted and extroverted character of the lunar phases translates into two general personality types—active and passive, yang or yin—analogous but not identical to those outlined by Jung as the extrovert and introvert. Where Jung's types are ontologically independent descriptions of personality derived from the poles of the astrological circle, the extroverted and introverted personality traits derived from the lunation cycle are purely dynamic and functionally integrated representations that rely on the precise admixtures of light and darkness in nature. How the personality functions, not what the personality consists in, can be clearly symbolized by the lunation phase. The individual responds both psychologically and physically but selectively on one or the other level, introvert or extrovert. In a given phase at a particular time, the balance of introspection and extroversion will shift in relation to the core level of comfort with the result that they will be relatively more focused on the psyche or on activity given a predisposing natal chart phase and later resonant real moon phase.

The moon symbolizes all those qualities that are associated with growth and transformation on earth. It regulates the infusion of solar energy into living forms. The substances on earth that perfectly mirror the moon's nature are chlorophyll and hemoglobin, which are necessary to the functions of photosynthesis and cellular respiration that sustain life on earth. The moon seems to act on human, animal, and plant homeostasis and consciousness through its subtle electromagnetic and gravitational effects on fluids in general and on blood, lymphatic fluid, and chlorophyll in particular.

The moon acts directly on the subtle senses and material bodies of all sentient beings by causing subtle changes in fluid pressure and osmolarity. Our poetic sense can better explain these feelings than rational science or my speculations. Traditional, naturalistic societies acknowledged both subtle and gross, psychic and physical changes induced by the moon, particularly by linking plant, animal, and human reproduction cycles and fertility patterns to the lunar cycle. The aim of this description is simply to identify two different and very general dominant patterns of response to the lunar cycle: one in which the waxing phase is the greater stimulus to physical activity, and the other in which the waning phase is the greater stimulus to physical activity. Implicit in this statement is the notion that individual activity in response to the waxing or waning cycle will be activity that is relatively more subtle or gross, introvert or extrovert, assertive or passive, yang or yin. Thus, visibly greater external activity means simultaneously greater internal/subtle passivity or potentiation.

The contemplative process is not externally in evidence but is concurrent; internal and external responses are always present, and whatever pattern shows up implies a counteracting invisible pattern that usually is of an opposing character. Simply put, the longer we jog, the greater the pressure for a rest phase. Jogging potentiates the rest phase; the capacity to jog is potentiated by rest. Such a simplistic dichotomy is not true to life where the astrologer finds rapid fluctuation, simultaneous manifestation or withdrawal of functions, and complex manifestations of correspondingly complicated underlying energies. Nevertheless, the illustration is useful in discriminating the broad patterns of cyclic response from relatively minor waves and eddies in energy, responsiveness, activity, fatigue, and other signs.

A template of patterns and qualitative standards may help in the understanding of some of the energetics and temperamental consequences of lunations and may compensate for the habitually extroverted portrait of lunations found in traditional astrology with its one-sided yang/waxing phase and yin/waning phase delineation. Responses to these cycles are incontrovertibly variable and individual. The two dominant responses to lunations are labeled arbitrarily:

1. type 1 as the physical, extrovert, yang, and male
2. type 2 as the psychic, introvert, yin, and female

The majority of men conform to type 1 and the majority of women to type 2, but there is considerable variation; at least 30 percent across gender and the patterns themselves vary asymmetrically with individuals. These patterns are rules of thumb and introduce the two levels and the oscillating pattern of relationship between them.

The Type 1 Masculine Personality

The type 1 physical response to the lunar cycle registers as a dramatic increase in gross physical strength, activity, or productive output during the waxing cycle of the moon. In our modern culture, this expresses as a can-do ethos or as organizational power, persuasiveness, or intellectual assertiveness. During the first half of the lunation, these people act rather than listen, assert rather than reflect. They feel ambitious, and they continually add to their list of tasks. By the full moon, as the pitch of activity increases, personalities responsive to this pattern begin to feel like they are stretched too thin.

Saturated with activities, they begin to lose control of events or details. Redoubling their efforts and increasing their efficiency, however, allows them to continue to improve performance until the full moon passes; at which point, their drive dissipates. Fatigue becomes a factor as they try to have it all at first by trying to grasp quality as they have grasped achievement. By turning to art, to music, to intimacy, or to family for renewal, they continue to implement ambitious goals set in motion early in the cycle. The full moon forces a reassessment. They cannot realistically complete all they intended; they must make sacrifices or reorder priorities and find time to think, to contemplate, and to anticipate. Of course, external activities continue beyond the full moon, and many are oblivious to the change. But even for these people, frustrations mount; things progress less smoothly or need unexpected revision. Attention is engage more fitfully and with less ease and effectiveness.

As energy works inwardly during the waning phase of the moon, type 1 people sometimes make unexpected associations and arrive at useful conclusions, which unceasing activity or aggressiveness could not have produced. When they assemble these insights and conduct research, they begin to build a reservoir of potential energy and ideas for future action. Now is the time for writing, for painting, for initiating activities with personal value even if sacrifices are made. If they are unresponsive or impatient with introspection, they will fitfully initiate new ideas prematurely and be unable to follow through effectively. Gears will spin, and frustrations will mount. External activities can become automatic, dry, and meaningless as the internal pressure for value, meaning, and solvency mounts. As the moon returns to the conjunction with the sun, the cycle ends, and the total cycle of energy completes itself by returning to its source. At the new moon, energy again releases outward into the new cycle. Rigid patterns of behavior are discarded. New patterns are forged with all the motivating enthusiasm accumulated from the stored potential energy and realizations from the previous cycle.

Type 2 Female Personality

For type 2, the waxing moon builds an internal potentiation while it subtly enforces a physical resting phase. For this type, to act takes an extra effort during the waxing phase, and to do so expresses personal need or internalized value. Energy to initiate and to do is muted or intermittent. The potential for action and release grows, however, to reach a peak near the full moon. The full moon is a release point for the externalization of internal physical and emotional processes that have gestated. The focus of this waxing phase is introspective. People are more reflective, cognitive, and thoughtful. Increased psychic potential, recalculation of aims, dream activity, emotion, and cognition build a charge that, if suppressed, leads to fits of exaggerated expression and exhaustion or a contrasting, gnawing anxiety or depression. Symbolized by menstruation, this powerful fluid-magnetic psychic charge begins to release into the outer world with the full moon.

During the waning phases of the moon, an increase of activity and external manifestation of the stored inner potential occurs. External action occurs more smoothly with less inner consideration. A feeling of freedom and decisiveness pervades. Important inner nonrational connections, conclusions, and insights that have gestated during the waxing moon now get rational form and implementation. The whole active catabolic process of life triumphs over the introspective anabolic tendencies, but in the type 2 person, meaning and feelings generally are more important than simple assertion and acquisition. The entire lunation takes on the character introduced during the waxing phase. Yet it should be remembered that no cycle stands alone; the reflective tendency of the waxing cycle of the individual focused on their psyche assimilates exposure to experience of the previous cycle.

During the waning phase, as the physical cycle uses energy and reaches its peak with the approach of the dark of the moon, the inner psychic potential of the type 2 personality is low. Actions become more automatic, compulsive, necessary, and sometimes meaningless. Nevertheless, pressure to turn inward in order to renew this reservoir of potentiality increases during the waning phase, and it reaches a dramatic peak of intensity at the new moon, manifesting as a desire for release of pent-up emotions or a yearning for a change for the better, for rest or withdrawal, or for meaning at the new moon. In the waning phase, external demands, pressures, and momentum frequently thwart introspective needs.

When the new lunation cycle begins, the process of externalization loses its momentum, and the energy turns inward. During the waxing moon period, external performance is more fitful. Will power is required to keep external activities moving, but this interferes with the rebuilding of psychic potential and leads over the course of the year to deep exhaustion psychically and physically.

VI. Medical Implications of the Number Seven

Based on the above intuitive outlook, the first contribution of the inclusion of lunations is to diagnose the basal pattern whether type 1 or type 2 or a variant. This can be determined by patterns of behavior and menstrual cycle patterns that are confirmed by tendencies either toward chi or blood deficiency. Many symptoms will fall into these patterns, and patients can be coached about when to push and when to wait, rest, and recover. They also can be treated in phase on meridians that will support the deficient phase, whether waxing or waning.

Knowing that a patient conforms or departs from gender norms can be important for determining intervention, for not all is as it seems to be. Women can have abundant chi and deficient blood, and men can have deficient chi and abundant blood; and the lunations provide a good insight into these relations.

The essential contribution of the number seven is its insights into harnessing the lunations in treatment. The number assumes the five-element diagnosis has been made and that the three-yin and three-yang meridians and their central, medial, and lateral planes and principles are known and understood. Seven simply employs the seasonal activation of elements and the full verticality of the six configurations in direct relation to the phases of the moon.

In the simplest terms, this means that during weeks when the moon is not visible or is in the last stages of disappearing or the first stages of reappearing, practitioners will get more mileage out of using meridians below the waist. Similarly, when the moon is prominent in the sky, half-moon, or larger, advantages accrue to treating above the waist.

The next stage is to focus on process. During the waxing moon, *shao yang* and *jue yin* can be treated to advantage; during the waning moon, *tai yin* and *yang ming* can be treated to advantage; and at the full moon and dark of the moon, *tai yang* and *shao yin* can be treated to advantage. Naturally, these advantages have to be weighed in relation to the seasonal activation of the elements and the time of day, which similarly stimulates certain elements. So a waxing moon at dawn and at the spring equinox will mutually reinforce each other through *shao yang* and *jue yin* meridians generally and *shao yang* wood in particular. A waning moon, at sunset, in the autumn will synergistically support *tai yin* and *yang ming* while counterbalancing *shao yang* and *jue yin*.

A more complicated picture would be an attempt to treat wood *shao yang* during the springtime but in the afternoon during a late waning moon phase. Here, the seasonal energies reinforce *shao yang* gallbladder, but the lunation is in the opposite waning phase below the horizon; and time of day counterbalances in waning phase above the horizon. The seasonal energies are strong to support gallbladder, and the structural oppositions give the opportunity to balance it with both metal and earth aspects of the waning cycle. This can be approached in various ways:

1. In a relatively healthy patient, treat the whole through the wood element. Assuming that GB. is the causative factor, the gallbladder can be reinforced directly through a source point GB. 40 or wood point GB. 41 to take advantage of the seasonal impulse. Local points on the level of the waist gallbladder and through would take advantage of the equinox. Liver 3, the

earth point, could treat the lunation phase by treating below the waist on a relatively internal partner meridian and on the source and earth point that links it to yin and to *tai yin* below the waist. The three-yin and three-yang partner meridians to liver and pericardium can be treated indirectly using a congruent principle above the waist like the metal point for the time of day. Metal involves yin within yang and is structurally metal activation, and it can be approached holographically through another meridian. This last would continue to treat the entire complex through the causative factor, through the activation of the spring equinox (horizon) and seasonal corollaries.

2. In a patient with more medical complications, the mixed portrait gives opportunities to support directly different struggling functions. At the time of treatment, the cosmos is providing energy in support of wood, earth, and metal. Knowing this is important both to diagnosis and understanding the pulse picture at the time of treatment, for treatment points can be selected synergistically to support any of these three or all three together. Where lung symptoms are to be treated, transfers from the abundant seasonal support of wood could be made to supplement metal; the moon phase supports *tai yin* earth, and this can be underlined and reinforced using spleen 3, the congruent earth point, or spleen 1 to use the abundant seasonal wood, then transferred through the tonification and earth point on *tai yin* lung 9. This kind of point selection moves available energy to a given treatment target that is selected based on symptoms but treated using orbs that have extra available chi from the cosmographical activation of the three-yin and three-yang blocks of meridians.

The lunar light cycle and phased position activates the blocks just as surely as the seasonal and daily cycle. It is important to remember that the moon is malleable, and its activation always is a refinement of the seasonal energy. In the above example, the moon's activation of *tai yin* earth owing to its late waning phase position was activation within the sun's position in *shao yang* wood at the beginning of the growth cycle. Just as a farmer would think, the waning moon-earth position in the spring is a good time to dig and till the soil preparatory to planting in the next waxing phase, a physician would think that using the earth point below or at the waist on the wood meridians would be an ideal selection of points. The springtime lunation qualifies the springtime and will, in this case, have only limited *tai yin* implications. However, given a causative factor or symptoms in *tai yin*, the lunar phase offers a way to access *tai yin* for treatment even though it is springtime, and the wood element dominates.

Just knowing when to select hand and foot meridians for treatment of a particular trajectory can be very valuable. With a causative factor in fire, an element dominated by hand meridians, the diminished moon phases offer opportunities in season to use points below the waist on either wood, water, or earth as appropriate in support of stabilizing, strengthening, or counterbalancing fire. You can work through the seasonal elements, the three-yin and three-yang partners either *jue yin/shao yang* or *tai yang/shao yin*, or theoretical oppositions such as water and fire, metal and fire, or earth and fire with their upper-to-lower and inside-to-outside opportunities for symmetrical distribution of needles.

Using the dark, half, and full moon phases in treatment are important. These are points with unique characteristics appropriate to the cardinal points. The dark of the moon is a time of intense focus, deeply internal; the seeds of a new cycle are born from the digested lessons of the previous cycle. The sun and moon are conjunct, which consolidates both of their powers in one location. This is a position of absolute yin and mysterious union, and yin can be strongly fortified in this phase. The half-moon phases like the east and west cardinal points create an atmosphere of tension that must be resolved either as increasing or decreasing light. This means that crisis points, including health crisis, will tend to lead in a certain direction and will show frank symptoms during these phases, for better or worse. Certain interventions can accelerate the intrinsic movement or trends of this period, and these phases also are helpful diagnostically about the extent and duration of the disease. Disease processes follow the seasons and lunations to some extent as noticed by Thomas Sydenham, who was one of the founders of truly empirical Western medicine. The full moon goes without saying as an important point for the support of yang. However, it also is the point where yin is able to connect with yang at its greatest point of extension, so it is a good time to regulate yang as well as a good point to support it.

Summary

Lunations detail the unfolding generative seasonal cycle. Each month, the moon traverses the whole cycle of the zodiac and in a way creates an asymmetrical electromagnetic wave motion that adds and subtracts to the relatively stable seasonal increase and decrease of light. This instability is reflected in the wavelike increases and decreases of temperature and moisture that develops not only every day but modulate either toward summer heat or toward winter cold. There may be no one-to-one relationship, but there are the tides; and during some periods, it rains every Thursday or every Tuesday, and agricultural societies could not help but notice these patterns. When we take the time and pay attention, these patterns are of subtle invigoration, and humans can notice lassitude internally. These patterns are important in personal and collective life, and people ignore them at their peril. Those who charge ahead when they should be resting make mistakes, injure themselves and others, and force people around them to live out of balance too. The ripple effect of following the chariot of the sun with all its pride and ambition while ignoring the times and phases of the moon can account for many cultural and personal maladies.

The distinct half-moon and full moon phases create a matrix not unlike the cardinal cross. This matrix, however, stands in relation to cardinal points, and the phases allow access to the principles of the cardinal points at every month and three times each season. That means that treatments based on the phases of the moon can harness the particular seasonal trajectories using a concordant cross mechanism when it is not specifically the spring or fall equinox. Moreover, using points above or below the waist can help the body adapt to the real locations and physiogical pressures instigated by season and lunar positions.

The larger eighteen-year cycles of the moon reflect a kind of hormonal and intellectual coming of age for humans; and if life has four stages 18, 36, 54, and 72, it means as much as any other years. They seem to indicate points of physical, emotional, mental, and spiritual maturity and completion. This is in keeping with gentle retrieval of experience shepherded by the lunations and the spiral developmental pattern that the lunations impose on the year. These are clearly more benign periods of life than those dictated by, say, the twenty-eight years of the Saturn cycle and more fulfilling than the tinny status of the twelve-year Jupiter cycles with which it coincides owing to the connection through three and six. This connection ties the lunations to elevations, and the elevations of the moon were very important to the ancients as established by Prof. Thom. Therefore, it is the moon that helps to establish the level of the body that deserves intervention. It is the number eight that fleshes out the vertical quadrants fully and allows for structural application of the dynamic process of the moon and the number seven.

Chapter 8 八

The Number Eight and the
Third Dimension of Depth

The number eight is a number that unpacks many dimensions of meaning that have made shadowy appearances in preceding numbers. The whole number sequence itself when applied to the body and cosmos may be viewed as a flower in a process of growth—where each new number displays a richer, more complete, and more sophisticated version of all that is contained in the greater whole or the one. By the number ten thousand, the complexity and density of meaning escapes human capacity using rational means, and we see in earlier numbers like eight an exponential increase in representational capacity beyond the preceding numbers—two, four, and six—and this confirms the fundamental notion in Chinese numerology of the generative character of numbers.

The preceding number seven offers some insights into the meaning of the number eight. Seven expanded on the cyclic divisions and meaning of the five planets and their corresponding earthly transformations, the elements, by adding the sun and moon. On the simplest level, the number seven divided any cycle into more detailed divisions of approximately seventy-two degrees, and these degrees rendered the cycle in greater detail in each segment and resulted in a graphic understanding of the anthropomorphic effects of seven. As a yang number, seven described a dynamic process, and that process had to do with the solilunar cycles. Generally, cosmologists do not use the seventy-two-degree divisions as seasonal markers or even as divisions of the day though it is entirely possible to do so, but the number seven plays out cosmographically as the four seasons, each of which contains seven divisions of the heavens. The cumulative annual total of these divisions or constellations is twenty-eight. The Chinese zodiac has twenty-eight designated constellations that range above and below the ecliptic along the path of the moon's orbit. This orbit was shown to describe a complicated generative purpose outlined by the union of male and female divisions of the circle that were uniquely connected to the division of seven.

The moon, which traverses approximately twelve degrees along the ecliptic each day, occupies one Chinese constellation each day over an approximate twenty-eight-day cycle. Based on the traditional notion that a smaller cycle symbolically equals a year, the twelve degrees reflects the nearly twelve months duration of the larger cycle. Thus, the divisions by seven of the Chinese heavens portray the heavens in terms of a day-by-day circuit of the moon in its monthly path through the zodiac. This cycle then generates the seven-day week and a calendar that is tied to the half and full landmarks in the waxing and waning phases of lunations. Moreover, this twenty-eight-day cycle fits roughly into the twelve lunations of the solar year as important landmarks in the progression of the seasons.

Because of the agricultural significance of these landmarks, the Chinese New Year and other seasonal celebrations are tied to this calendar. Clearly, in Chinese culture and medicine, after determining the five-element bias in a patient, the position of the moon in a lunar mansion and the phase of the moon in its solilunar cycle became important for diagnosis and treatment. The number seven explains the pattern of calendrical movement in heaven having to do with time and is a derivative of the heavenly stems seasonal pattern determined by the five elements.

What are implicit but not stated in the number seven are the structural ramifications in space and its three dimensions of this lunar cycle. All cycles we learned from the number two portray both absolute dichotomies such as light and dark (or any derivative polarity) and processes of change or tendencies either toward increase of light or increase of darkness. The landmarks of the lunations, the dark of the moon, the half-moon, the full moon, and the declining half-moon, which were organized by the seven-day week and the seven constellations in each quadrant of heaven establish four spatial divisions in heaven vertically and which match the four divisions of the cardinal points on earth horizontally. This generates the number eight. The number eight is the spatial consequence of the number seven in a very literal way. It may productively be seen as extending the three-dimensional model that emerged with the number six and the three-yin and three-yang divisions and processes of the body and world by yielding spatial divisions of four-yin and four-yang segments (fig. 1). As with the number six, these divisions have complex functional relations to the vertical and to the horizontal planes and express degrees of relationship either with structure of function of one or the other plane. Moreover, it is clear that the number eight introduces a concept of spherical depth that differentiates interior from exterior, and this differentiation is perfectly expressed in two additional special organs, the gallbladder and the spleen, in their quasi-interior to exterior relationship as duct structures related both to digestion and immune functions that protect the body from exterior pathogens but have no contact to the materials on the exterior. These both express unique interior-to-exterior relationships that have to do with the horizon and middle, and these act in counterpart to the brain and sexual organs. These complex organs symbolize the mediation of relations between the surfaces of the sphere *tai yang* to the point in the center *shao yin*. In terms of element affiliations, these special organs relate to their waxing and waning phase functions, wood and earth, while they also manage interior-to-exterior balance and create equilibrium between the fu (GB, fat metabolism and intestinal immunity) and *zhang* (sp, blood purifying) organs.

The link between the number seven and eight, with their emphasis on the lunations, suggests that the divisions of eight, when applied to the human body, will have something to do with finding structures and points that are responsive to peak points of the lunations and that use the geometrical structure of the solilunar cycle to augment treatment or to affect whole systems or structures that are related to linked functions. Thus, for instance, during the early waxing phase of the moon before the half-moon, it might be appropriate to select a point on a foot yang meridian because it is below the waist, but yang in nature is increasing below the surface and at night as the moon makes its first appearance. This idea is enshrined, as we will see in the placement of the yang chiao master point UB 62, that allows all the yang meridians to be activated from a yang meridian below the waist. If these were done during the early waxing phase as implied by the point, the effects could be augmented, and all the yang and chi functions that are activated and sustained by kidney yang might be effectively supported. The following paragraphs will further examine the structural implications of eight based on the assumption that eight has been generated from the number seven with its emphasis on the solilunar relationship.

I. The Number Eight

The character for the Chinese number eight, [八] Ba, means "to divide, to partake." It is a primitive representation of the division in two parts, the separation;

this figure now mean eight, this number being easily divided into two equal parts (note that four, a square is a kind of unity in the Chinese reckoning).[95] This suggests that eight is easily divided into two squares and this conforms with the widely understood geometrical representation for eight as the cross and the X or in three dimensions, two pyramids with four faces—the dodecahedron. The essential meaning of the divide recalls the number two that divides yin and yang, which, in space, produces up and down, back to front, side to side, and, now with eight, interior to exterior, the three dimensions. The symbol is in the shape of a triangle with an abbreviated line above (heaven). This recalls Timaeus's assertion that the basis of all forms is the triangle, and here, the apex of the figure rests in heaven. Eight then is somehow a result of the emergence from heaven to form downward into three dimensions. The dodecahedron is constructed out of eight triangles divided in half, creating two squares or matter in its two dimensions of structure and process, the four directions, and the four interactions between their qualities.

II. Cosmographic Structure

Figure 1 shows the simple division of the circle by eight, and this image can be found as the major landmarks of almost any compass.

Eight

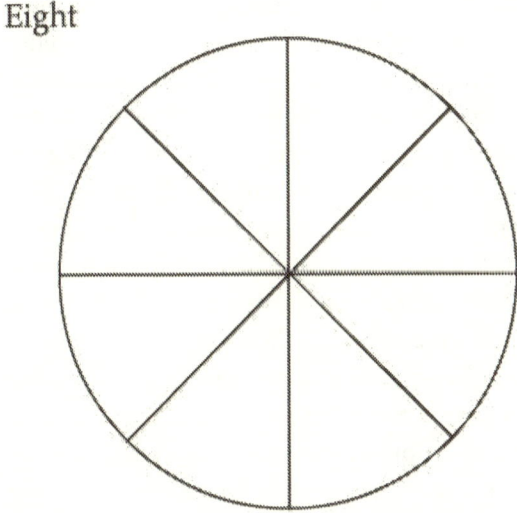

Figure 1. The Eight Divisions of the Daily, Monthly, or Seasonal Cycle of Light

Eight
With Squares

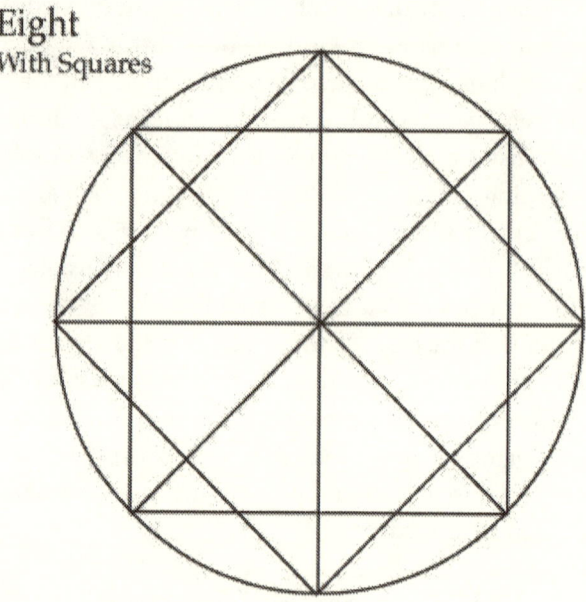

Figure 2. Eight Divisions with Geometrical Squares That Represent the Domains of Structure and Process Respectively

Viewed as though from above, figure 2 shows two pyramids; and as the square is uniquely composed of a series of equilateral triangles, this is an image for the cube of the number two or the three-dimensional octahedron of Plato.[96] On part of the octahedron treats above, the other below, and the middle is a square representing the cardinal directions. This is an intriguing start but hardly a dynamic image or one that yet approximates the sphere.

The sense for the dynamic meaning starts and ends with the circle of the year and sphere of the cosmos. On it, the midpoints of the cardinal points always have been considered important and useful landmarks in geography and astrology (fig. 3). In every case, the number eight has mundane practical uses and meanings. In astrology, they are thought to consolidate or blend the characteristics of the two adjacent cardinal points, and this conception develops directly and immediately from the consideration of the seasonal transitions between the prime qualities associated with the cardinal points and division by four. The four humors of Western astrology are simply names applied to the behavioral and medical traits that result from combinations of qualities (fig. 4).

Eight
With Squares

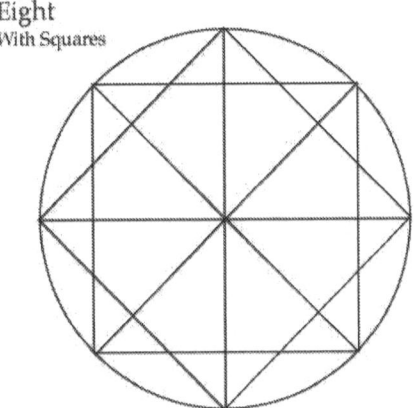

Figure 3. The Planes Formed with the Midpoints between the Cardinal Directions

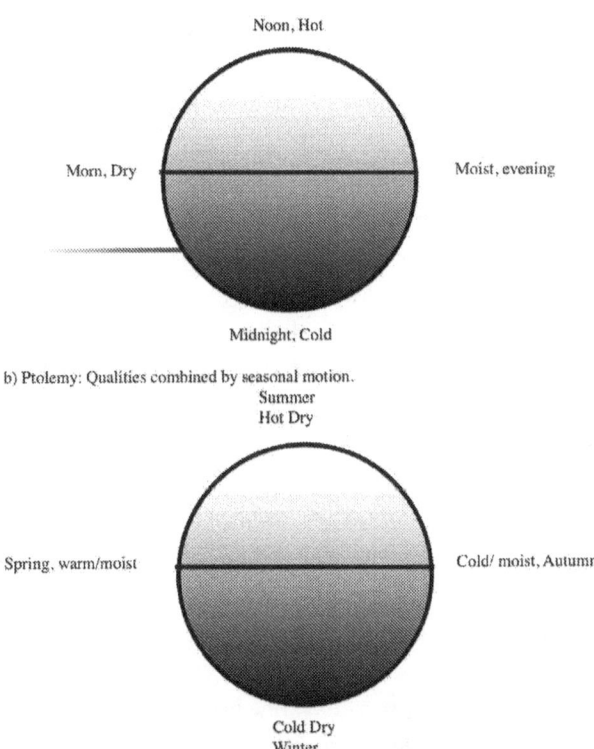

a) Ptolemy: By diurnal motion.

Noon, Hot

Morn, Dry

Moist, evening

Midnight, Cold

b) Ptolemy: Qualities combined by seasonal motion.
Summer
Hot Dry

Spring, warm/moist

Cold/ moist, Autumn

Cold Dry
Winter

Figure 4. Prime Qualities

The division of eight implies a four-directional division in heaven and earth, so it is the simple embodiment of the yin and yang (earth and heaven) aspects of the four directions and the four seasons. Thus, in this three-dimensional portrait, when we look to the east, we now can distinguish the meteorological and astronomical features of heaven and the state of vegetation conditions on earth, where with the number four, both were compressed together as a single undistinguished whole. Clearly, the object under observation is the same, the conditions are the same, but the distinction allows for useful differentiation, which can ultimately give a better sense for the future or what is in the process of becoming.

The two-dimensional drawing of figure 1 may also be seen as two distinct crosses or as two interlocking squares as in figure 2. One cross represents the spatial organization of the vertical and horizontal planes now fully established since the number six, which conform to the structural absolutes of prior heaven: heaven, earth, and horizon. The *X* represents the dynamic processes of change and interaction that occur within the structure or the principles of posterior heaven. Both are necessary, and both are fully articulated in contrast to the divisions of the number four that could only articulate one of these processes at a time or combine them in a hybrid form, where the north-to-south divisions, with their implied vertical, represented prior heaven, and the east-to-west division represented the changes of posterior heaven. Individually, each produces a geometric square, and this double square contains all of life in a symbolically stable structure, both the absolute and the relative. There is a way in which the processes of life symbolically are contained in the structure of the great *tai yang* sphere just as human vitality and sensation are contained by the body. The image also produces an octagon, which in a more limited rational way imitates the larger circular structure of the whole, and this fundamentally is why the number eight and the *ba gua* representations that use the octagon have such power in people's thinking.

Timaeus describes to Socrates the formation of the *X* as a pair of interlocking planes:

> He (the creator) then took the whole fabric and cut it down the middle into two strips, which he placed crosswise at their middle points to form a shape like the letter X; he then bent the ends round in a circle and fastened them to each other opposite the point at which the strips crossed, to make two circles, one inner and one outer. And he endowed them with uniform motion in the same place, and named the movement of the outer circle after the nature of the Same, of the inner after the nature of the Different. The circle of the Same he cause to revolve from left to right, and the circle of the Different from right

to left on an axis inclined to it; and made the master revolution that of the Same. For he left the circle of the Same whole and undivided, but slit the inner circle six times to make seven unequal circles, whose intervals were double or triple, three of each; and he made these circles revolve in contrary senses relative to each other, three of them at a similar speed, and four at speeds different from each other and from that of the first three but related proportionally.[97]

This passage describes in Greek terms the geminating power of number when placed in its originating cosmological context. The number eight contains the lunar motions of seven and the cyclic divisions of six inherent in the two motions. The two rotating planes recollect the two planes formed by the major and minor stations of moon's Metonic cycle and the anticlockwise motion on the ground of a calendric circle defined by the clockwise motion of the sun. The language is esoteric but is understood by the cosmologer-mathematicians. The division of inner and outer, with the inner having a counterclockwise motion, reflects the same fundamental split between the clockwise motion of the solar wave and the counterclockwise and yet subordinate lunar phase process and earthly rotation. One circle has to do with motion in vertical quadrants, the other horizontal, but the X describes the weaving motion of the two.

If these geometrical squares are applied directly to the body as an object in space, clearly, one square represents the vertical axis between the earth and the midheaven the spine ren/du and the other axis or plane of the waist/horizon *dai mo*. The second cross represents structurally the four limbs, the two arms, outstretched above the legs below. These relations reflect the recognition in discussions of the number four above that the X of the four limbs uniquely represent the human ability to move and adapt to our environment and environment defined by the four cardinal points of the horizon. The two crosses then differentiate what was one process for the number four and usefully find a way to show what is structurally absolute and stable versus what is mobile and adaptive in the human body. This distinction establishes a relation between self and the world between inner and outer, heaven and earth, and core and adaptation that was not differentiated either in the number four or six.

Any two-dimensional drawing of eight suppresses some relations and components; thus, figure 1 implies several different kinds of relationships that cannot be fully expressed by a single drawing. Figure 1 when viewed directly as the division of four, as in figure 5a shows a human body and world divided at the horizon/waist between heaven and earth and bilaterally east and west at the anatomical midline. Figure 5b shows the view of the body and world from above, looking down toward the plane of the waist or horizon. This defines the cosmos in terms of the four divisions on the horizon—north, east, south, and west. When divided in four quadrants, side to side, and back to front, the figure produces a clear and important anatomical differentiation of four different from the front view. Together, the images produce eight, which combines four divisions of the prime vertical and four divisions of the horizon—two divisions of four. These structures are somehow connected to the eight phases.

Moreover, either half of figure 5a or 5b multiplied by the two sides of the plane upon which the divisions are made produces eight divisions—that is, the upright division divided into back and front yields eight, and the top-down view divided by up and down produces eight. These drawing makes it clear that figure 1 attempts to display in a temporal sequence of eight divisions what in a spatial drawing would be simply the four quadrants visible from any direction and the hidden invisible additional four that they imply.

The visible is yang, the invisible yin. The figure of eight implies, but does not state directly, the notion of depth in addition to width and height. In short, the divisions of eight with the various squares imply, but don't display, the three-dimensional application to the body, which always is about the quadrants of the body's back and front or up and down, depending on the vantage point. The number eight magically reveals the mystery of the yin and yang of any particular perspective and promises an explanation of the relation between a person and the world as an extension of the relationship between the whole and part and the outer and inner directionality.

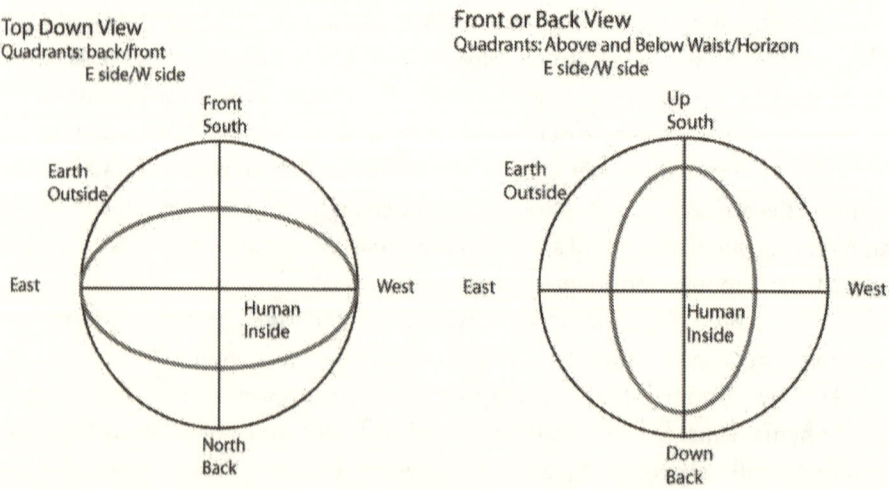

Figure 5.A, The Human Body and World Viewed from the Front with the Prime Vertical Defining the Quadrants; B, the Human Body and World Viewed from above with the Horizon/Waist Plane Defining the Quadrants

These eight quadrants are defined by specific planes, which following the law of two, yin and yang—which divides left and right, back and front, above and below. Figure 6 shows the planes involved in the a three-dimensional representation.

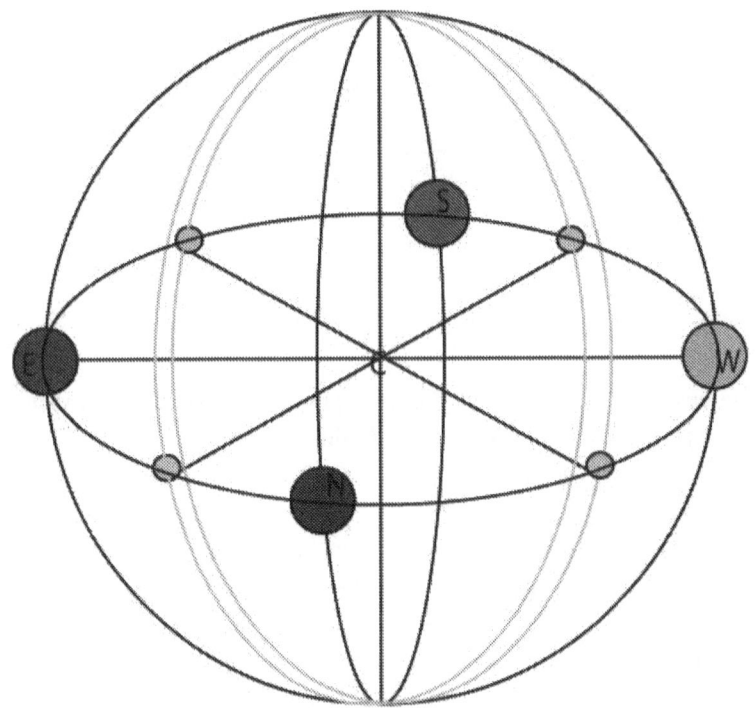

Figure 6. Three-Dimensional Representation of the Planes Behind Eight

This image gives a glimpse of the front-to-back dynamic as well as the more familiar up-to-down and side-to-side relationships. Owing to the infinite loop characteristic of energy flow and structural housing in the number eight, any representation of a human figure must be superimposed on the organizing planes of the world in a way that the figure can be seen as lying on the horizon facing up or down with their head in any direction or standing up vertically facing front or back with their head facing in any direction (fig. 7). The human figure in eight achieves a high degree of physical definition, but the more structured it becomes, the more elusive and mobile it becomes as well. More and more, the structure imitates the complexity and multidimensionality of the larger world and adapts to it with the result that the human being realizes an exponential increase in mobility and possibity.

Eight

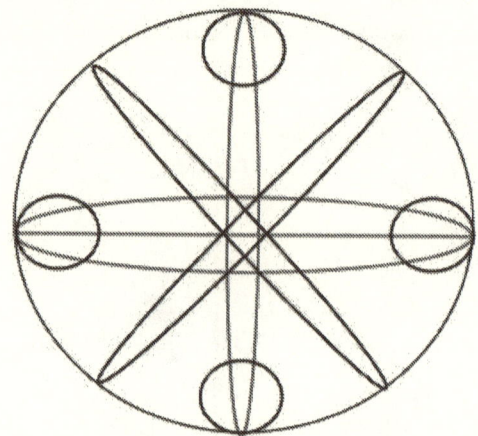

Figure 7. The Human Stick Figure Expressing the Relation Between Body and World in the Number Eight

The dynamic complexity of eight sometimes obscures its anatomical complexity, which reduces to the various quadrants times two. It is these quadrants—four above and four below—and their defining planes that are linked by tradition to the eight extra meridians both physically and functionally. What is unique about these planes is that they define equally the quadrants of the body and world; and therefore, the great implicit in the number eight is the full exposure of the magical relation between the body and the world. When the body moves, perspectives change, and the world moves or changes with it, ever-retaining the organizing quadrants and their four hidden counterparts (8). Whether day or night, dawn or dusk, there always remain four quadrants horizontally and four quadrants vertically (8) to identify locations in space. Here, the spatial feature of the one-to-one relationship is very fully expressed, and space is everything.

There is an implicit differentiation between body and world, inner and outer; but the differentiation is so tenuous as to be just emerging, just beginning to be apparent with the number eight, because, like the birth of a child or the appearance of the new moon, it is first discernible in outline (as within or part of something larger) with the structural wholeness of the number eight that shows the motions of consciousness more fully according to the motions of the moon.

III. Medical Practice—Theory

To understand better the traditional connections between lunar cycles to body structures and processes and to physical change, it is necessary to turn to oracular writings and astrology.

The *I Ching* is an oracular text that fully details phase relationships based on hexagrams, hexagrams being the two sets of three solid or broken lines into six lines organized as expressions of the three yins and three yangs.[98] The trigrams are a more basic set of numeric standards having to do directly with the relations between yin and yang and the verticality of heaven, horizon, and earth in all their permutations and combinations. It just happens that the permutations and combinations yield eight principle trigrams. This relationship perfectly expresses the relationship in acupuncture between the numbers six and eight, which devolve to the three-yin and three-yang meridians and the system of eight quadrants. The eight quadrants (the limit of combinations of three-yin and three-yang lines) contain and give the physical framework in which the six meridians and their functions unfold. The *I Ching* examines the three yins and yangs in all their permutations and combinations and contemplates humans and the world together. Nevertheless, everything that occurs within the matrix of four horizontal and four vertical directions are established by planes that define eight.

Six, as developed above, was the first number to combine the four directions (horizontal) with the vertical implications of the number three (heaven, horizon, and earth). The sixty-four combinations are, in a way, the detailed examination of phase relationships within the prevue of the three-yin and three-yang conception of any cycle and relationships in nature and the body as defined by six, the three-yin and three-yang meridians in their relations to the midline, back and front, and the horizon. The framework in which they function are defined at the outset to the arrangement of eight trigrams, which, for oracular and interpretive uses, defined the seasonal cycle in two models: (1) Fu Hsi, which is the prior heaven, absolutist, and spatial model; and (2) the King Wen, which is the posterior heaven temporal model. These eight-sided figures fully articulate the seasonal implications of the division of eight in space and time respectively.

The Fu Hsi trigrams were first organized visually in a circle that depicted the absolute cycle of nature in its pure oppositions—summer-winter, sun-moon, marsh-mountain, spring-fall, cold-heat, heaven-earth—that details the dualism of the *ba gua* (fig. 8). The system is a purely rational and visual set of relations that reflect the homeostasis in a permeating yet timeless geometric visual totality.

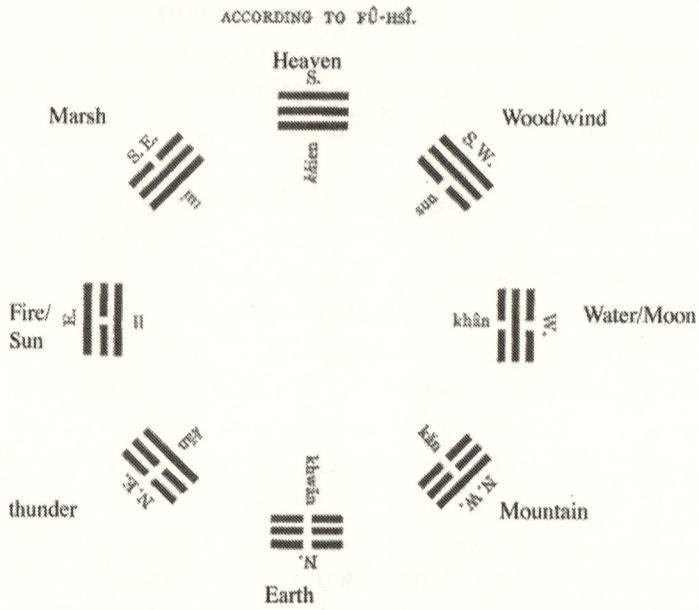

Figure 8. The *Fu Hsi* Prior Heaven Model for the Eight Trigrams

In contrast, King Wen's model reflects the posterior heaven reality of seasonal fluctuation more than the anterior heaven world of ideal relations. The model is horizontal and terrestrial, mirroring the horizon better by replacing heaven in the dominant south position with the element fire. Fire and water still anchor the principles of midsummer and midwinter on the north-south axis, but the cycle is all about evolution and involution as nature emerges through wood in the spring and retreats inwardly through metal and heaven with the autumnal retreat. Mountain and thunder describe the source of water and the springtime rousing of nature that precedes growth. Conceivably, the ramifications of this innovation are among the great inventions of Chinese philosophy, for the model in figure 7 breaks the confines of a static oppositional conception of change and transformation, a stasis imposed by the geometric idealism of the cardinal points. The interesting thing is that this dynamic yang process derived from five-element conceptions and in it the number seven with its lunar cycle implications are applied directly through the trigrams to an even number. This is made possible by the way eight allows the application of the number four both to earth and heaven, which in turn allows the application of the vertical conceptions of heaven and earth (three in the form of trigrams vertically disposed as hexagrams) to a horizon-based model with a perspective organized by the cardinal points.

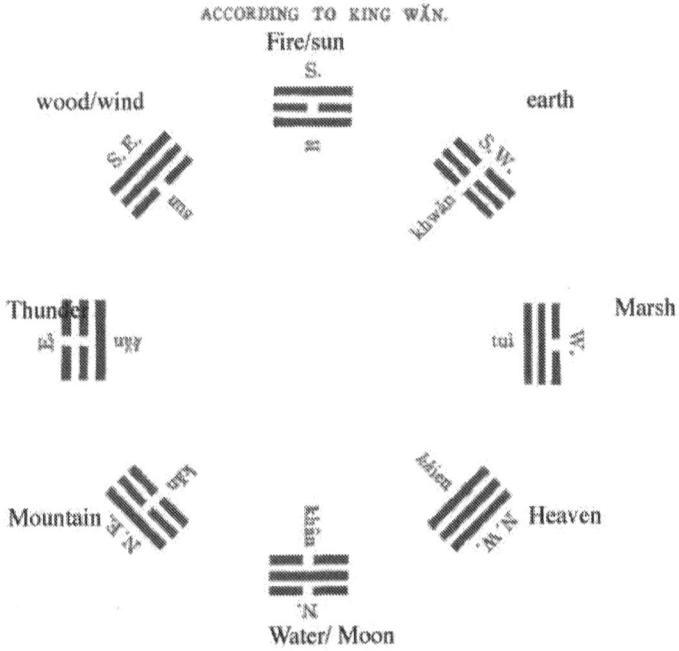

Figure 9. King Wen's Posterior Heaven Functional Time-Based Model

Lu and Needham recognized the *Ling Shu* and *Nei Jing* linked the three yins and three yangs to the cardinal points in the pattern beginning with *tai yin* metal west, central *tai yang* fire south, and *shao yin* water north; and *shao yang* wood springs east:

> Thus we have Thai-yin, Thai yang, shao yin, Shao Yang, Chueh yin and yang ming. Although the first four of these were regarded as the chi of the four seasons.[99]

Lu and Needham have affirmed that the principles of Greek astrology were applied to Chinese astrology in the house systems and traces of all the numbered divisions of the season percolate along the silk road.[100] Comparing the eight house divisions of heaven from the roman Maternus are instructive to understanding the heaven versus mundane perspective. Maternus's eight houses were thought to predate the twelve-house system, but I think that is unlikely. Rather, the eight divisions represent a description of heaven that is earlier in the numbered sequence of celestial portraits in numerology.

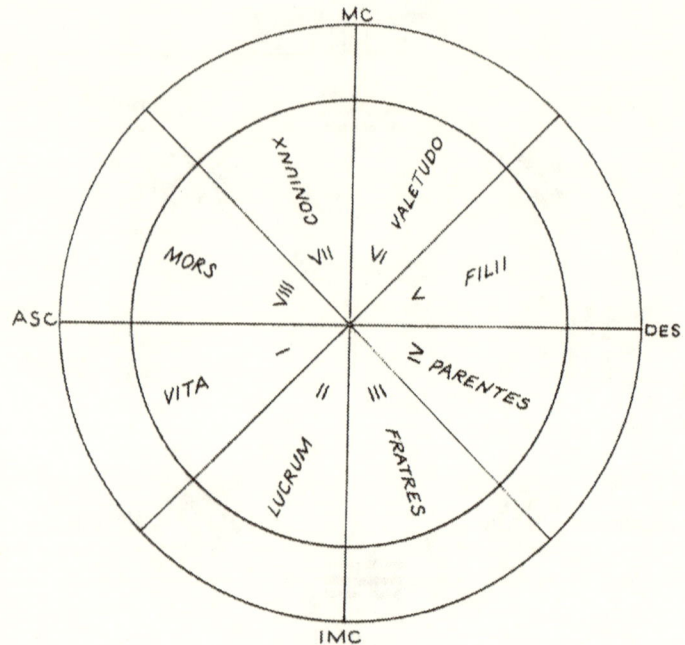

Figure 10. Firmicus Maternus's Eight Mundane Divisions of Heaven with Their Localized Meanings[101]

Table 1. Translation of House Meanings

I	Vita Life;	Identity
II	Lucrum	Wealth
III	Fratres	Brothers and Sisters
IV	Parentes	Parents
V	Filii	Children
VI	Valetudo	Health
VII	Coniunx	Spouse
VIII	Mors	Death

Figure 10 shows the roman house meanings distributed in a division by eight of space at a given time of birth. The template for these divisions assumes a start point of the cycle to the east that progresses counterclockwise in harmony with the movement of movement along the zodiac rather than clockwise in harmony with the diurnal rotation; north and south also are reversed in the drawing to reflect this. But the key to this chart is to understand that the eight divisions of the day or year depict mundane circumstances and processes as distinct from heavenly or zodiac dispositions of bodies. This chart begins with life and progresses around the circle to death in the same way the King Wen diagram begins the living process in the stillness of winter and follows it with a springtime birth and ends with a return to heaven. The Greeks merely chose a different starting point for the growth cycle, which they focused at the east and spring equinox rather than the north and winter solstice. This distinction between mundane process and celestial position is shared across cultures and sheds some light on the distinction that King Wen is making, a distinction that may have come from the Greeks.

Maternus's division of eight precedes the typical astrological chart with its twelve divisions, just as the King Wen breakdown precedes the twelve-division horoscope shown by Needham (chapter 12). The eight-division chart is earlier in the number sequence, not necessarily earlier in time. What is really interesting for Western astrologers is to note the numerical rigidity in the application of the same sequence, 1-8, in the division by twelve shown in figure 10b. The result is that both the structural and cyclic meanings employed by Manilius with the characterization of beginning /life, followed by relations with the world that is completed with an ending in death, is destroyed in the twelve-division sequence. From the cyclic perspective, late winter in the division of eight is suddenly switched to late fall, but all the meanings stay with the numbers instead of the seasonal phase. These changes nullify the value of the present meanings of the twelve-house divisions by undermining the relation between house meanings and the cardinal points or seasonal sequence.

The Roman model is somewhat thrown together as a sequence and does not represent true oppositions of meaning spatially or a true generative sequence in time. Thus, by reapplying to the Romans the very useful space and time/ structure and process distinctions of the *I Ching*, the original house meanings can be improved in their conformity of the cosmographical processes as in figure 10c and 10d.

Both models place life and identity at the midheaven position associated with the sun, not the transitional eastern horizon. Death is placed in the midnight position, and the beginning of the cycle can start anywhere owing to the congruence of house meanings with the cardinal points. However, if death and midnight are considered the end, the best point to start the cycle would be with parents, the source of life below the horizon, in the dark preceding individual life or bodily condition. An idealized chart would place life opposite to death, body and health opposite spouse, parents opposite children, or wealth opposite siblings. These oppositional relations refer to practical conditions (fig. 11C). The positions in both models respect the basic division between the individual in the east and social relations in the west and honors, activity or life above the horizon, death, and duty located below the horizon.

b. Standard Western 12 House Meanings

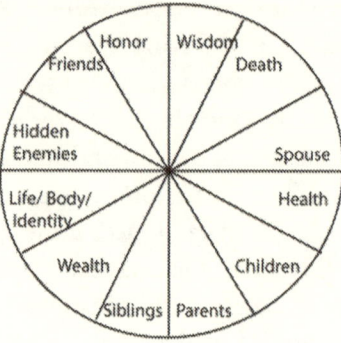

c. Alternative Spatial Arrangement; Anterior Heaven

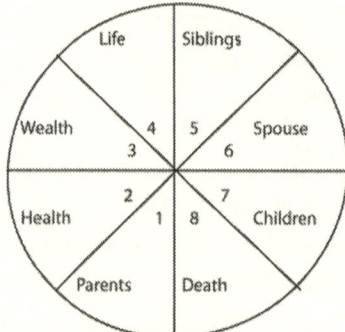

d. Alternative Temporal Arrangement; Posterior heaven

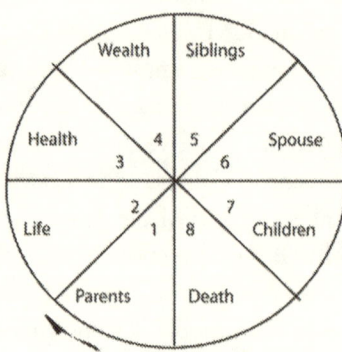

Figure 11. *A*, Reconfigured Roman Houses; *B*, The Original Eight-House Sequence Shown in the First Eight Houses of the Twelve-House System (Counterclockwise Direction); *C*, The Cardinal-Point Spatial Distribution of House Meanings as Oppositions; *D*, The Seasons of Life Generative Distribution of House Meanings

However, that organized meanings should align with the seasonal and diurnal phases determined by the cardinal points and should be reframed when the issue is revisited in discussions based on the monthly divisions by twelve. The object, nature, and the body remain the same for every number window.

The divisions of eight, both east and west, have the capacity to emphasize the full range of human life whether viewed from the perspective of heaven or earth and space or time. In Chinese thought, the memory of an earlier complete number system governing cosmology and medicine is kept intact within the vocabulary of traditional Chinese medicine. The immediacy of application to body structures shows the mimetic tendencies of a thoroughly realistic philosophy. We have seen that the distribution of the trigrams (heaven, horizon, earth) around the divisions of eight in two models was based on principles developed in "two," which led to "four" and the distinction between yin and yang and seasonal changes around the cardinal points. Four leds to "six" with the introduction of the vertical and the possibility of locating cosmographical zones of influence on the three-dimensional human body. The three-yin and three-yang (6) meridian structures with their associated points, then, are very revealing about how the ancients conceived of the division of the body in which the three yins and the three yangs operate. These divisions rely on the combination of horizontal and vertical quadrants into a system of eight topographical and functional zones because certain three-yin and three-yang acupuncture points were selected as influential points or control points for these quadrants. These patterns of resonant points, known as the eight extra meridians, explain implicitly how the quadrants were viewed and how they can be mobilized for treatment in alignment with the phases of the moon.

IV. The Eight Extra Meridians and the Eight Divisions

Acupuncture theory describes trajectories for eight extra meridians and assigns points located on the three-yin and three-yang meridians either as resonant points or confluence/control points thought to be associated with them. The location of the meridians, the functions they possess, and the relations with control points say much about the theory behind the meridians. In general, it is obvious that the concept of the eight extra meridians derive from the geometrical conceptions behind the division of eight. Unraveling the way that these relations developed and how they should be used in treatment is slightly complicated by the differences between the Fu Hsi and King Wen models—which distinguish between structure and function, space and time. The meridian applications clearly were an attempt at compromise to allocate some meridians to structure and some to function in a way that shows their basic interdependency. Some compression of points and implied symmetrical points was involved. Similarly, instead of creating sixteen functions and meridians based on the same set of planes or great circles applied to the body, they differentiated some planes into halves, while some were treated as wholes; and the number of control points was compressed as well. Figure 11 shows the basic distribution of confluence points according to the planes they supposedly control. Even a cursory glance shows a displacement of KI 6 and UB 67 from the centerline and an overlapping of wood and earth meridian points on transitions that suggests an attempt to handle the rising-east and declining-west aspects of the horizon in a living process.

Eight Extra -Confluence Points by Plane Of Influence

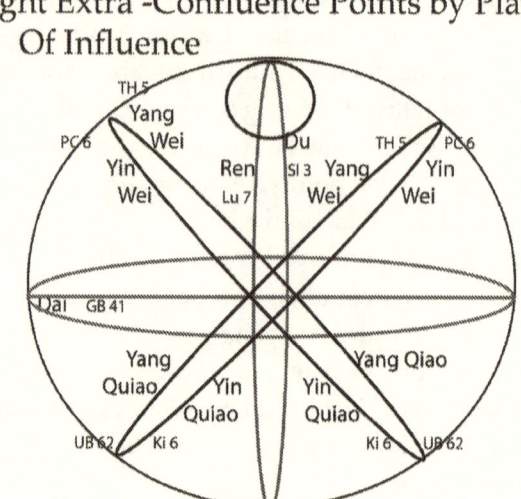

Figure 12. The Planes and Their Associated Eight Extra Meridians with Confluence Points Associated with Them

In addition to their being a link between the basic body planes and divisions of eight as in figures 5 and 6, we learn from study of the eight extra control points and the pathways that the ancients did conceive the unique character of eight to be the ability to distinguish interior from exterior. This is particularly evident in the creation of the chong or penetrating vessel and the incongruities created in the application of *tai yang* and *shao yin* control points. Of all the dimensions, the distinction of self, not interior function from exterior function, is the first expression completion that encompasses all the steps necessary to the creation of a completely individualized form apart from the totality. This completion of form matches the seasonal attributions of the number eight and explains the concept of decline as a necessary accompaniment of any form that emerges from the Tao into some kind of fullness of appearance. It means that an exponential hardening and complexity are replacing the impulse. Thus, the eight principles is an important conception that completes the list of component parts necessary to human existence and provides a comprehensive structure for analyzing a medical condition in its mental (inner) and outer aspects.

The important point is that medical theory, even materialistic medical theory, emerged first from cosmographical fact and thinking. These principles were yin and yang (2) with the experience of hot and cold, moisture and dryness (four cardinal points and four prime qualities), up and down (6), interior exterior (8).

Figure 12 applies a cosmographical order by showing the literal arrangement of control points on the human body in its relation to the eight spatial divisions in figures 5 and 6, where the arms and legs are distinguished. Figure 12 makes a basic distinction between the fixed and the mobile parts of the division of eight as a kind of template for inner to outer relations. Here too we see that points are placed symmetrically on the limbs and at points where the interior and exterior of the limbs are roughly parallel. With the arms and legs extended to match the forty-five-degree angles between cardinal points viewed one way and the vertical to horizon at ninety degrees to another, we can see that the physical symmetries are designed to match exactly the cosmographical symmetries. What those symmetries designate does register on the three-yin and three-yang grid of acupuncture points as well. This is particularly evident in the placement of the linking vessel points, PC 6 and TH 5, which are placed in the middle of the three-yin and the middle of the three-yang parallel meridians.

The leg meridians deviate from this pattern, and the reason for that too has to do with physiology and the extension of the legs well below the torso. In that case, the symmetrical linking of partner meridians *tai yang* and *shao yin* on the feet at the lowest point of contact with the earth roots the transformative processes of the earth, horizon, and terrestrial branches to a dominant *shao yin* kidney at the bottom of the structural and vertical midline plane. This in effect creates a macrocosmic orbit or surrounding mandala from du 20 at the top of the head (also above and outside the torso) ruled by *tai yang* fire to the interior partner *shao yin* kidney at the bottom of the foot well below the torso that stands in an exteriorly and interiorly position relative to the torso.

This exterior-to-interior conception is reinforced intellectually by the opposition of du and ren as halves of the same whole symbolized by du 20 and ren 6 or du 2 to ren 1 on the meridian and *tai yang* small intestine 3 as du confluent point with *tai yin* lung 7 as ren confluent point. Both the great yins are in oppositional relationship, du having reference to the macrocosmic *tai yang* of the back (north-south opposition) and ren to the generative power microcosmic *tai yin* or that yin that has the greatest visible capacity and scope of manifestation in form (as opposed to *shao yin*, which is the least visible and most minute and most thoroughly unusable, unknowable yin) on the front. Giving lung 7 the most interior three-yin meridian position with the hands above and the most exterior when the hands are below at rest (a fact that reflects the reciprocal inhale and exhale of the breath), confluent relations to ren establishes the lung as the key to understanding torso relationships; and it speaks volumes about the relationship between lung and kidney and the whole balance of organ relations in the torso.

The way the control points are placed confirms this understanding that the eight divisions are imposed by planes and quadrants of relationship that are associated either with structure or process but complicates it by making sure that three-yin and three-yang control points, which normally have allegiance to predominantly structural relations based on the solstices cross over and are woven together with process relations, are based on the terrestrial horizon and equinoxes; but the truly interesting thing about this application of control points is that it derives from the actual structures of the body, which themselves dictate and explain the meanings that are symbolized by the points.

Only two of the eight extra meridians have points that stand outside the three-yin and three-yang meridian complex. Those points fall legitimately on what can only be considered as two sides of a single geometrical plane that runs vertically down the center of the body, back to front—thereby, dividing the two symmetrical sides of the body. Du and ren meridians in effect represent a single planar division with points located on it according to a combination of organ location and geometrical proportion. This is important because it implies that the eight other meridians with their borrowed points (from three-yin and three-yang meridians) may also represent similar planes or functions defined by or organizing the quadrants created by the planes, yet they must somehow be derivative or dependent on the vertical meridian that defines the symmetrical halves of the torso.

This notion of a link between geometrical planes and eight extra meridians is reinforced by the rough trajectory of the borrowed points of the *dai mo*, which circles the waist at ninety degrees to the ren/du plane. These two taken together appear to link the eight extra meridians to the Fu Hsi model of eight divisions, which emphasizes a vertical structure of absolutes with heaven above and earth below in their relation to the horizon, and this is structurally mapped anatomically. We can see that the two of the eight extra meridians distinguish back-from-front aspects of the single plane that divides left from right. Two points on the meridians, du 20 and ren 6, seem perfectly to embody the polarities of the two halves; but the two halves are associated with confluent points *tai yang*, si 3 for du and *tai yin*, lu 7 for ren. If this approach were applied to the waist/horizon, it too would have yang and yin front and back or left to right feature; but instead, it has a single point applied to it, *shao yang* GB 41. What would its opposite counterpart be and why doesn't it have a second point attributed to it in conformity with the pattern of spatial divisions by eight in the model of Fu His? If du to ren is heaven to earth, then *shao yang* GB would be associated with fire sun. Instead, GB seems to derive from the King Wen model for thunder; and its opposite would be Mars or Stomach, probably stomach 36, the earth point, or 42, the source.

These relationships cannot be fully realized because of the limitations imposed by the number eight and its geometrical forms, but clearly, there are more meaningful yin yang relations and confluence points than are evident in the traditional ones. Those extra confluence points demanded by symmetry between yin and yang portions of a plane are placed in brackets.

By relying on the basic horizontal and vertical planes that these two grand meridians represent, we can clearly see the imposition of a cosmographical idea on the body, perhaps even more clearly than was possible when considering the three yins and three yangs alone. In actual fact, the three yins and three yangs make no reference to the centerline or the *dai mo*; but the plane is assumed, perhaps more by instinct than by justification operative in the division by six. We saw in that arena that there was either a fixed pole on the vertical axis or a fixed pole on the horizontal axis, but both could not be accommodated by any representation of six. It is only in the division of eight that both the vertical and horizontal can be simultaneously represented, and it is very significant that the vertical in its yang and yin aspects has prior heaven notions built in, owing perhaps to its inclusion of the head/heaven in its structure; and the horizontal is assigned to a posterior heaven construct.

Similar constraints must apply to the divisions by eight, constraints only made explicit in the number ten (the eight directions with interior and exterior made explicit) or twelve (the principal of four on the three levels of heaven, horizon, earth), not merely on the levels of heaven and earth). As a hypothesis, we can say that the ancients anchored the divisions of the eight when applied to the body in these two conceptions—horizontal and vertical or process and structure. The vertical conception is given the greater weight by the body itself because the midline dictates a new meridian.

The invention of du mo to represent a geometric plane in turn calls into question the typical notions of what a meridian is. A point clearly can have functional relations to its zone and through the brain to other zones, but the idea that a meridian really consists in a geometrical relationship and that treatment depends on the brain's geometrical or proportional construction may be an ancient notion that needs revisiting by modern science based on clinical effects.

We saw that figure 3 imposed a body that shows the arms and legs in the positions of transition between fixed anatomical points of reference, so we would expect both arm and leg to show reference to King Wen model transitions appropriate to seasonal midpoints. As mentioned above and true to form, the symbolic points—inner and outer frontier gate on the *shao yang* and *jue yin* meridians, which are located in the middle of the arm, back, and front—symbolically links all yin and all yang (back and front) aspects of the upper quadrants bilaterally (treated as one to account for the body facing north or south). These points are compatible with wood-wind transitioning to fire and sun. Essentially, these points symbolically refer to the upper half of the body in the front and the upper half of the body in back first, then by quadrant left and right of the mobile and, therefore, terrestrial King Wen midpoints. That the confluence points refer to body length meridians themselves symbolically proclaim that at certain key times, these points act as a fulcrum point or activation point to mobilize all yang or all yin capacities above and below in service of a particular quadrant or trajectory.

We already have explained the relations between *tai yang* and *shao yin* of the foot and the regulation of the *chiao mai* and lower quadrants. Normally, we would have expected them to rule the ren mo because of the close proximity of *shao yin* kidney in function and location to the ren centerline including the conception of water and yin, midnight, and midwinter with its implied hinge effect. However, within the tight constraints of the division of eight and the real extension of the legs all the way to touching the earth whereas above a least when the arms are at rest and humans are not exceeding their station, the arms are below the head; and the head itself and the du meridian rises above the torso and touches heaven, not the *tai yang* confluence point. Nevertheless, the attribution is initially a surprise but one that is dictated by the body. The *chiao mo* and *shao yin* have more to do with the activity of the activation of yin and yang in the lower quadrants than lung, spleen, or liver; and that may be the reason why in addition to structure, this link was made.

The fact that one of the eight extra meridians is called the penetrating vessel and its trajectory closely parallels ren and follows the trajectory of the kidney meridian accomplishes two important tasks:

1. It reconfirms that the number eight is closely tied to the concept of generation, conception, birth, and form and that this meridian symbolize clearly the interior-to-exterior relations unique to the number eight.
2. It salvages the connections between *shao yin* kidney and the centerline that were sacrificed by putting the confluence points that control the lower lateral planes of the leg in *shao yin* kidney and *tai yang* bladder when normally, the affinity to these quadrants would have fallen to liver and gallbladder.

Overall, the choices dictated by the model force, the model to visually and structurally emphasize the interior to exterior dimension of the plane of the ren/du, and, therefore, its relationship functionally more in a front to back than side-to-side relationship. Rhythms of intercourse, for instance, generally are rhythmic front-to-back motions more than side-to-side motions, and this may be what the *chong mo* ren (conception vessel) is emphasizing. However, all three-yin meridians are involved in reproduction and connect with the ren pathway and centerline so we have to see the *chong mo* as somehow being a mediator, perhaps referring to the interior mediator between the three yins in the front of the legs and body and the three yangs at the back of the legs and body. When the feet are pressed together, spleen 4 is near the point where the feet actually touch at the centerline. The other point is the interior maleolus and kidney 6. This again reinforces that interior mediating role of the spleen and kidney for organizing and connecting between the yang left and yin right sides of the body on either side of the ren-du plane.

Eight Extra -Confluence Points by Plane Of Influence

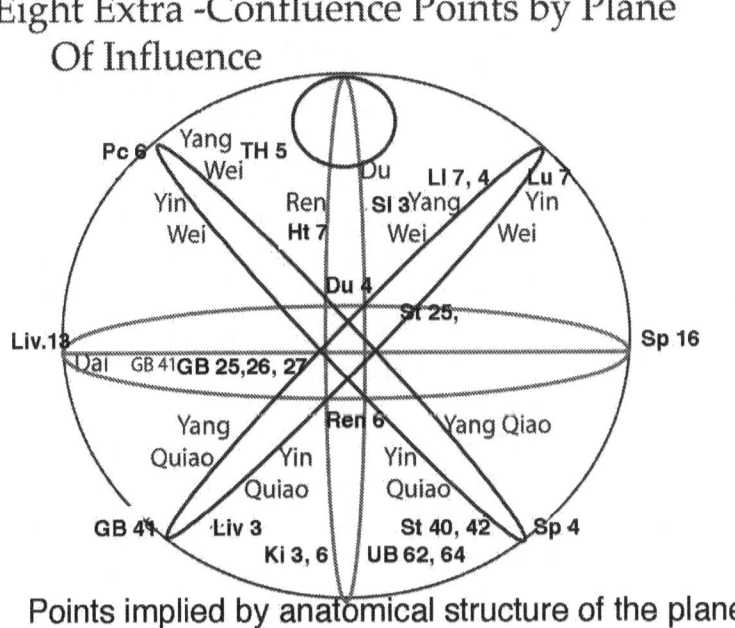

Points implied by anatomical structure of the planes.

Figure 13. The Unpacked Relations between Confluence Points for the Eight Extra Meridians and Their Implied Counterparts

Eight Extra Meridian and Eight Divisions
Time Phases and Structural Location of Confluence Points

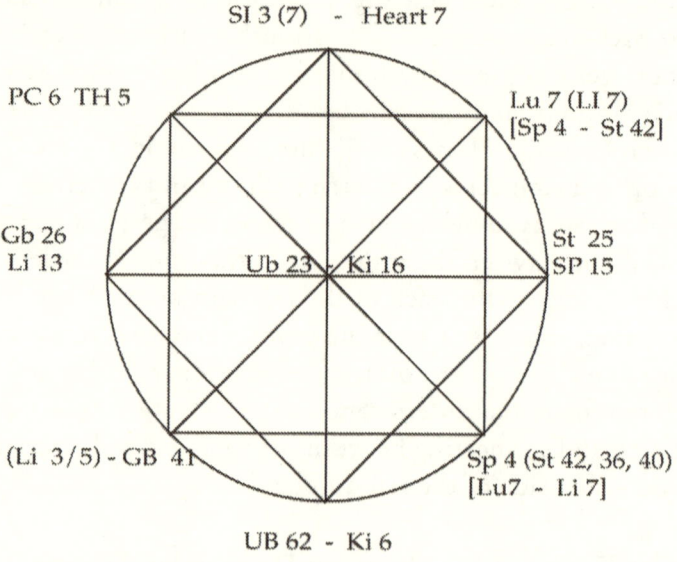

SI 3 (7) - Heart 7

PC 6 TH 5

Lu 7 (LI 7)
[Sp 4 - St 42]

Gb 26
Li 13

Ub 23 Ki 16

St 25
SP 15

(Li 3/5) - GB 41

Sp 4 (St 42, 36, 40)
[Lu7 - Li 7]

UB 62 - Ki 6

Figure 14. Simplified Figure 13, a Two-Dimensional Drawing of the Two Crosses with the Time-Phase Distribution of Confluence Points

Figure 13 lays out the confluence points structurally, and they have been distributed with their implicit yin and yang partners using comparable connecting luo or element points. They have been distributed symmetrically in oppositional contrast to their cosmographical opposites, which reflect the intention of the yin yang divisions of the various planes. Figure 14 lays this out two-dimensionally as in figure 2 showing the structural and temporal order of eight extra meridian phases harmoniously with the three-yin and three-yang cycle whose points conform topographically with those phases. The pattern in figures 13 and 14 shows that in the waxing cycle, *shao yang* and *jue yin* points above will regulate the linking vessels and quadrants and below will regulate the motility while during the waning cycle, *tai yin* and *yang ming* above will regulate linking quadrants and below quio quadrants.

In greater detail, the figures show that the luo points of great yin lung and spleen are both used, the former for ren and the latter for chong—two related interior functions. As illustrated in the descriptions of the five-element depictions of the waning cycle, lung and spleen are anatomically interchangeable as above or below depending on whether space or time are being treated. If great yang spleen and lung symbolically regulate the ren and chong vessels with their interior and back-to-front dimensions, it is interesting that their partner meridian *yang ming* stomach and key points 40 (luo) or 36 (earth) points with large intestine 1 (element, source) or 7 (luo) regulate the declining western aspect of the horizon. They operate opposite *shao yang* gallbladder 41 (wood) or 37 (luo) to act as confluent points for yang aspect of the horizontal *dai mo*, which refers cosmographically to connection with dawn and dusk and the outermost manifest world of action. The internal partner to the gallbladder is the liver; and therefore, liver 1, 3, or 8 acts as an oppositional counterpart of stomach spleen in the regulation of the chong mo. These symmetries point to the confluent points and eight extra meridians as abbreviations of a set of counterbalancing forces that keep equilibrium on the yin or yang aspects of the planes that they supposedly represent. Moreover, their partners assist in this equilibrium by treating interior to exterior, side to side, back to front, and up and down depending on whether the aim focuses on symmetry (space) or process (time) when you fill in the picture.

Ren is regulated by lu 7. Lung 7 will treat the symmetrical quadrant on the side the needle is inserted. It will activate the relatively interior, upper, and front aspect structurally and the lower interior front aspect in relation to process; the yang partner (li 4, 7) treats upper and outer and back quadrants on the side where the needle is inserted and has more to do with external defense, and it affects the *dai mo* by changing the quadrant equilibrium. *Tai yin* spleen spleen 4 (also 1 or 6) treats the ren below the navel in its interior or descending aspect; its partner, stomach 40 (36), treats the lower outer quadrant in the front and in so doing affects the equilibrium of the *dai mo*.

This description shows the intimate relationships proposed by the traditional system between the vertical centerline plane and the horizontal plane. The same intimacy can be expected between the horizontal plane and the centerline in its aspect as the du mo. The one affects the other

Du is regulated by *tai yang* small intestine 3, the tonification point. *Tai yang* is the meridian closest to the centerline, and that parallels the du mo so this makes sense. The activating point could also be the luo, element, or source point depending on the situation. The small intestine regulates from the most exterior or interior position depending on the position of the hand. This structural contradiction is contained in the polarity in partnership of *tai yang*, the outermost, and *shao yin*, the innermost meridians; and the reversals are interesting. Still, the heart is not emphasized in the system; and it is lung, not heart, that regulates the upper ren pathway, the counterpart to the upper du pathway.

The implicit partner to the small intestine is heart and probably does work to regulate the du mo and to equilibrate the ren and du in the upper body with some effect on the *dai mo*. If weak, there will be gas and upward pressure against the diaphragm for instance; and if relatively strong, there will be stagnation or heartburn. Heart 8 or 5 could theoretically regulate the inner aspects of the upper body quadrants, and it probably does. However, the heart so dominates the upper body and is so central that in the Chinese mind, it equally regulates the interior aspect to the du and the ren and that automatically. The choice of the tonification point of small intestine then emphasizes that the only problems that really will show up in the upper quadrants are problems of deficiency and that these will show up in the quadrants or near the du mo as back and costal pain. The selection of the tonification point will take some pressure of the *dai mo* and diaphragm while it augments the chi in the chest, helping the heart indirectly. The unique passive role of the heart and the importance of the rhythmic function of the lungs in their relation to the heart particularly through ren 17 and the zhong qi may fundamentally explain the allocation of ren mo to lung 7 and the regulatory control of *tai yin*.

The balance between pericardium and three heater to regulate the upper divisions through the plane that runs side to side, dividing back from front, has been explained. However, *shao yang* three heater is the upper body partner to *shao yang* gallbladder, and it regulates the *dai mo*. Thus, it appears that the exterior aspect and the upper back quadrant is regulated through three heater and gallbladder while the interior aspect upper front is treated by pericardium and the interior of the *dai mo* by the [liver]. This explains the uses of the liver points on the *dai mo*. On one side, we have the *dai mo* being explicitly regulated by the gallbladder; and having the liver and three-yin and three-ang partners from above and on the other hand, we have it being influenced implicitly from lung in the front upper aspect and from quadrants below through the spleen and stomach counterparts. There are spleen points and the earth *mu* point on the *dai mo* to confirm that these relations are patent. What we are seeing, then, is that the same five elements and prime qualities are operating in the eight extra meridian system; the eight extra meridians simply define the planes and quadrants that outline the human frame and, in particular, define the interior-to-exterior relations within those quadrants along with the above-to-below, back-to-front, and side-to-side planes.

The eight regulatory points that supposedly affect the meridians or activate them really does no such thing. They are cosmographical signposts that flag the systems of checks and balances within the three yins and three yangs that likely will affect the quadrants defined by the planes involved. Most of the points on the meridians establish above-to-below or inner-to-outer connections that justify the theoretical model by linking relevant blocks of meridians on the front and back or above and below to compel advantage in a quadrant.

The luo points, generally, are used to confirm inner-to-outer connections between inner *zhang* organs and their outer counterparts and possibly their tendino-muscular counterparts to treat muscular pain in particular quadrant. The emphasis on luo points, particularly for those wood and earth elements, that dominate the horizon emphasizes the inner-to-outer dimension of the divisions by eight. The yang organs tend to use element or tonification points to reinforce or strengthen linking or outer defenses and to mobilize chi around an area. Clearly, once the cyclic principle is understood the luo or element point can be selected according to need so long as the trajectory *tai yin* or *yang ming* is selected in the declining phase. Similarly, despite the distribution of confluent points the system clearly is talking about proportionality and equilibrium in bodily zones that are defined by planes but that already have infrastructure from the three yins and three yangs. The confluence points are flags that speak to the system of three-yin and three-yang, interior-to-exterior points that need to be brought into equilibrium in order to balance the quadrants.

V. Medical Practice—Treatment

How then to best use the system in its original cosmographical sense? Eight is the child of seven, so it gives us the structural counterpart to implement and respond to the day-to-day ebb and flow of twelve seasonal lunations and their phases. Each day has a characteristic energy associated with the phase of the day in combination with the phase of the lunation in a particular season. This means that in the springtime generally, we can best access the whole body through the *shao yang* and *jue yin*; and within that framework, we can refine treatments by taking into account the waxing and waning phases of the moon and the diurnal cycle. If the moon is waxing, we get further support for treating these same trajectories, and if daylight is treating through pericardium and three heater. The only difference with the number eight is that it defines the quadrants of the body that are most active or that are interactive in phase. An early spring phase will justify treating wood a later spring phase pericardium three heater. An early spring phase with a late waxing moon would suggest treating both wood meridians and ministerial fire together with an early waning moon-wood meridians with metal, such as liver 3/li 4 for a characterological or structural treatment, liver 3 with spleen 1 for a process or digestive treatment. With the early spring, waning moon and morning treatment including pericardium 6 or 7 or liver 1 with pericardium 8 and spleen 1 or the alternative for metal liver 4, pericardium 5 (which as a union point includes lung), and lung 9 would generate a congruent treatment for all the phases involved.

These blending of treatments to include phase relations in the general mix of points that support the treatment target are powerful.

The power of these kinds of point selections is that they severely focus treatments toward particular aims using the maximum available active energy of the phases. This has the potential to harness more chi and create equilibrium in the whole through the functions that are in play. It is difficult to sleep after dawn and stay awake in the evening, and the same functional characteristics of periods of time can be harnessed. This energy can be harnessed in several ways:

1. The person's energy can be harmonized with the world through this treatment alone.
2. If the causative factor is being treated, the season, month, or day phase that is harmonious with the causative factor can be used to support that weak link with good effects on the whole.
3. Particular TCM syndrome treatments and needle patterns can be refined and harmonized by shaping and selecting points on the meridians selected that have to do with the phases or the causative factor or both. This provides a way to focus the lens of a treatment toward treating the syndrome. It also gives a way to help a patient that goes beyond the syndrome diagnosis when their condition has improved. A treatment that harmonizes an individual with the state of things generally is supportive.
4. Diagnosis and prognosis are improved by following the moon phases and seasonal trends. Symptoms can be synchronistic with these phases as can be their resolution.
5. Using the pulses, symptoms and signs within the cosmic perspective can show where the extra chi is to apply to treatment.

Regardless of method, the direct application of points above and below the waist and back to front according to the relative yin or yang condition of the phases is the legacy of the divisions of eight, which establish quadrants vertically and horizontally that are linked to the movements in the greater cosmos.

The system of eight divisions can use figures 12 and 13 above as a basis for selecting master points that structurally are resonant with whole quadrants. Since they largely are luo points when in phase, they allow for the opening of the interior-to-exterior relationship between the chi of humans and their world, which results in an equilibrium, either resulting from the release of an interior excess of pent-up emotions and/or the infusion of cosmic chi into the associated deficient quadrant. These points can be seen as an improvement on the standard rote use of confluence points as if they had a magical association with the so-called eight extra meridians. Instead, we see that these meridians simply define planes that link the existing three-yin and three-yang meridians because their X or forty-five-degree trajectory crosses the vertical planes and topography of the three yins and three yangs. Selecting these points is not magical, rather it makes explicit with the addition of implicit meridian partner points to what before was a shorthand system for recollecting the geometrical and structural balances between quadrants of the body.

Like the five-element and six configurations before it, this system has to be understood in its Fu Hsi and King Wen aspects.

Fu Hsi

We treat the quadrants of the body as Fu Hsi structures knowing that there is a one-to-one relationship between the body and the world. In this set of timeless relations, we can structure effective treatments by balancing quadrants above and below the waist or alternatively back to front by using the three-yin or three-yang blocks through the confluent points and their counterparts. We've learned to treat upper body and back pain and deficiency through *tai yang* small intestine and its lower body partner UB. And we've learned that we can similarly treat it through the front-to-back relationship using the partner meridians *shao yin* heart and kidney (particularly the kidney points on the front at the level of the heart). We also can treat back problems bilaterally using meridians bilaterally, by supporting the deficient side, or by doing X treatments that use the quadrant opposite side to side and above to below together. Focusing on the bilateral process between heart and lung, small intestine and large intestine can support this bilateral structure. Finally, the eight divisions allow us to treat interior to exterior, which is handled to some extent by the heart as interior small intestine as exterior relationship; but this can be augmented by treating ren and du points together, which in some cases can be the sole cure for back problems particularly those that derive from extreme compensations expressed as hyper control and hysteria.

Structural treatments on the three-yin and three-yang meridians can refine the oppositional character of quadrant treatments. If you treat on the far outside of a quadrant, you can treat the opposite side on the far inside of that quadrant, which in some cases above the waist will be treating *tai yang* small intestine to *tai yin* lung or spleen, others as *yang ming* large intestine to *shao yin* heart or kidney. These are strictly structural oppositions. The eight divisions associated that are structurally disposed points with metal element meridians above the horizon can be used in these cases to augment any points in the front or back and above or below quadrants of the body.

King Wen

Treatment of the process makes the quadrant treatments fall more in line with the five-element phase treatments than those of the six configurations. The link between a standard five-element treatment of wood and the eight divisions, for instance, is simply that the five-element treatment can be augmented by (1) treating in season augmented by the lunation phase or time of day which we already knew and (2) using the process distribution confluence points from figure 13 according to the phase to augment the normal five-element treatments.

The most important thing to grasp is that each seasonal phase is a world unto itself when it manifests. You cannot convert autumn into spring. You have to access the qualities of spring through the vehicle of the autumn during the autumn. During the autumn, qualities of spring are largely absent, below the surface, and we know that they never are eliminated entirely. This cosmographical illusion is that the present spatial vista is all there is even though latent in it are all other times and circumstances. This understanding is so important to treatment, and it is an understanding that arises spontaneously from the interior to exterior relationships of the divisions of eight. We have to treat the interior spring through what is manifest, namely autumn. This understanding of literal physiological connections with the world process profoundly conditions point selection. The process view of the divisions of eight requires this recognition more than the purely spatial Fu Hsi configuration.

Owing to the nature of seasons—vertical spring, horizontal fall—I've included more source and element points in the *shao yang* and *jue yin* rising aspect of the cycle than in the declining phases because the opportunities for chi augmentation are as important as the achievement of equilibrium. However, the division of eight explains on an entirely different level, the interior to exterior significance of luo points, which on this level refer to the interior-to-exterior relationships between quadrants of the body as well as between five-element orb and yin and yang partners. Clearly, the luo points function to create equilibrium above to below, side to side, and back to front for whole quadrants particularly when the three-yin and three-yang luo points are in phase or in an opposition to phase.

In the autumn or afternoon, the body can be visualized as facing head down with feet above, and this naturally torques the treatment by using some important functional oppositions. Taking the example used above of an early spring wood treatment, the functional opposition is earth *tai yin* spleen and *yang ming* stomach rather than metal, and an early waning moon would reinforce this. Such a model creates treatments that balance wood and earth, the two functions that govern the middle. This configuration allows for the balancing of functional oppositions of the relatively moist and the relatively dry, the functions of early decline and early emergence from the seasonal model. This provides lots of ways to control dampness, dryness, heat, and cold as they relate to the digestive process and the normal day to day as opposed to extreme equilibriums of life. This particular relationship is all about the horizon and trends up to it. Gallbladder is about rapidly rising chi; the liver gives a gentle downward counterforce. *Yang ming* is about rapidly declining chi, while the spleen is about a gently compensating rise in energy. Thus, this functional treatment is about seeing that these two processes work. The liver and spleen together represent the reservoir of yin that restrains fire in the spring and restrains cold in the fall. Gallbladder mobilizes fluids toward upward growth and form; *yang ming* mobilizes fluids for downward elimination and decline. The local points along the horizon combined with the element points that resonate with whole quadrants can strongly support one or the other of these tightly coordinated functions.

Based on cosmology, it is clinically more effective and appropriate to view the eight extra meridian points and command points as quadrant treatments than meridian treatments. They define trajectories in empty space more than substances or pathways. The following comments are made from that perspective. To treat the *dai mo*, we treat the middle according to the quadrant location of the problem. If we need to mobilize, reinforce, and strengthen, we turn to thunder and GB. If we need to smoothen, nourish, and lubricate dry conditions, we stimulate the liver. If we need to calm and promote digestion, we treat the spleen; and if we want to accelerate the removal of dampness, we treat the stomach and large intestine. If we need to treat stamina, reproduction, balance, strength, ambition, or supplement or drain an excess below the waist, we treat the lower body quadrants through the confluence points and partners. We can do the confluence points for all the planes that surround or demarcate a zone. In the back lower zone, we can treat UB 62 for the centerline and inner and outer equilibrium GB 37 or 41 and partners or stomach 40 or 36 and partners for *dai mo* and outer connection with these real-world circumstances.

Chapter 9 九

The Completed Body in Three Dimensions

With the establishment of a comprehensive framework for describing any space in two dimensions, the number eight when applied to the body shows the beginnings of future developments that would include the third dimension of depth. Depth is described for Plato's cube of the plane of the prime meridian that is represented by the meridians of du, and ren has a trajectory that penetrates the body back to front and, therefore, contains the notion of interior to exterior. This idea is reinforced by the creation of the chong mo that with its links to the uterus reinforces the importance of this inward versus outward distinction. The succeeding yang number nine like all yang numbers has to do more with process than with structure; and because of this, it compresses the three dimensionality emerging in the number eight into action, rendering a very powerful image of fullness and completion when drawn. Where eight was the outcome of the inclusion of the lunations and the details of monthly and weekly cycles to the annual progression of four seasons, the number nine sets the stage for the complete three-dimensional landscape symbolized by the number ten. This landscape includes heaven, earth, and horizon and not as a mere vertical possibility for life but fully rendered in three dimensions as a window that nearly approximates the one. This is the mystery of ten that it restores the whole in an image that allows for the infinite expansion of particulars in a fully rendered whole. It is like the one but not the one. It is a prototype human ready to learn from experience. The number nine displays the processes and conditions that require three dimensions.

What is required in the world of three dimensions, which is spatial recreation of the conditions in which life can flourish, is a detailed understanding of how the light cycles affect every aspect of life—in heaven, in the earth, and on the horizon between. As demonstrated in discussions of the three yins and three yangs (six), we saw that the three phases of the cycle of light had a vertical pole having to do with space and a horizontal pole having to do with time. In that construct, dawn and dusk were key transitional points on the horizon defining the directionality of the cycle. The axis of midday and midnight forms a neutral center, the kind of center a person standing on the earth would have when standing in a field to view the whole process. In addition to the three above and three below the number six added the yin number meanings of space and of verticality (up and down 2) to the four cardinal points with their spatial (north and south) meanings and process (east and west) meanings. Now, in the number nine, there is the possibility of fullness where the perspective on the horizon can be fully amplified for the first time. It is no longer the three yangs of heaven and the three yins of earth but the horizon and the geographical terrain of life that can be fully understood with the number nine. The same laws formerly applied to heaven and earth now can be applied to the horizon, to the individual body, and to all of life.

I. The Number Nine

The number nine, [九] jiu (3),(chiu) (3), has "as its primary I (1) or germination of lesson 9 in Wieger, but it has a support oblique line like P'ieh (1) and a horizontal line at the level of heaven. The germination is appended or hangs off the line for heaven while the other line is an upright slash that penetrates heaven but is predominantly below. The character has three strokes. Chiu 3 has variants that include the character for smoky quartz and for duration. Chiu 4 variants have Ju (4) or root images. The two verticals have an empty space between which suggests three zones horizontally and three zones vertically or nine."[102]

In my view, the graphic for nine is a complex expression of the human capability. The symbol for germination is accompanied by what could be a slash or upright pole of support, which implies an action that leads to germination underground or a support for sustaining growth upright. Alternatively, the character implies something upright with a perspective from above. The extension of the line above the horizontal and the roots below are a bow to the three levels—heaven, earth, and human—and a bow to the number three. The three levels are supported by two vertical strokes, yin and yang; and one line is masculine, the other feminine and containing. The number nine is the template for the human form with all its verticality and capacities for freedom of motion, consciousness, and generativity. It is consistent with ancient philosophy that the human form is the unique end product and highest outcome of the cosmic order, and this symbol contains that implication.

Gestation of the human foetus is nine months, a fact that would not have been lost on the sages. The image of germination and human reference combined with the growth cycle's ninth month work synergistically as a catalogue of integrated meanings.

II. Cosmographic Structure

Implicit in the number nine is the application of the three modes of astrology which are the mutable, cardinal and fixed; the three comparable gunas of Hindu astrology which are the sat, raj, and tam; or in more modern terms the three states of energy which are the potential energy, kinetic energy, and inertia—all of which described the three phases of any natural light cycle, the beginning with rising light, the middle or transition at the noon/midnight neutral phase, and the declining light phase. These processes can be used to divide space or isolate phases in time. Within this framework, most events and experiences can be identified in a meaningful causative setting.

That form itself emerges from the application of the three-yin and three-yang conceptions on the plane of the horizon as well as in heaven as phases of bodies and weather and earth as transformations of the growth cycle. When applied to the body, it simply means that the visible divisions of the body can be used as a matrix in which to see the effect of the light cycles and the generative relations of positive neutral and negative. For instance, the three chiao or divisions of the torso wherein each are divided to left, center, and right; and it is understood that the left-center-right divisions contain the three physiological processes in all their ramifications such as chi blood and balance, yin yang and balance, that each relate to the relevant upper, middle, or lower zone in the body. This is congruent with the symmetrical divisions of cells and development of the foetus.

The vertical and horizontal quadrants now considered functionally in relation to the different, more detailed levels of the body have a functional balance between yang and yin in each zone, and these zones are organized in humans by the number nine (fig. 1). Practically speaking, the three zones of the head reflect the three chiao on the body, which reflect the three divisions of the legs, making the nine zones in the structures of the body in which a unique equilibrium of yin and yang must be maintained. The equilibrium is different in each zone, and each zone reflects the nature of heaven, horizon, and earth in the functions contained in them. The self-evident relations are (1) heaven, breathing, air; (2) heart and fire, which are the higher centers of the brain and in fetal position the higher functional capacities of hands and feet; (3) earth, feces, and urine reflect the rivers and material substances of the earth and the jaw of the head, which is designed to consume them; (4) horizon, where the dynamic processes and experiences and transformations of life occur as with the digestive process and where the sensory apertures are in the head.

Nine

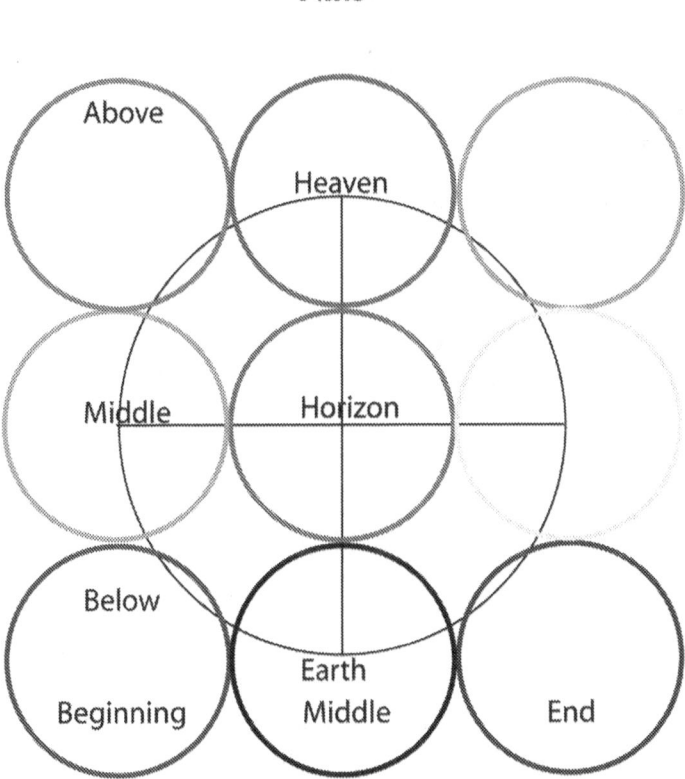

Figure 1. The Standard Image of Nine with Three in Each of Heaven, Earth, and Horizon

Figure 1 shows the consequence of applying the principle of three both horizontally and vertically in the cosmos as it is distributed in the real terrain and structures of the body. This is the famous magical square of the Arabs. The visual magic it contains is that the formless cyclical processes of nature (yang) when understood horizontally and vertically creates a fully rendered three-dimensional structure, a square. This is the principle of the pyramids, the mystery of the relation between prior heaven and posterior earth between the basic vertical organization of life—heaven, horizon, earth—and the basic horizontal division defined by four directions. The magic is geometrical and somewhat surprising that the number three times three a number of completion for the principle of three achieves stability and completion. The reason for this, however, has been established by earlier numbers as far back as two and four, where it became clear that yin and yang require that everything be viewed as both structure and function in space and time; and the odd to even oscillation of numbers fully examines this necessity. However, it is at the number nine that the square symbolically is completed that looking at the human body face from any perspective, the whole thing can be encompassed both structurally and functionally.

Nine

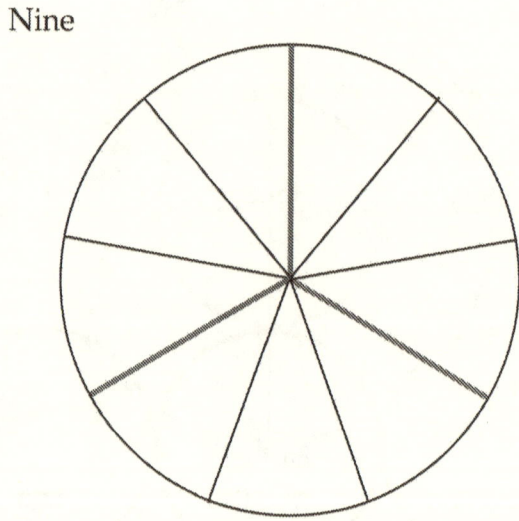

Figure 2. Nine Divisions of the Circle, the Result of Plato's Square of Three

The simplest two-dimensional model for nine is nine divisions of the annual circle, which creates nine forty-degree divisions known as noviles (fig. 2). In keeping with yang numbers, the apex is pointed at the noon position in heaven. There is no symmetrical connection with the horizon, and there is no symmetrical link with the three-yin and three-yang divisions of six, which is unexpected. Unlike six, nine has no evident link to the root yin number two. This is in keeping with the thoroughly yang and mysterious, unmanifest character of the number. Forty degrees as a multiple of four is a gratuitous confirmation of the square, but this symbolically reflects only the evolution of nine from eight. The addition of a single stroke compresses the squares of the number eight, creating a dynamic wheel whose base structure is three not two. What is distinctly symmetrical are the three 120-degree equilateral triangles generated by nine (red, 3 x 40 degrees = 120). These triangular structures can be created using any spoke as a starting point, and this reflects a revolving structure that nevertheless has a stable internal organization that allows for adaptation. It also confirms the unique position of nine as a pure extension of three and its principles but now presenting in a postnatal real world of forms, change, and circumstance.

The traditional philosophical implications of three and, hence, of nine are the differentiation of the three spirits of rational, animal, and vegetable; Chinese shen, hun, and po; the three gunas of sat, raj, and tam; the three Greek modes cardinal, fixed, and mutable. Eventually in the twelve divisions of the heavens, these three principles were each assigned to one of the three seasonal lunations of thirty degrees days each. Here, we see them purely in their own self-sustaining, self-regulating context infinitely reflecting each other as ramifying the number three in the form of the beginning, middle, and end of the beginning, middle, and end of the beginning middle and end and so on: 1.333333333 . . .

Each incremental geometrical division of the circle gets ever closer to the circle without the possibility of achieving it but achieving nonetheless an approximate complexity. The novile in astrology is an irrational aspect that implies this by describing a movement from harmony to the tension in full manifestation and completion. The horizontal and the vertical are in a state of tensive balance in which a nudge can move the whole outwardly unstable but inwardly strong structure. It is a position, therefore, of illusion—the illusion of form; the illusion of time, space and direction; the illusion of balance, depth, of options and choice—where in actual fact, nothing is fully manifest. Everything is temporary and elusive; there is only the imprisonment of consciousness in a dynamic prism, a mere yet near approximation of wholeness and identity.

Figure 3 shows the three horizontal planes from the Chinese number three that form the foundation to the left, center, and right in nine. The upper and lower planes are defined by the conditions set by the geometry of the number eight. The figure shows the implicit memory of the preceding yang number seven. Compared to seven, everything about nine is a mystery. It fits asymmetrically with every system except three. Even three thirties, which equals to ninety days, and three times ninety days completing the symbolic year is eerily imperfect. It refers in a replicative way to the principle of three, three divisions of the year, which ultimately is equal to three months. It represents to the mystics among us the five visible planets, the sun and moon, and the two invisible planetary giants—Neptune and Uranus. Nine has the feel and quality of the epic and moment-to-moment tension between the resistless tide of the spirit of evolution and the relentless pull of gravity held together in some balance by our instincts and actions.

Figure 3 contains an image of the planes involved in the number nine including those of the preceding numbers seven and eight.

Nine

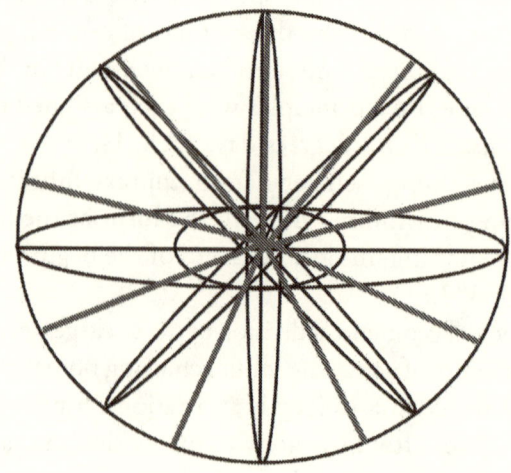

Figure 3. The Planes behind the System of Nine

The three horizontal planes that connect the active reach of the *X* pattern from eight also serves to link the three yins and three yangs on each of their different levels of function in the body, not only above and below, but also their capacity for combined and integrated action. Figure 4 shows the vertical human figure superimposed on the model. This is a very complex figure because it can dynamically face in any direction and has multiple capacities for functioning—spiritually, mentally/emotionally, and physically. It is able to initiate actions on all three levels and pursue them by spiritual or intellectual journeys with beginnings, middles, and ends; by socially interacting and producing food and livelihood; and at the most basic levels by genetically reproducing and applying mammalian rules to raising offspring.

Nine: Three Dimensions

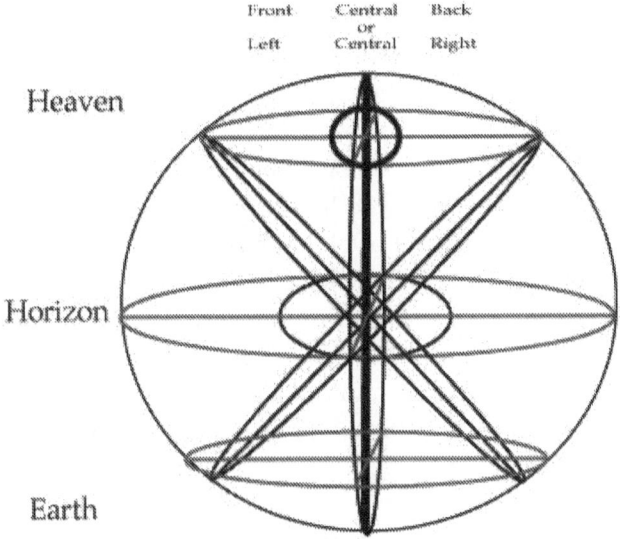

Figure 4. The Vertical Stick Figure of the Cosmic Man Superimposed on the Planes

Figure 4 shows that the number nine views the body as a unified structure, but it does not accommodate the arms very well. The arms have to be at the sides or if above, included as mere passive extensions of the head or torso. This is unexpected because the arm and legs together were so basic to understanding the differentiation between form and function on the basis of the body's quadrants as defined by the eight extra meridians. Clearly, the number nine for all its structural evidence in the body is not fully integrated either as a description of structure or of process. The lateral movement and functionality evident in the human being is constrained somehow; and it is the square that constrains it. The body itself, now fully rendered is in a sense the prison of the soul, the prisoner of our capacity to identify with the whole. Nine in a way creates the body itself as a prison as a limit or confinement that is incomplete. It is a confinement of time defined by the beginning, middle, and end of every completed structure. It is a whole, a creation, complete in itself; but it is not the one, and while all its functions work, it has a discrete and limited identity. Number nine compresses the vision of the whole to discrete objects and to how the laws of heaven and earth apply to the horizon, to a life on the horizon as a distinct creature. Thus, it is in the purview of number nine that we are able to have conditions that allow for the generation, growth, completion and death of form, the growth and completion of crops, the harvesting of grain, and for the annual cycle the nine-month gestation and the birth of children. But this very capacity, the capacity for life and form, is intrinsically requiring separation, requiring identity and autonomy that somehow limits a person, separating him from complete whole and function.

It is only with the number ten that the inhibition is lifted on the completed human; and the arms can be raised to heaven, the human can explore freely the world of interaction, the world of philosophy—the heaven principle. In a sense, the illusion of wholeness is complete; but the human does have almost infinite capabilities for adaptation and movement, virtually godlike powers over other creatures owing to its fulfillment of and understanding of numbers. Nevertheless, the functional and anatomical structures and the number nine are very basic and important in the differentiation and selection for use of specific acupuncture points.

As above, so below is a timeworn truth. Each level is replicated in miniature, holographically in heaven, in earth, and in between. Similarly, each action has an implication on all three levels and a bias toward one or the other level. These actions can be identified now as trends or as a bias toward one level or another in an individual, and these trends can be broken down into nine or eighteen subtypes of personality. This capability is similar to the way that the general principle of five elements is extended to twenty-five different primary diagnostic combinations and treatment interventions through the access provided by the symbolism of five elements point on each of five meridians. Here, the standard is not associated with the horizon only. In this model, the entire knowledge and associative power of the five elements is contained within the spectrum of the horizon with its more localized heaven, earth, and horizon system with its external and internal causes of disease and natural element circulation. Now each of the three primary zones of existence also are included—the celestial with its planetary motions and the earth with all its lunations and preconditions for life in the form of natural laws. This means that the three zones can each be comprehensively studied, not horizontally only, but also in their vertical connections appropriate to the number three.

The verticality of the divisions of three and nine explain some of the mystery and sense of confinement in the number. Each person is imprisoned in his or her own equilibrium. Nor does life with all its levels play out on a circle as cleanly as the cardinal points and five elements. Rather, it holistically describes in three dimensions and in association with the prime vertical the impact of the horizontal cycles on three distinct levels, which yet operate interdependently. This is less of a cyclic model than a study of the structural locations and dynamics of a given limited period of approximately two and a half hours.

Some further cosmographical projections of three and nine on the circle may help to refine an understanding of the numbers.

The natural breakdown of any division or partial division of three is nine because of the homogeneity of three—each part having a beginning, middle, and end as well as an upper, middle, and lower part, ad infinitum. This yields figures 5a and 5b, where adding two hexagons for a total of three creates three phases for each of the three yins and three yangs projected on the cosmos. Figure 5 shows the day, month, or year divided into eighteen periods, each of 1.33333 to infinity hours. Once again, the assymetricality and mystery of three creeps into the mix of the merely material. Three is in a way a stable structure that can encompass the range of conditions. Three describes the necessary motion intrinsic to a whole comprised of more than two parts that necessarily must interact to retain wholeness. Nine phases are yang, nine are yin; and they further break down the three yins and three yangs into components, which necessarily can be viewed once again as polarized, either as structural or operational.

Just as with the six divisions, the model can only satisfy the requirements of the vertical or the horizontal. The blue line shows the empty unaligned and asymmetrical phase that happens to the model. The number nine is always present but it can only connect as process (beginning, middle, end) or as vertical (heaven, horizon, earth), not both together with any coherence. If space is modeled, there is no beginning, middle, or end. If time is modeled, there is no up and down. The number nine is schizophrenic in this way and adapts to one perspective or another but not both. Even so, one implies the other, and together they beg for the number ten to give a fully realized structure that will express the full dimensional ramifications of space and time simultaneously. Indeed the sequence may go 2 = back to front, 4 = side to side, 6 = up to down, 8 = interior to exterior, 10 = space and time.

The division of the circle by eighteen still retains the division between form and function as a recollection of the mysterious self containment of the number nine as portrayed in the figure of nine divisions on the circle (fig. 2). The disconnection continues until the division of the circle by thirty-six, where full connection is established with both the horizon and the prime meridian.

Nine: Space and Structure
Divisons of the Circle by 18
Divisions of Three Yin
and Three Yang by three

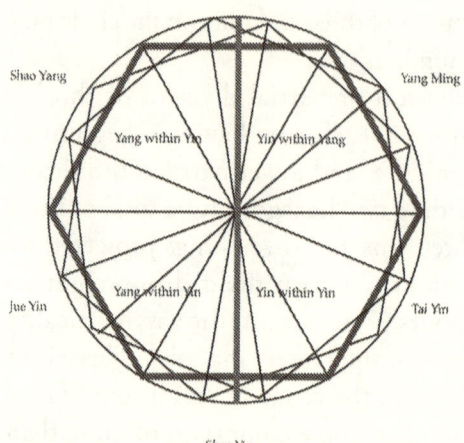

Nine: Vertical Time and Process
Divisons of the Circle by 18
Divisions of Three Yin
and Three Yang by three

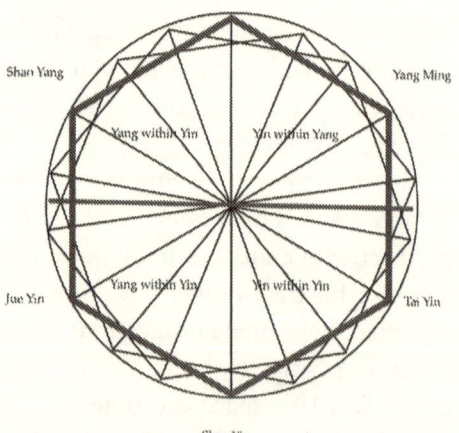

Figure 5. The Application of the Principle of Nine to the Hexagon Formed by the Three Yins and Three Yangs: *A*, Plane of the Horizon Contacted by the Figure Dividing Heaven and Earth Vertically in the Standard Three-Yin and Three-Yang Configuration; *B*, Plane of the Prime Meridian Contacted by the Figure Dividing Heaven and Earth and East and West Emphasizing Time and Process

Figure 6 extends the division of nine to each of four quadrants showing how the whole concept of nine applies in each seasonal distribution in a complete circle of thirty-six divisions. Symbolically this shows the nine divisions of China and the nine divisions of the body, complete and whole in their adaptation to each of the four seasons.[103] By multiplying nine by two or four, connections between the mystery of nine and the yin and yang or the four quadrants become possible. The black hexagons show the links to the vertical and horizontal planes, while the orange show the structural links and the ongoing motion implied by yang. Hexagons form stable sixty-degree equilateral triangles in geometry, and these complex representations combine the structural stability of six in relation to the anatomy of the human body.

The lesser division of the circle by twenty-four develops from the extension of the concept of six divisions (three yins and three yangs) to each phase of six divisions (fig. 7). This image highlights some of the differences between six and nine while retaining shared links to the number three. Of first importance is recognizing that the number two in six allows the image to connect with the horizon and the midheaven. This gives the image a decidedly structural feature that contrasts markedly with cosmographic models related to nine up until fig. six with its four(cardinal points) divisions of nine. In the annual cycle, figure 7 shows the twenty-four periods identical to same half phases of the moon (Chapt. 11). The connection with the structures of the lunations is again dependent on the yin number two, which allows for the consideration of form and change.

Based on discussions of the three yins and three yangs and the later ones about lunar cyclic processes of number seven, it is clear that this image allows for the choice of treating the upper or lower body meridians of a particular three-yin or three-yang phase. The discrimination of what point to use was based on the seasonal element, the consequent six configuration meridian partners, and on lunar phases above or below the horizon. This new presentation of six divisions within six (fig. 7) goes further, and it allows for the detailing of element within element in the selection of points for treatment, depending on whether the month or day is early or late in the phase. This principle can be used to treat temperaments that conform to impairment of the particular phase combination, be it fire within wood or water within fire.

When this background dynamic is carried further with the division of quadrants by nine (fig. 6), it allows the acupuncturist to refine the choice of acupuncture point selection beyond hand or foot meridian and element points on it to the targeting of specific zones on the meridian—if heaven-above, if earth-below, if horizon which is the middle of the pathway on the arm, leg, or torso.

Nine
Division of the Circle by 24
Divison of quadrants by 6
Division of Three yin and three yang by 2

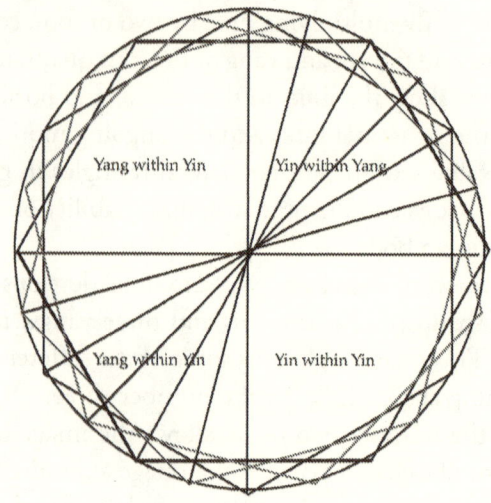

Figure 7. The Division of Twenty-Four, Six Divisions per Quadrant

These decisions about zones can be made independently by considering symptoms topographically or according to the phase activate at the time of treatment or indicated by periodic symptoms.

In his appendix, Unschuld sites *Hou Han Shu* to explain the concept of nine mansions, which I quote here in its entirety with Unschuld's diagram. This number sequence is a hybrid of heaven and earth conceptions based on the eight trigrams in the King Wen sequence. It shows how the number nine is generated from eight by pointing to the implicit center of the circle in a cosmographic map of eight divisions. Five at the center implies the five agents and stillness at the center with movement around the heavenly pivot. The model creates the magic square. "Its numbers always form a sum of 15, regardless of whether they are summed horizontally, vertically or diagonally."[104] This square is created by the transposition of two and eight and the retention of nine at the midheaven. The magical effect of numeric balance is more important as an expression of balance in the postnatal model than it is as a basis for oracular projections. This contrivance is more understandable when it is viewed as a development from earlier models involving the trigrams and the three-yin and three-yang compromise in the declining cycle between vertical and horizontal models. It also is helpful when viewed as a form depected in two dimensions that is moving toward a three-dimensional sphere. These two considerations can justify two above and in back showing the direction of the rising cycle yet containing it, and eight is down and in front showing the direction of the declining cycle yet ensuring that yin will not triumph. Yang is counterbalanced at its peak and yin at its apotheosis respectively. The concordence of nature and number in such an image can make metaphysicians giddy with delight. However, the natural sequence of seasonal functions and numbers on the circle would be as shown in figure 8b.

Hence when Taiyi comes down to the nine mansions, he starts [his movement]from the kan mansion. From there he continues to the kun mansion. From there again he continues to the chen mension. From there again he continues to the sun mansion. This way, he has [completed] one half [of his movement]. He turns to the mansion the center for rest. From there again he continues to the quian mansion. From there again he continues to the dui mansion. From there again he continues to the gen mansion. From there again he continues to the li mansion. His movement has now reached the end of the circle. He rises to take rest in the star of tai yang and returns to the purple mansion.[105]

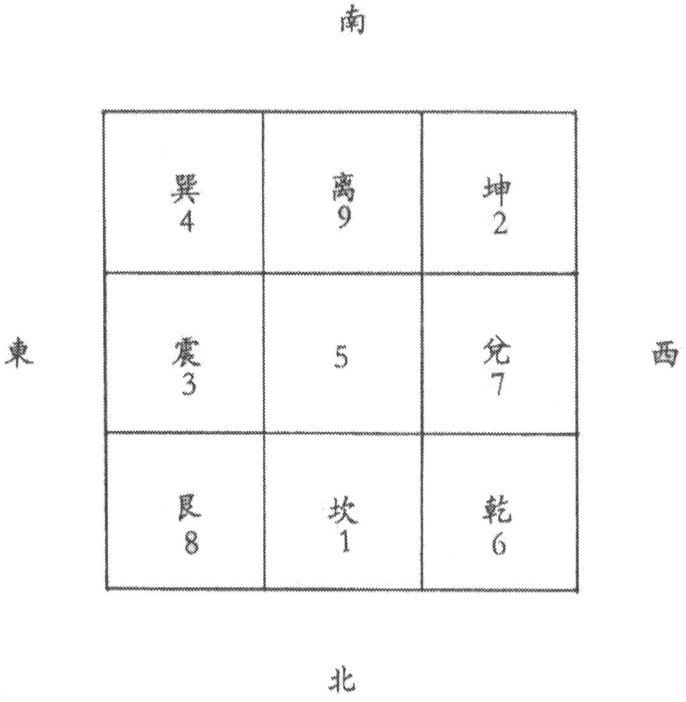

南

巽 4	离 9	坤 2
震 3	5	兌 7
艮 8	坎 1	乾 6

東 西

北

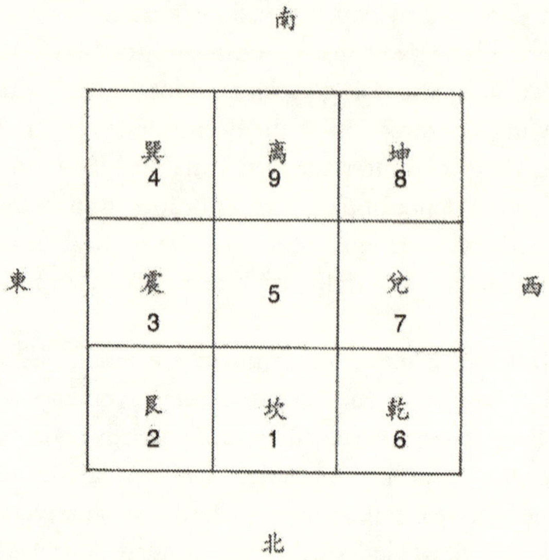

南

巽 4	离 9	坤 8
震 3	5	兑 7
艮 2	坎 1	乾 6

東   西

北

Figure 8. A, The Nine Mansions (diagram 18) from Unschuld, a hybrid posterior heaven, time-phase model where the pole star has come to earth; B, Unmagical Posterior Heaven Sequence with the Cosmic Man Facing Downward on the Declining Cycle as with the Three Yins and Three Yangs

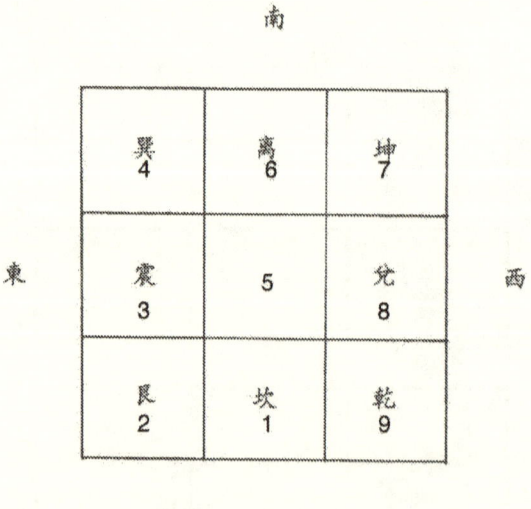

南

巽 4	离 6	坤 7
震 3	5	兑 8
艮 2	坎 1	乾 9

東   西

北

Figure 9. An Anterior Heaven Model for Nine with Five at the Center as the Mediator or as the Yang Number and, Hence, Unmanifest Vertical Axis

Figure 9 shows the unmanifest balance of the Fu Hsi spatial sequence.

Looking ahead a bit in figure 10, the two interlocking hexagons show the three yins and three yangs doubled (2). This brings the form of one hexagon into alignment with the horizon the other, with the prime meridian accomplishing in one step what took six hexagons to accomplish using the divisions based on three and nine. Here in figure 10, one hexagon provides the structure for the other's process reciprocally. Behind the four divisions of three is a unique union of the principles of three and four/two. This harmony in the division of twelve highlights by contrast the unique yang extension of three in the relatively directionless and unstable number nine and its derivatives. There is no intrinsic link to the cycles, only an endless possibility created by the completion in three dimensions of a stable relationship between potential energy and time potential energy and space. In nine months, the baby can be born and take its first independent breath in ten.

Twelve
Divisons of Three Yin
and Three Yang by Two

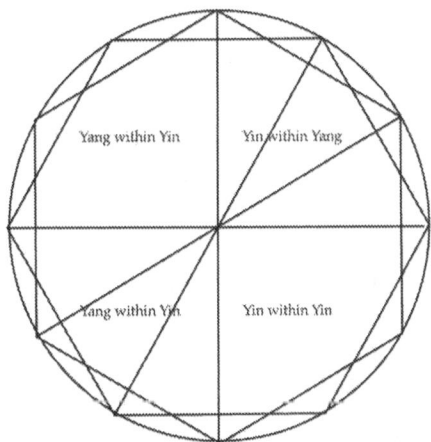

Figure 10. The Division of the Circle by Twelve, Two Aspects of the Three Yins and Three Yangs, Form and Function

III. Medical Practice—Theory and Treatment

The cosmographical figures for the number nine describe a system that is every bit as important as the five elements or eight principles. The principle of three has an infinite regress that allows for precise detailing of relations between congruent body zones. In this model, heaven or the head has not only three divisions but also nine—same with the torso, legs, and arms (fig. 6). If you treat reproductive problems, you are treating in the zone of earth; but if there is a rational or psychological component, you can treat the heaven in earth (the feet). If there is a nutritive component, you can treat the horizon in earth (the leg below the knee, stomach 36 and 40, etc.), and if there is a hormonal or blood component, you treat the earth in earth (the thigh). The zones profoundly dictate the appropriate intervention. These relations quickly become nuanced. You can treat higher or lower according to the degree or relative yin (down) or yang (up) of the condition; and these differentiations will show up in the color, tongue, and pulse. Moreover, you can augment your basic treatment of emotionally and psychologically driven (heaven; and foot) problems, for instance, by treating one of two ways:

1. Treat the heaven in heaven itself for a strongly augmented psychological affect. Selecting resonant points on the hand or head with the foot points will emphasize and augment the stress reduction aims of the treatment.
2. Treating the earth (lower torso, jaw/neck, and upper arms) either in the horizon or heaven will significantly reinforce the hormonal and physical side but through resonant higher functions.

These nuanced uses of heaven, earth, and human/horizon conceptions in nine divisions provides a way to treat literally the body, emotions/mind, and spirit levels and to synergistically support these levels while intervening in any zone of the body for alleviation of symptoms.

The nine basic divisions can be subdivided into nine more into infinity. This means that point selection can be further nuanced by recognizing in head pain, for instance, the upper, middle, or lower zones; and if it is on the upper, it has three and nine zones in a vertical tier that complement the vertical zones of *shao yang/jue yin, yang ming/tai yin,* and *tai yang/shao yin.* This can establish precisely if it is a *shao yang* headache whether it is upper, middle, or lower; and then if it is, upper—body points on the torso and points on the hands and feet can be selected over other typical alternatives. Some standard points are clinically effective, precisely because they are structurally resonant with the target of treatment.

The divisions of nine suggest that if we are treating conditions related to heaven, we would use points on the hands, feet, cranium, and the upper chest. Heaven is the origin, and it takes precedence over earth and the horizon. These points would have a causal, psychological, mental, and systemwide effect that differentiates them from other points. Their metier is potential energy. The number of nerve endings, the proximity to the brain, and vital organs having to do with speech, emotion, and thought confirm this connection between brain, hands, and feet.

Treatment of conditions related to common sense, experience, the balance between activity and fatigue, digestion, and transformation relate to the horizon. This is the world of the cardinal points and five elements and six configurations; the world of human exchange, information, fact and action, and conditions that affect pure functionality, energy, and nutritive strength would best be treated anatomically in middle burner—the arm just proximal to the hand the leg just proximal to the foot and the area of the head associated with the sense organs.

Treatment of conditions related to genetic weakness, hormone imbalance, nervous system, joints and bones, reproductive disorders, bowels and urine, blood building and immune disorders, and the root conditions of the body not directly involved in digestion or supporting consciousness and interaction would, in this model, be amenable to treatment from the upper arms, the jaw, the lower *jiao*, and the thighs. These problems often result from habits and inertia.

These generalizations help to identify gross terrain for intervention. Strangely, these differentiations were not automatically present in the three-yin and three-yang structures but had to be inferred. Thus, the divisions of nine go much further. They go beyond treating bilaterally as with five elements and six configurations. They offer a way to extend differentiation of points and treatments not only into the vertical planes of the quadrant structures but also the horizontal planes at different levels. This creates a grid of horizontal and vertical zones on the body topography that allows discrimination at the level, for instance, have the waist and middle *jiao* in a way that was unavailable before. If we have an emotion of sadness and considerable confusion and some arm or shoulder pain on the right side, we can determine that we are treating the mental and spiritual aspect of heaven through the monarch fire element, *tai yang* fire, in its mental aspect; and we can treat by opening the zone with *tai yang* small intestine 3 for the central aspect and *shao yang* three heater 6 for the lateral aspect of the upper-back zone and then on the right side and counterbalance it either with its partner *tai yang* UB 62 and lateral motility vessel point on the opposite leg or in combination with its yin partner heart 9 tonification or 6 junction above or below *shao yin* kidney 4, the luo, or 6, the motility vessel control point. If you want to keep all the focus above the waist but maintain some balance below, then GB 41 the *shao yang* leg counterpart to PC 6 and the point that regulates the lateral aspect of the body generally along with the *dai mo* could be selected.

The waist defines the middle *jiao* and the horizon with the basic quadrant definitions derived from the prime vertical back from front versus prime meridian side versus side. Yang structurally is the back and for time or phase the left, and yin structurally is front and for time right or west. As a rule of thumb, then, for qi and yang, support the back and for yin and blood, the front. Similarly, the times of day when the treatments would be most effective would be the period just before and after dawn at which time you could treat chi and yang with gallbladder or yang points on liver bilaterally or unilaterally on the left. Tonificaiton would work well at dawn or in the morning. To treat the yin, you would treat liver bilaterally or on the right. Sedation would work well in the evening with liver points since they will already be at a low ebb, or the opposite stomach spleen points could be supported to support and counterbalance the damp or dry condition of the liver.

All the great vertical planes pass through the center waist of the body, so both heaven and earth energies can be treated there. This is where yang penetrates into the earth to give fertility, and it is also where yin rises above the earth with exuberant growth. The matrix can be treated in its rising cycle phase gallbladder and liver or declining cycle phase spleen and stomach on the physical structure where needed or at the time phase with the appropriate meridian points—morning for wood, evening for spleen. The opposites can also be treated to get leverage on the counterpart.

However, what is critical about the number nine is that it defines the locations and times for treating the functions associated with rising light or declining light in the daily, monthly, and annual cycles. Thus, treating wood and using the *shao yang/jue yin* self-regulating complex is good in the spring, the waxing moon phases, and the morning near dawn—particularly for supporting chi and yang but also for creating a balance in the whole system when the yang aspect of the hun spirit equilibrium is deficient. Earlier in the season, moon phase or day, the yang aspect of water would be activated; later in the day, the yang aspect of fire *shao yang/jue yin* would be activated. And finally at midsummer, midday, *tai yang, shao yin*; but in each increment of time with the increase in light and heat, the locus of treatment will move up the body in the appropriate meridians just as the declining cycle will move down the body with each phase of the declining cycle.

The literal one-to-one relationship between time of day and structural location of the intervention is a powerful way to consider the selection of points. Points selected in this way will activate body zones that correspond to the proportional zones and functions being activated by the solilunar positions.

The middle *jiao* is the point where heaven and earth both intersect, but only leg meridian points are visible at that level. This reflects the obvious fact that all forms are an extension of the physical substance of the earth and that there is an invisible agency that descends into the earth to generate and support life. The arm meridians have internal connections even when they do not have points of their own at the level of the waist that are accessible through certain points—in particular ren 4, 6, and du 4, UB 23—which connects with shen on the kidney level. The yin meridians extend up to the shoulders but not up to the head. This has esoteric implications, not the subject of this essay.

The significance of the number nine on the level of the waist is that nine completes the assemblage of conditions and structures necessary to post natal life. The navel is the locus of prior birth nutrition, and it is at the same level that the horizon is represented. When yang penetrates below the horizon, it goes symbolically into the earth in a generative way, and that is symbolized in ren 4 and 6; when yin rises above the horizon, its visible rise is limited at the top of the tree canopy, the diaphragm, hence, *jue yin* liver 13 as the uniting point of all yin organs. The point can be used on the left side in the morning and the right side in the evening to support liver yin; and it can effectively be complemented with ren 12, zone point li 1, li 3, GB 41, then by the liver points on the shins to treat the middle, digestive, and horizon-based problems.

For a diagnosis of stagnant liver qi and dampness, a history of social problems, including aggressive behavior, thoughtlessness, and some nausea when emotionally disturbed, the model of nine helps make some useful distinctions. If the portrait includes soft chi deficient pulses or kidney chi deficiency from energy burn and damp stagnation owing to chi deficiency, then activating the chi yang spectrum will be useful. To do this, the needles would activate the *shao yang* gallbladder and three heater; and the lateral points on the left of the body, the *shu* points in the back to mobilize chi, move dampness. The distal points will be left above and right below or one-sided left or right depending on the weak side. If points are used bilaterally, they will be activated. If the stagnation is deeper owing to deficiency of blood and yin, lack of blood storage capacity, blood pressure, or stress on the system, the emphasis would be more on the right yin side and below. Points like liver 13 and ren 8-12 and du 10 create links between left and right and front and back by using the leverage of the *dai mo* horizon on the appropriate zones, left or right. Thus, GB 41 in combination with *tai yang* UB 62 left for chi right, for blood would be effective with the proper local support points GB 25, for yang liver 13, for yin UB 19 or secondarily 18 for yang.

The lower *jiao* is the home of kidney yin and kidney yang. It is the area of the body that differentiates between solids, liquids, and gas; its functions are basic and preliminary to life. In the lower *jiao*, warmth is life; cold is death. The lower *jiao* reflects those basic preconditions in matters that are necessary to life; and they have much to do with minerals, hormones, fluid equilibrium, and structural stability. This is the root area of reproductive desire, the unconscious, and the automatic functions of life; therefore, habit is everything. Good habits preserve life; bad habits destroy life. Lower body problems usually means the owner is out of touch with basic needs and drives or is forcing a pattern of living that the body or nature cannot sustain. Understanding the simple root requirements and communicating them to patients can be very important.

In reproductive problems, you have to make a root discrimination as to whether the problem is from poor circulation (heaven), poor nutrition and transport (horizon), or poor habits and root resources (earth). Any one of these three can inhibit conception in a woman and healthy resources in a man, and all three do. So while doing standard points on the back to augment chi and yang and front to augment blood and yin and local points to improve local organ function, check the pulses closely for deficient blood and for the nerve and lifestyle-driven tight pulses. Based on these diagnosis, the practitioner can augment both the upper or middle *jiao* but steer needles to the upper, middle, or lower zones on the torso appropriately in conformity with the bias.

Chapter 10 十

Wholeness and Completion

Ten completes the initial whole number sequence by returning to zero. The zero of ten is not the unformed unnamable primordial Tao implied by zero, the something that is nothing, rather it is a portrait of any birthed fully human manifest form in three dimensions that possesses a unified field of time and space to operate in. The body viewed unilaterally with five fingers and five toes emphasize heaven and earth and together generate ten, and it is the body that best expresses the symbolism of ten. The fingers and toes also generate ten in the bilateral mirror above and below with at total of twenty. However, the bilateral replication simply implies that the principle of ten operates in the waxing phase, the waning phase, and the center—divided into above and below. This is the literal roots and branches of human life to which Chinese medicine refers. These pure manifestations at the tips of ultimate yang and ultimate yin of the number five encompass all the integral-numbered vistas on the body and the world that preceded it. Nevertheless, in between, the body is organized around the five senses and the nine body outlets in men and the ten body outlets in women. Ten represents equilibrium between heaven and earth, yet it differentiates itself from the representation of the yang number nine by completing its capacity for physical expression.

I. The Number Ten

Ten, [十] *shih* (2) Wieger, "The number that contains all the simple numbers. Symbol of extent in two dimensions and have the five cardinal points—east, west, south, north, and center. It is the twenty-fourth radical. Derivatives include terrestrial cycle shen (the ninth in series); the plus includes all things and implies the minus, the simple line for absence. All things contained between them. Ten is a symbol for manifestation."[106]

The primordial vertical cross was the number five indicating the four directions and center, but it also was transformed into an *X*. These became the arms and legs of the cosmic human in the cosmographic representations for the number eight. For the number ten, the vertical cross is restored because the number ten contains both function and direction, the cross and the *X*; it has five yang and five yin. Five and ten are linked closely through the ten celestial stems, which when paired form the yin and yang aspects of the five divisions of any cycle. As the whole number that contains all simple numbers, it represents completion of a cycle and a new beginning. It also reasserts the horizon line of "I" or one indicating the whole. Ten was the jumping off place for the ten or ten thousand things of Chinese philosophy, because ten is a kind of cosmographical completion for the whole number sequence through which all the basic conditions and relations in nature are mapped. The numbers one through ten explain the structure of relationships in the cosmos—encompassing the heaven earth and man, the four directions, the five planets and their corollaries, the three yins and three yangs—which explain the light phases and introduce the vertical dimension, the process of germination, and dynamic change indicated by seven and dominated by the Sun, Moon and planets, the eight-sided *ba gua* of the *I Ching* that details the comprehensive spatial structure of nature in three dimensions and nine, which introduces the unique human form and consciousness into the setting of the creation. After ten, you have simply the enumeration of particulars. Once you have man, you have ten thousand men; once you have growth in time and space, you have the famous ten thousand things of Chinese philosophy.

II. Cosmographic Structure

Cosmologically, ten divisions of the circle of the year, the solar cycle, also has the unique characteristic of uniting heaven and earth. The figure of ten can be formed simply by extending the radius lines for each of the divisions for the number five (fig. 5-1). The five divisions of heaven resulted in a figure with five radii at regular seventy-two-degree intervals. When these phases are divided into their yin and yang aspects by the extensions of the radii or by the division by number two, each of the paired celestial stems associated with the five phases becomes responsibility for a single phase, and the result is figure 1. The figure implicitly consists two pentagon, a pentagon facing south combined with a pentagon facing north.

Ten

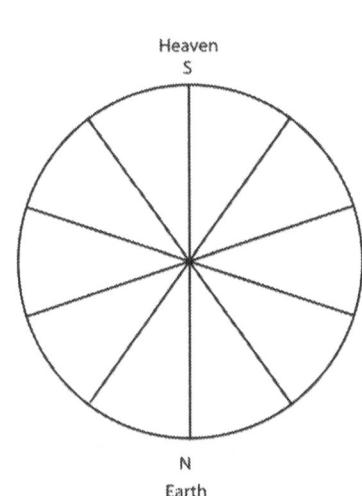

Figure 1. The Ten-Phase Portrait of Heaven

The typical five-element mandala described in chapter 5 was an extension of the cardinal points and showed a portrait of the seasonal transformations that resulted from the annual solar cycle. It gave a set of associated standards that allowed for noticing deviations from the standard motion. Figure 1 fundamentally describes in a two-dimensional drawing the original horizontal five-element cycle of the horizon and the recreation of the principle of five for the vertical line (prime meridian) and plane (prime vertical), yielding a dynamic yet stable image for five elements unfolding in both time and space. However, whereas the image for five elements was oriented toward the south in timid acknowledgement of the midheaven, the figure for ten in acknowledging the vertical pole directly points to a system that primarily has to do with action, yang, and heaven rather than the horizon. For these reasons, the ten celestial stems were developed as symbols for the phases of the growth cycle as it pertained to heaven, the sun, and the five planets in their day and night operations. Day and night are implicit in the figure but not explicit because this is a purely solar image having to do with the circulation of the five planets in five seasons in yang and yin aspects. It is only in the terrestrial figure of twelve that connects the east and west horizon to the vertical and includes the symbolism of sun and moon, yin and yang, that day and night are fully differentiated and applied.

Figure 2 extends the concept of ten to the division of the circle by twenty and figure 3 extends it to forty, which diagrammatically shows the application of the principle of ten to the quadrants, then to the quadrants horizontally and vertically.

Ten:
Five Divisions/Elements in Each Quadrant

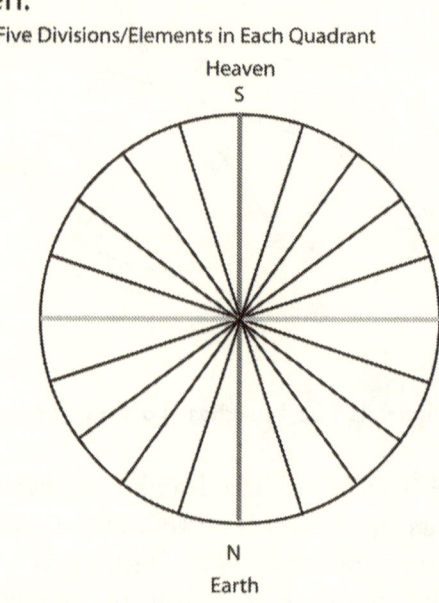

Figure 2. The Principle of Ten with Five Divisions (Elements) Applied to Each Quadrant to Show the Presence of All Elements in Each Season

When figure 2 is divided in half by the horizon, there are ten above and ten below; when it is divided on the centerline, there are ten to the east and ten to the west. Figure 2, therefore, indicates the facts of the body, namely, that unilaterally the five fingers on each hand apply in space to above time and process when extended horizontally and similarly with the leg below and to the sides. This notion extended bilaterally generates figure 3 with its forty divisions, ten in each quadrant, which result on the body from taking into account all the combinations of roots and branches available when applied to the quadrants, lying down on the horizon or standing up against the prime vertical.

Ten: As Forty Divisions Extending on Five
Ten Divisions/Elements in Each Quadrant

Heaven
S

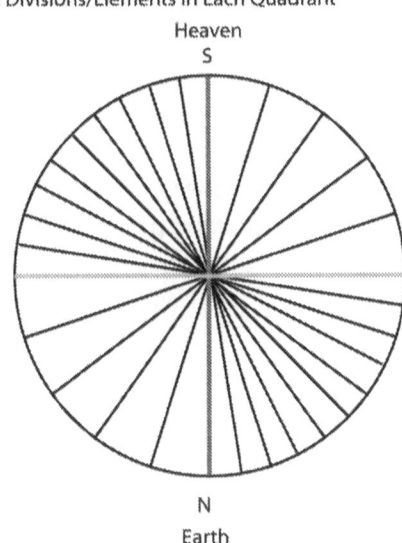

N
Earth

Figure 3. The Principle of Ten with Ten Divisions Applied to Each Quadrant

The divisions of the body suggested by the number ten imply ten vertical planes that complement the nine horizontal divisions of the body. On the simplest level, this simply implies that the five elements are active holistically in each of the body divisions. These meridians theoretically would supersede the three-yin and three-yang meridians without nullifying their value but have never been developed into a system, perhaps owing to their complexity. Nevertheless, the fingers and toes suggest two possibilities.

First, that the principle of ten can be applied best by treating the hands and feet exclusively in reference to heaven and earth respectively. Earth in a way is simply an extension or reflection of heaven and the hands and feet as we have seen in relation to the number nine with the same heaven—upper body, upper head, brain affinity. Profoundly then, ten in a system that treats hands and feet exclusively as reference points for the whole system and works because of the close connection with the brain or seat of the soul. The hands would treat the whole through the window of relationship, heart, and consciousness; the feet would treat the whole through the window of intrinsic nature, instincts, drives, and issues of survival and sustenance. The system suggests a specialized kind of access to the energies of heaven and earth appropriate to these extreme extensions, which may explain the effectiveness of these techniques being developed for the most part in Korea.

Second, they indicate the presence of twenty meridians—ten yins on the front inside and ten on the back, five on each side of the midline and prime vertical—each of which organizes the body in biomagnetic fields and topographical lines that form a grid with the nine horizontally stacked zones. This creates a more refined grid than the three yins and three yangs, one that differentiates exclusively one element to each of the five meridians; and it suggests as well a more direct relationship between fingers, toes, meridians, and the five senses and their related organs. Whatever pattern emerges, it will respect what has been learned about the same cycles through earlier numbered divisions, and it will conform uniquely to the meanings associated with the five elements and ten stems.

The five planets traditionally were the key to understanding the relationships between heaven earth and the five elements. If these basic element associations are maintained and the lessons of the three yins and three yangs also are retained, the riddle of meridian associations in a system of ten becomes clear as tabulated in table 1.

Table 1. The System of Planetary Associations and Element Contrasts between above and Below

Heaven	Fire	Earth	Metal	Water	Ministerial Fire
	Sun/Mars	Saturn	Venus	Mercury	Jupiter
Horizon/ Implicit					
	Moon/Mercury	Venus	Saturn	Mars	Jupiter
Earth	Water	Metal	Earth	Fire	Wood

The chart shows the clockwise celestial sheng cycle rotating above and the reverse terrestrial sequence below. Although they reverse in every case, there is a leg meridian below complementing a hand meridian above or the reverse. Moreover, this suggests a crossover effect from above to below that may explain the X Chinese character and the X factor in treating above to below and left to right. The relationships are complex, and their dynamics need further study, but the principles are sound.

The multifaceted and reflective relations between above and below are reflected in the pattern of element affiliations that are consistent in sequence with what has been studied before and confirmed by the fact that each meridian in sequence reverses its direction in a regular order. This regular oscillation and balance between qualities and elements is all tied to the planetary associations, and they work cosmographically. What the system does show that anticipates the wide-ranging discussion of the Chinese clock of the twelve divisions is the importance of understanding the clockwise solar motion with the counterclockwise rotation of the earth and lunations as the basis for the generation of chi that supports life. The planets circulate in heaven and activate these complex dynamics in a systematic and structural way.

For a clear sense for the meaning of the ten phases in the natural cycle, see appendix 2 that intuitively explains the implications of the seasonal ten stems. If the year is divided by ten-day weeks, there are thirty-six weeks in the year nine in each season, which hints at the generative connection between nine and ten. Each month based on the cycle of ten divisions has thirty-six days or four divisions of nine days. These are perfectly legitimate divisions for determining treatments based on a division of the body by ten meridians, five per side above and below the waist. That system has not been developed, but its shorthand will allow the use of ten clinically as an extension of the number nine with its nine vertical divisions. Including depth and a sense of daily and seasonal time entirely driven by the sun, the nine horizontal divisions on the body can be divided in segments and treated in yin and yang segments just like the three-yin and three-yang planes are.

Chapter 11 十一

Heaven to Earth—Introduction to Twelve Terrestrial Branches and Basic Cyclic Correlations

Eleven is not a linear number only. Cosmographically, eleven begins a spiral; it completes the circle and returns to the number one without achieving it. It has become something else for having gone through the original whole number sequence. Eleven captures all the formless mystery of yang, the creative energy, the one that sustains the whole of existence; yet like an infant, it contains it as a recent memory but lives in a present world of form and limitation. Figure 1 shows the spiral return of eleven horizontally and vertically and in depth in keeping with the complexity of the number. This display shows a kind of return horizontally, a rise and descent vertically, and an asymmetrical displacement and movement in space and time in depth. These combined motions capture the notions of a creative movement that is profoundly contradictory and irrational. That it is a movement forward but of return, outward and inward, etc. In this sense, the fundamental nature of eleven is one of ambient evolution based on the amoebalike equilibrium. The annual cycle is dynamic and stands in relation to thousands of years of dynamic equilibrium. The present year never starts with one or even eleven, more like eleven billion, and it is this timeless and spaceless movement of chi in an ongoing evolutionary equilibrium that is merely symbolized by the number eleven.

The consequence of eleven as a transition from ten to twelve is that it introduces the notion of return and reincarnation. This backward motion from heaven to earth, a motion that is generally toward involution and incarnation, acts as a counterpoint to the realization of body and form in three dimensions of ten. Ten was based on the generative principle of heaven. The contrarian nature of eleven is to bring the conception down to earth to require breath and action and life. It is a transition from heaven to earth and from celestial stems to terrestrial branches that are forever identified with the number twelve owing to the terrestrial lunar cycles.

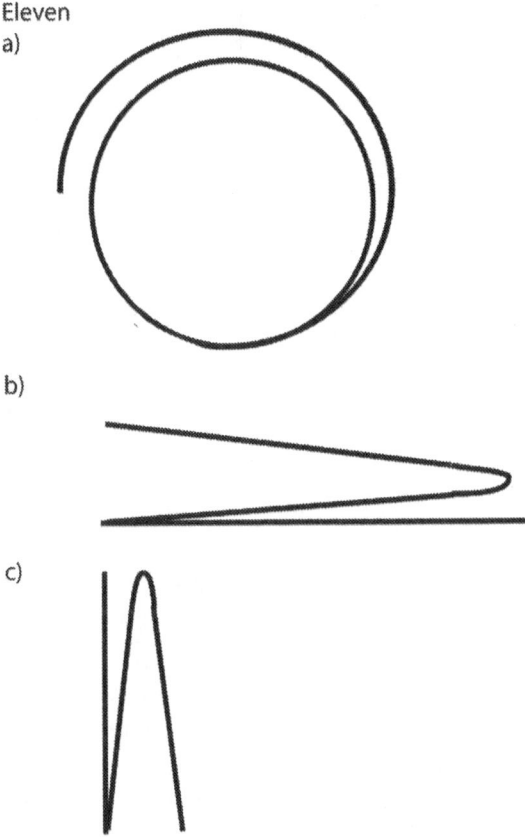

Figure 1. Three Versions of Eleven

Figure 2 shows the standard two-dimensional rendering of the number eleven. The figure is similar in its yang orientation to heaven, and it again shows the lack of connection to earth or horizon. As in figure 1, it could equally be pointed horizontally or downward, which would show its actionable and directional character more than affiliation with one or the other domain. Divisions of the circle by twenty-two, forty-four, and eighty-eight bring the principle of eleven to the cardinal points and the great celestial sphere but do not much change its limited bodily or physical reference.

Eleven

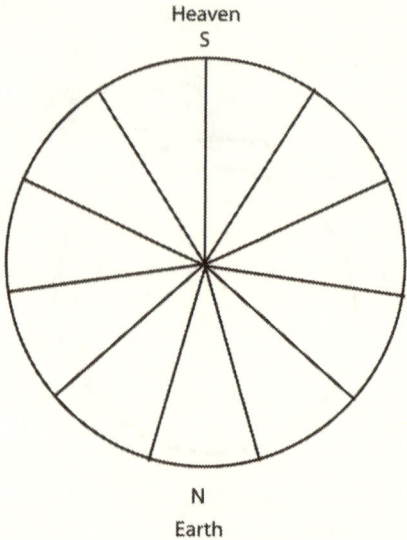

Figure 2. **The Division of the Circle by Eleven**

From here on, eleven will be interpreted as the implicit yang partner of the even number twelve and as a transitional link between the yang celestial stems and yin terrestrial branches. Its role simply stated is to explain the mystery of *yang ming*. It symbolizes the capacity in the autumn season and the return phase of the year to represent the involutionary and downward motion of yang into the earth when the natural association of yang is above. Perhaps, the mystery of generation has to do precisely with the necessary downward action of yang in its search for a consort that will reflect what it is and define to some extent what it is not. For this larger reason, I think the proper disposition of eleven in a two-dimensional cosmographical drawing is as in figure 3, where the line of the horizon is added.

Every year can be divided into twelve periods that roughly conform to twelve complete lunations, and this was the basis of most early calendars. The sun's complete circuit through the heavens with accompanying seasonal changes defines the year, but time was best measured through the changing shape of the moon as the perpetual cycle of days were differentiated into weeks and months. Concurrently, each lunation transited through roughly thirty degrees of the zodiac, and this resulted in twelve spatial divisions for the months and days. These spaces translated into twelve two-hour periods and twelve months. Thus,

Eleven: The Return

Descent to the Terrestrial Branches

Heaven

S

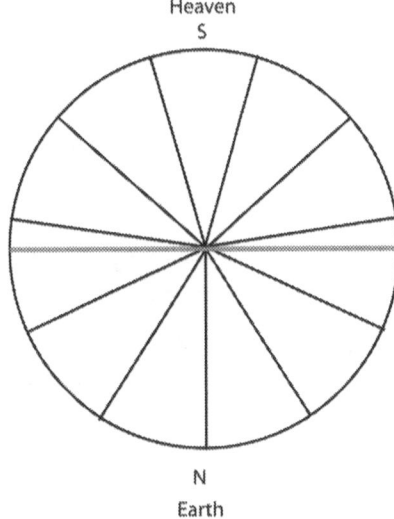

N

Earth

Figure 3. The Downward Necessities of Yang and the Role of Eleven as a Transition from Celestial Stems to the Derivative Terrestrial Branches

traditional crude measurements of space and time conspired to elevate twelve to a unique status. As an even number divisible by three (odd) and four (even) numbers, twelve allowed for the unique inclusion in a single image both of the principles of heaven, earth, and human (3) and the cardinal points (4). The model is both dynamic and stable; it explains both movement and form, heaven and earth, in manageable increments that retain the contributions of root numbers (fig. 1). This chapter will examine the traditional Chinese characters known as the twelve branches and their uses and meanings as a prelude to creating a twelve-division clock and system of diagnosis and treatment that is congruent with the discoveries in earlier chapters of cosmographical connections between the body and the world.

The division by twelve shows the cardinal cross in the center, which must be conceived simultaneously as lying flat on the horizon and standing vertically against the prime vertical. It is a two-dimensional representation of a whole system of relations that were introduced in chapter 3. The two-dimensional drawing is significant because the ninety-degree angles of the cardinal cross are complemented by 30-, 60-, and 120-degree triangular configurations also shown by the way a few spokes are drawn. Thus, the meaning of three divisions of the

Twelve
Divisons of Three Yin
and Three Yang by Two

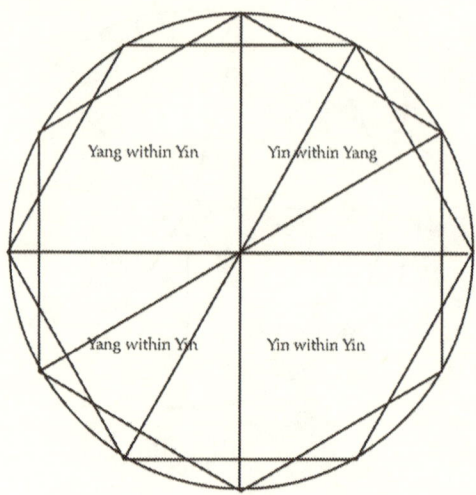

Figure 4. The Twelve Divisions of Time and Space

circle (120 degrees), four divisions (90 degrees), and six divisions (60 degrees) is subsumed in twelve divisions. The principle of three can be applied either to quadrants or to the cardinal points to qualify seasonal meanings. The principle of five, seven, eight, nine, and ten are implicit; and that is a limitation in the number twelve. Eight can be implied by the principle of four applied horizontally and vertically. Ten is an extension of five and is responsible directly for the generation of twelve. This leaves the numbers five, seven, nine as the odd numbers, which, in this system, must be viewed through the lens of their preceding and succeeding numbers, as dynamic processes seamlessly unfolding within the rather stiff and arbitrary frame of four and twelve. Symbolically, that limitation is the limitation of matter and form applied to the reservoirs of energy in the universe, and that combined with the twelve annual lunations is why twelve is associated with the earth and its forms. The root values of twelve branches from the same growth cycle enshrined in ten celestial stems.

The number eleven is a number of transition that can best be used by sensing the trajectory of health and disease in the body zones. It is not enough to select a zone and spatial location. Treatment requires the directing of chi and using the movement evident in subtle symptoms and signs to augment that intention. Eleven is about observation of movement in the whole view of the body and honoring that when selecting points in a zone.

The eventual framework for eleven is the twelve-month year based on the twelve terrestrial branches. The phase meanings of the branch symbols are described in appendix 2. The following discussion depends on these basic cyclic meanings, but the number eleven describes the transtion into this framework; and there are variable possibilities in how the twelve divisions could be set up, and the following discussion clarifies the choices that were made.

I. Variations in Twelve Branches and Derivative Divisions

The twelve branch characters roughly display the movement of the year, but their starting point is at the beginning of winter, not the beginning of spring. In my view, this is a bow-to-the-orbital motion of the moon and the rotation of the earth, which run counterclockwise to the circuit of bodies through the zodiac. The implication of the placement is that there is a counteracting, reverse cycle that retains balance and which nevertheless bows to the overriding solar will and seasonal progression. The start point of the terrestrial branches and their associated animals and, eventually, in Chinese medicine, the meridians in *tai yin* lung reflect this acknowledgement of earth, the reverse direction of the rotation of the earth and the unfolding lunar phases in the waxing cycle. The breath with it, inhale and exhale, perfectly express the waxing and waning cycles of light cycles and the movement from above (air) to below (body/earth). The start point of the twelve terrestrial branches reflects this dualistic foundation that found a necessary prelude in the mysterious number eleven.

Following are several variations in the modeling of the twelve branches and the assignment of elements to them. This variation represents a dialogue carried on from different perspectives that nevertheless shows a general allegiance to the root seasonal conception of the branches.

The Yellow Emperor's advisor Chi Po in the *Nei Jing* describes the action of the twelve astrological animals associated with the twelve lunar months, which in turn were correlated with the twelve terrestrial branches. He beings his description at the New Year but with the third branch character and animal:

> Yellow Emperor: How are the nine fields applied to the body?
> Chi Po: The left leg is according to the starting of Spring, the date is Earth Yang/Tiger and Earth Yin/Ox. The left side of the Thorax is according to the dividing of Spring the date is Wood Yin, Rabbit. The left arm is according to the starting of summer, the date is Earth Yang, Dragon and the Earth yin, Snake. The upper chest throat,

neck and head is according to Midsummer, the date is fire Yang, horse. The right arm is according to the start of fall, the date is Earth Yang, Monkey and Earth Yin, Sheep. The right side of the Thorax is according to the dividing of fall, the date is Metal Yin, Rooster. The right leg is according to the beginning of winter, the date is Earth Yang, Dog and Earth Yin, Boar. The lumbar, backside, anus and urethral openings are according to absolute winter, the date is Water Yang, Rat. The Six yang organs and the three lower Yin organs are according to the center.[107]

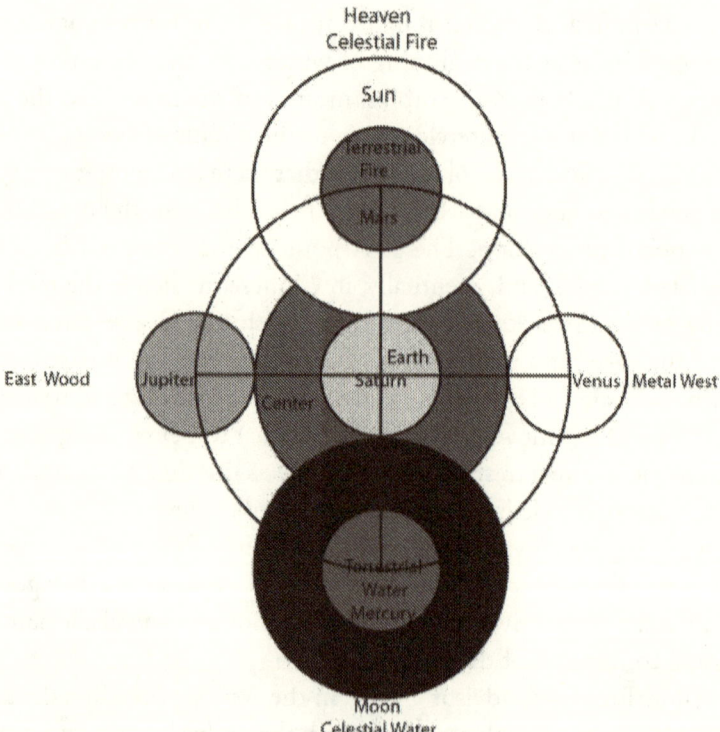

Figure 5 (also fig. 4-5). The Circular Eastern Model for the Twelve Seasons and Animals

Su Wen makes direct connections between body parts and the phases of the year. The sequence begins with the earth-yang-tiger attribution for the first lunation of the New Year. This is the first phase moving out of winter but the third phase in heaven in the clockwise motion of the seasons from tzu, the first branch that is linked to the rat.

The model is circular rather than from top to bottom as in the anatomical attributions of Western astrology. Of particular interest is the correlation of the first two complete lunations of each season to earth—the first being yang and positive, the second being yin and negative—while the cardinal points retain the conventional seasonal attributions associated with the four seasons and their associated elements. This pattern follows figure 2 above, and it designates the function of change and movement to the earth element as follows:

Table 1. Cardinal-Point Distribution in the Twelve-Division Model

Terrestrial Daily Branch	Animal	Fixed Element	Seasonal
+	Tiger	Earth	Spring
-	Rabbit	Earth	Spring
+	Dragon	Wood	Spring
-	Snake	Earth	Summer
+	Horse	Earth	Summer
-	Sheep	Fire	Summer
+	Monkey	Earth	Fall
-	Rooster	Earth	Fall
+	Dog	Metal	Fall
-	Boar	Earth	Fall
+	Rat	Earth	Winter
-	Ox	Water	Winter

The importance of the Su Wen portrait is that it treats earth as the place of transformation and the ruler of periods of transition between elements and qualities that in their turn dominate at the cardinal points of the year. This is a clear extension of the earth-centered cardinal-point model discussed in chapter 4. Earth as the nexus of the prime meridian and prime vertical at the level of the horizon is implicitly throughout the cycle as the locus of change.

From another point in the *Nei Jing*, Needham extracts a table showing a slightly different emphasis. Su Wen applies the cardinal points to derivative five-element correlations and, ultimately, to underlying stem and branches relations as shown in table 2: from Lu and Needham's table 12.

Table 2. Needham's Table 12[108]

Element Yun	Season Shih	Cardinal Fang	Stem (denary Cyclic signs Kan	Branches (duodenary cyclic signs and the animals Chih
Wood	Spring	East	Chia I	Yin(tiger) mao (hare)
Fire	Summer	South	Ping ting	Wu (horse) Ssu (serpant)
Earth	Late Summer/ transitions	Center	Wu chi	Hsu (dog)chhou (0x) Wei (sheep) Chhen (dragon)
Metal	Autumn	West	Keng yun	Yu(cock) Shen (monkey)
Water	Winter	North	Jen kuei	Hai (boar) Tzu (rat)

This table shows clearly that the wood element commencing in the spring with the New Year is associated with *chia* and *I*, numbers one and six of the celestial stems, but with yin (tiger) and mao (hare) three and four of the terrestrial branches. This means that the celestial stems that treat the solar year commence at the first lunation of spring after the winter phases of winter solstice and apply as well to the first phases of the emerging day at 3:00 to 5:00 a.m. The first terrestrial branch *tzu* (rat) occupies the point appropriate to the yin, earth's terrestrial branch origination point, that just prior to and during the winter solstice.

Later, when applied in the twelve two-hour periods of the daily cycle, this placement is just prior to and during the midnight phase.

Where the celestial stems move clockwise with the advancing sun, parts of Chinese astrology consider the terrestrial branches to move counterclockwise in harmony with the characteristic of the lunations. This is the basis of much of the entry-exit point sequence in acupuncture.

Also of interest in this table is the application of four terrestrial branches to the earth or center, not eight as in the figure above, which gives one each for the four transitional points between elements at the cardinal points. This conforms to Porkert's model of cardinal-point transitions from chapter 5 figure 4.

Present-day Chinese astrology practice starts descriptions of the annual lunations in the month proceeding the winter solstice, which also conforms with the diurnal cycle of twelve two-hour periods that begin with tzu for the period between 11:00 p.m and 1:00 a.m. Table 12 shows a direct link to the cardinal points and the number four of the horizon-based system. The rat as the root phase of the cycle is associated with water/winter and the winter solstice. The reverse motion from the winter solstice leads to the metal fall lung/metal.

Table 3. The Twelve Terrestrial Branches with the Twelve Associated Animals and Their Fixed Elements[109]

Terrestrial Daily Branch	Animal	Fixed Element	Seasonal
+	Rat	Water	Winter
-	Ox	Water	Winter
+	Tiger	Wood	Spring
-	Rabbit	Wood	Spring
+	Dragon	Wood	Spring
-	Snake	Fire	Summer
+	Horse	Fire	Summer
-	Sheep	Fire	Summer
+	Monkey	Metal	Fall
-	Rooster	Metal	Fall
+	Dog	Metal	Fall
-	Boar	Water	Fall

The element sequence of modern astrology shows a clear dominance of a single element for all three lunations of a given season. This cardinal-point model sacrifices the earth, which now is fully subordinated and implicit, and earth attributes in a chart derived solely from vestigial assignments from the celestial stem or element assigned in the central stem/branch pair for the year, month, or hour. The cardinal point emphasis on the opposition of wood and metal rather than wood and earth derives from the wide view of opposition between springtime growth and moisture and the autumnal decline and hardening of plants (chapter 4).

The following table 4 includes the corresponding two-hour periods for each animal. Joseph Needham makes it clear that the application of branch/animal to horary periods was a late development in Chinese astrology occurring after the ninth century.[110] This reflects a simultaneous attempt to give greater precision to interpreting charts for individuals; and the diurnal cycle offered the best means to differentiate one person born, for example, in the same rat year and boar month from another. Individual chart interpretation was fully developed in the ninth century and later.

Table 4 (11). Astrological Cardinal-Point Distribution of the Terrestrial Animals (assigned to daily two-hour periods with implied elements)

1.	11:00 p.m. to 1:00 a.m.	Rat	Water
2.	1:00 a.m. to 3:00 a.m.	Hours ruled by the Ox	Water
3.	3:00 a.m. to 5:00 a.m.	Tiger	Wood
4.	5:00 a.m. to 7:00 a.m.	Rabbit	Wood
5.	7:00 a.m. to 9:00 a.m.	Dragon	Wood
6.	9:00 a.m. to 11:00 a.m.	Snake	Fire
7.	11:00 a.m. to 1:00 p.m.	Horse	Fire
8.	1:00 p.m. to 3:00 p.m.	Sheep	Fire
9.	3:00 p.m. to 5:00 p.m.	Monkey	Metal/air
10.	5:00 p.m. to 7:00 p.m.	Rooster	Metal/air
11.	7:00 p.m. to 9:00 p.m.	Dog	Metal/air
12.	9:00 p.m. to 11:00 p.m.	Boar	Water

This model conforms to modern astrology where the earth is implicit. The elements associated with the twelve branches as diurnal periods follow the seasonal order of the cardinal points. The day in this terrestrial branch conception, a conception that implies the counterclockwise motion appropriate to the branch divisions, begins at the most yin portion of the day/year just before midnight.

During the first millennium BC China, each of the twelve lunar months were divided into a yang-positive waxing phases and yin-negative waning phases producing twenty-four shorter chieh-ch'i periods with the following translated names.[111] The names attribute precise meteorological and agricultural properties to each division of the year starting with the New Year. These strongly reinforce and qualify the basic meanings associated with the twelve branches. The precision of the periods and their dates also are interesting because the lunar dates are placed in the context of the solar year. The dates of the combined periods correspond roughly with the symmetrical cardinal-point basis of the Western zodiac as seen above in Lao's table.[112] However, the start point like that of the Su Wen cycle is at the New Year rather than the month prior to the winter solstice.

Division 1 of this ancient sequence for the first waxing moon of the Chinese New Year starts before the typical New Year, so it is important to understand that the Chinese New Year is variable and that the dates are variable. The solstices and equinoxes are in the right place, however, so the variations may be only a few days.

Table 5 (12). Twenty-four Divisions (chieh-ch'i)

1.	Great cold	January 21 + or -
2.	Beginning of Spring	February 5
3.	The Rains	February 20
4.	Awakening from Hibernation	March 7
5.	Spring equinox equinox GB 6	March 22
6.	Clear and bright (weather on GB	April 6 Pt.
7.	Rain on the Grains	April 21
8.	Beginning of Summer	May 6
9.	Lesser Fullness (of Grain)	May 22
10.	Grain in Ear	June 7
11.	Summer solstice	June 22
12.	Lesser Heat	July 8
13.	Great Heat	July 24
14.	Beginning of Autumn	August 8
15.	End of Heat	August 24
16.	White Dew	September 8
17.	Autumn Equinox	September 24
18.	Cold Dew	October 9
19.	Hoar Frost	October 24
20.	Beginning of Winter	November 8
21.	Lesser Snow	November 23
22.	Great Snow	December 7
23.	Winter Solstice	December 22
24.	Lesser Cold	January 6

If the above half lunation divisions are reapplied to full lunations and the periods of the twelve animals with the start point at the beginning of the wood element, the first branch of the lunation for the New Year is the wood or earth (Neil Ching) tiger. The terrestrial seasonal details apply then to the waxing and waning phases of the terrestrial lunations/months with their implicit element correspondence as follows:

Table 6. Consolidation of Chinese Lunar Months and Zodiac Solar Sign, Dates and Qualities from Porkert and Needham[113]

1	Tiger	Aquarius	January 21-February 19 Great Cold Beginning of Spring Wood Liver Yang stirring underground
2	Rabbit	Pisces	February 20-March 20 The Rains Awakening from Hibernation Wood liver gallbladder Sprouting
3	Dragon	Aries	March 21-April 19 Spring equinox Clear and Bright Wood gallbladder Ten thousand things beginning to grow, awakening to life
4	Snake	Taurus	April 20-May 21 Lesser Heat Rain on the Grains Beginning of Summer Ministerial Fire Pericardium or Earth Transition Ten Thousand things begin to flourish: plants begin to break through the soil.

5	Horse	Gemini	May 21-June 21 Lesser Fullness Grain in Ear Ministerial Fire/Monarch Fire Three Heater/Tai Yang Yin is exhausting, Supremacy of Yang
6	Sheep	Cancer	June 22-July 23 Summer Solstice Lesser heat Monarch Fire Heart Meeting between Yin and Yang
7	Monkey	Leo	July 24-August 23 Great heat Beginning of Autumn Metal or Earth Transition Ten Thousand things are flavored, Taste of Fruit
8	Rooster	Virgo	August 24-September 23 End of Heat White Dew Metal Yin gains control, suppresses the ten thousand things
9	Dog	Libra	September 24-October 22 Cold Dew Hoar Frost Metal Ten thousand things are growing old, completion.
10	Boar	Scorpio	October 23-November 22 Hoar Frost Beginning of Winter Water or Earth transition

Ten thousand things turn to ashes,
yang withdraws; yang chi is small,
goes underground, earth becomes
fertile.

11	Rat	Sagittarius	November 23-December 21
			Lesser Snow
			Great Snow
			Water
			Yang energy is hiding below, yang
			in touch with yin underground.

12	Ox	Capricorn	December 22-January 20
			Winter Solstice
			Lesser Cold
			Water
			Ten Thousand things are being
			nourished below in preparation
			for growth.

Although the rat is given, the start point of the animal cycle tradition, the New Year always starts with the tiger and third branch. Alternatively, there is a forward-moving celestial and solar cycle from the perspective of the winter solstice; the start point is in a reverse direction for the terrestrial cycle that begins on the back side of the midwinter-midnight position. Even if the terrestrial cycle moves forward with the solar wave, the heart of the terrestrial system is symbolized with the end of the cycle, not the beginning and with an implicit reverse, counterbalancing motion to the solar system. This clearly is a vestige of a more elaborate lunar system that lingers in traditional astrological and medical systems. However, all the systems have as their basis the seasonal conceptions, which then were applied to the diurnal and monthly cycles as well.

The consolidated table of the lunations provides a singular insight into the Chinese view of cycles and number, so that when yang reaches its maximum expression, it encounters yin at the summer solstice and that when yin predominates at the winter solstice, yang begins to mobilize. Thus, any positive appearance implies its opposite operating below the threshold and out of the limelight. Each half of an opposition generates and strengthens the other either inwardly or outwardly depending on the pressure for manifestation. Clearly as heat reaches its peak expression in summer, nature already has moved past the solstice, and the days are becoming shorter; yin already is gaining ascendancy below the surface appearances.

Only the momentum of the waxing light cycles is pushing heat to its extreme. The more inward or the greater the pressure for expression, the more outward or the more its foundation is weakened allowing the potential of the opposite to develop. Implicit in the seasonal expressions of yin and yang is a profound psychology and social psychology that was and now is employed politically. The Wu yun and branch cycles became a metaphor for identity and a rational system for predictions of outcomes. We see the same tendency in the *I Ching*, where a positive line implies it's opposite, a broken line and the reverse. Systematized, these relations theoretically express a profound set of natural and psychological states of equilibrium between varying degrees of suppression and expression. This is a system that shows, unlike rationalist systems in the West, the weakness inherent in extremist positions and rigid principles, which of necessity empower their counterparts in various subversive ways. The critical point in these determinations, always, is to determine the directionality of flow between expression and repression, in conformity with natural cycles, and inevitable cycles of social and individual action and consequence.

One more set of relations projected from the lunations is worth considering at this point. The division of twelve branches refers fundamentally to the twelve lunations. The moon returns to the original zodiac point of the first union (conjunction) in twenty-eight days; but owing to the sun's motion, the moon catches up after twenty-nine and a half days, which ancients generally rounded to thirty days. Owing to the emphasis on lunations in agricultural societies, each day of a lunation cycle could be differentiated first by its quadrant and second by the effect of the degree of light associated with it. The lunar cycle around the zodiac was projected onto the continuously visible constellations above the horizon as twenty-eight divisions or as four quadrants of the heaven, each containing seven divisions each of a span that matches the distance traveled by the moon in a single day or nearly thirteen degrees. These lunar mansions called *hsiu* added interpretive value to the segments of heaven and the phases of the moon transiting them. The seven hsiu in each of the four quadrants of the sky added distinct meanings to the cardinal points using a familiar language. Moreover, the placement of the moon, rahu, or ketu (the nodes of the moon where the plane of the moon's orbit intersects the ecliptic) in a division associated with a particular division had additional interpretive value particularly in a night chart.

Chapter 12 will examine the uses of the division of twelve in traditional Chinese medicine. However, that discussion depends on the assumptions developed in the involutionary assumptions of the number eleven and how that plays out in the terrestrial growth cycle. Any system, whose foundation is the twelve branches, must honor in its presentation the generative growth cycle and the implicit reverse motion of the moon that drives it in twelve annual lunations.

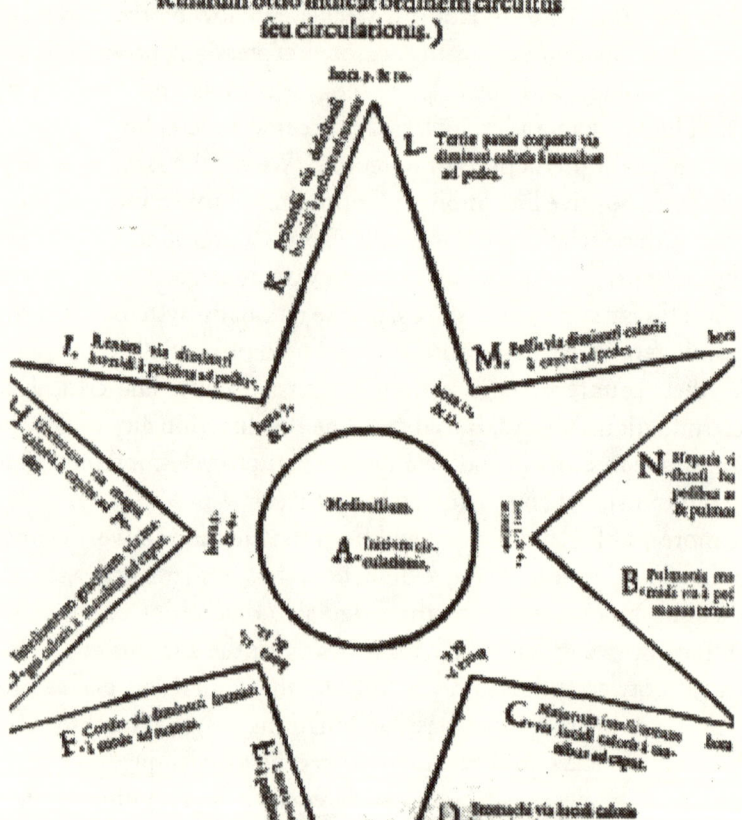

Chapter 12 十二

The American Clock in Chinese Medicine

The number twelve is the culmination of the principal geometrical numbers of the early cosomographical order. It was thought to contain all the combined meanings of prior numbers that were applied to understanding nature. The figure above shows the twelve-phase organization of the day, or year with meridian affiliations. It is a six-pointed star, and each point has two sides, a yin and a yang aspect. The twelve divisions of the circle can be seen as the three yins and three yang s (6), each divided into two aspects, or as two interlocking hexagons as in figure one1. Figure one 1 shows the perfect alignment of the twelve divisions with the four phases of the year defined by the cardinal points.

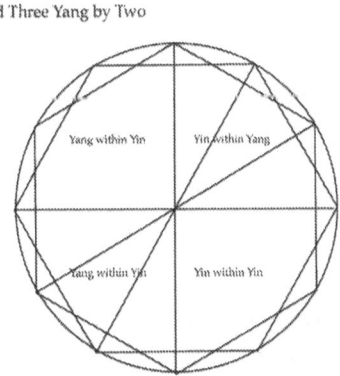

Twelve
Divisons of Three Yin
and Three Yang by Two

Figure 1. The Geometrical Division of Twelve Showing the Interlocking Hexagons and Connection to the Cardinal Points, the Horizon, and the Prime Meridian.

The direct approach to understanding the number twelve in cosmography is to see it as an extension of six. Each of the three-yin and threeyang phases of six now are divided into yang and yin aspects creating a more detailed differentiation. Instead of a hexagon, you have two interlocking hexagons- one organized by space, the other by process.

The Greeks were good at geometry; it it was on the basis of geometry that their twelve-division astrological model with its foundation in the four elements and four humors developed. Figure two shows the cosmic man placed in the sequence of twelve. Agrippa's figure is standing in a square indicating the four directions, and the arms are in positions that indicate transistions and humors that together create a model for eight divisions. This deliberate organization places the meaning of the twelve zodiac signs on the foundation of simpler geometrical relations, ultimately to yin and yang and the triangular consequences in all other geometrical forms.

Figure 2. Agrippa's Cosmic Man Showing the Underlying Cosmographic Organization of Eight with the Square and Four Limbs that in Turn Sustained the Meanings of the Twelve Signs.

The division of twelve in the context of Chinese medicine describes the commonly known twelve meridians that link acupuncture points. These are well established at this point to be simply the planes of the three yins and three yangs differentiated into heaven and earth or above and below at the waist or shoulders depending on the model. The six-yang meridians with their four *fu* or hollow organs with the special organ gall bladder, and the whole conception of hollow spaces symbolized by "three heater" link to the hands and represent the hollow of heaven and sky. The six yin meridians with their associated five *tsang* or solid organs with the additional symbol for life, heart contained in earth or the pericardium, link to the feet and to form. Figure three displays the twelve divisions in a purely five-element sequence in which a yin organ is immediately followed by a yang organ in an alternating sequence, and the five-element partners are linked to congruent five meteorological phases.

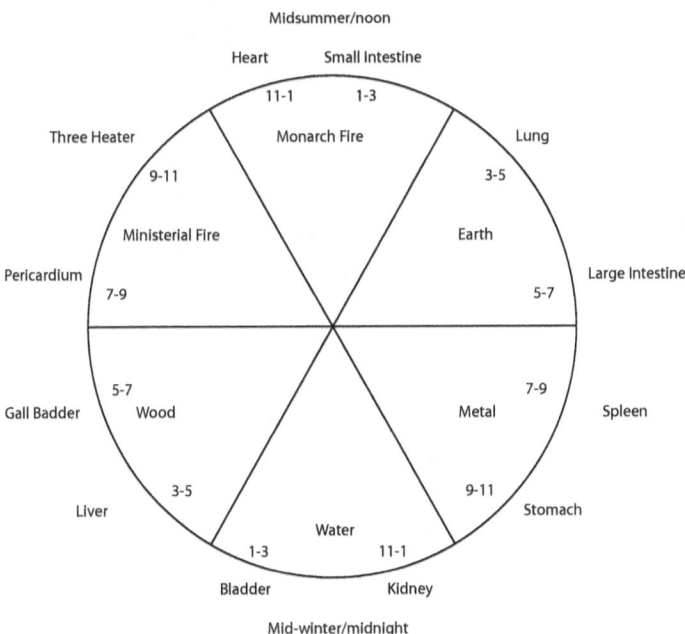

Figure 3. The Division of Twelve Showing the Five-Element Sequence of Solar Activation and Acupuncture Meridian Affiliations with Alternating Yin and Yang Meridians.

It is important to remember what has been commonly forgotten in this five elements model,—namely that the twelve divisions were conceptually the same as the traditional twelve divisions of the Chinese astrological chart of the heavens.

Figure four 4 reproduces a twelfth-century Chinese natal chart displayed by Lu and Needham. It shows a chart anchored by the twelve branches on the second circle defining the twelve divisions of the heavens and the time of day. The typical twelve-house divisions are shown fourth circle with house meanings identical to those found in the writings of the Roman Maternus (Fig. 8-10): life, health, siblings, parents, children, health, spouse, death, beliefs, honors, friends, enemies, endings." Instead of zodiac signs, it shows in the sixth circle the twenty-eight lunar constellations. The ascendant is not fixed to the east, since the Chinese did not detail stellar ascensions, rather, the chart emphasizes the position of the sun in the sky, and the ascendant falls where it may in relation to that. The planetary positions fall in their real dispositions in relation to the unique placement of the divisions lending their bias. The twelve divisions have fundamentally to do with day and night, winter and summer, considerations that allow humans to plan and to live wisely. The Chinese clock is nothing but an extension of astrology and an old symbol for integrated living.

PLATE XVII

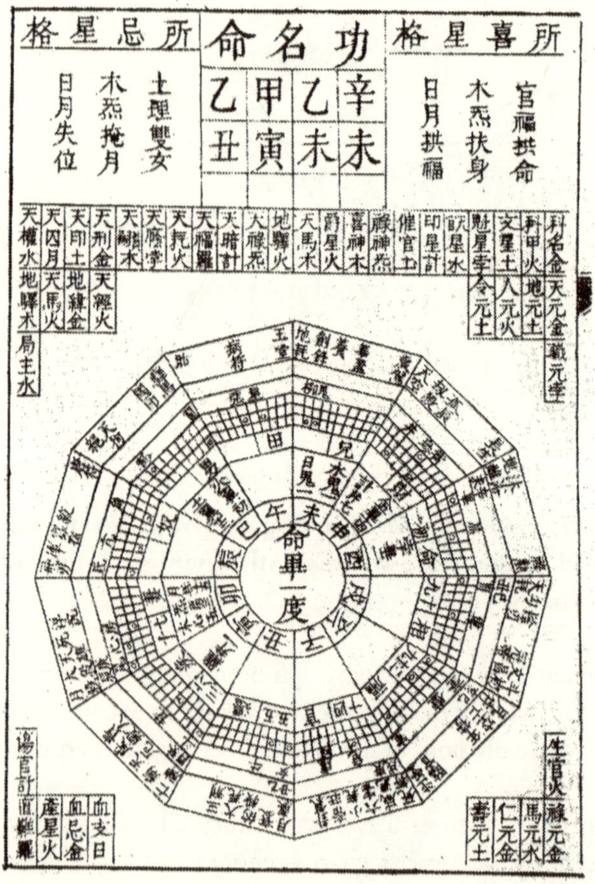

Figure 4. The Traditional Chinese Horoscope from Lu and Needham.

I. Cosmographical Structure

Tradition gives numerous different ways to distribute the three yins and three yangs, and that is complicated by the need in cosmography to show the generative mother-to-child relationship between the numbers five and six. We expect congruence because the overall pattern in nature being described is the same, even though the two were shown to operate on two different principles, celestial stems and terrestrial branches. Each model tells a different story—one of time and process, the other of space and anatomy.

Figure five 5 shows the traditional clock with its twelve divisions, and these initially seem to reverse or nearly contradict the cyclic principles of the five elements shown in figure two2.

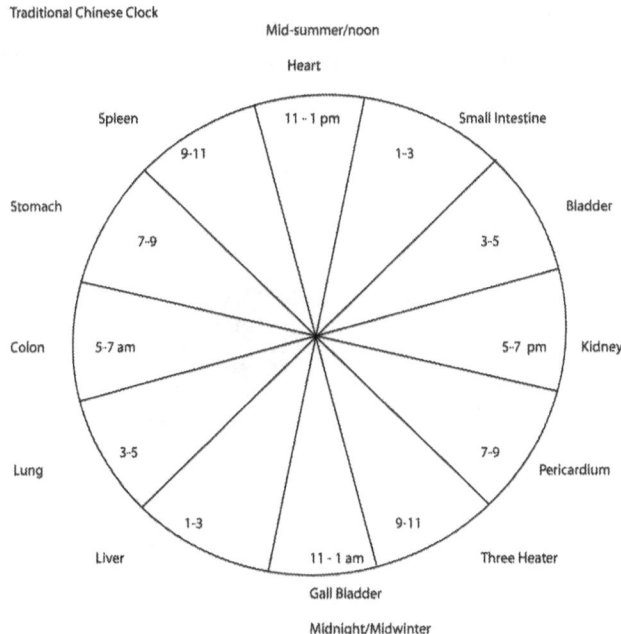

Figure 5. The Standard Chinese Clock.

The present commonly used system of twelve two-hourly periods used in Ttraditional Chinese Mmedicine follows the order shown in figure five. This can be represented in three-dimensions as in figure six. The obvious incongruity and asymmetry of this distribution of meridians and functions is owing to several factors.

1. The problem raised by contrasting circulation of the internal and environmental motion of the terrestrial branches shown in the lower two levels of figure five. The blocks show a reverse direction in their entry—exit sequencing, but they are forced into compliance with a clockwise motion. This ambivalence is complicated by the two-dimensional representation, but also shows a weak attempt at resolving a direct conflict between the poles of heaven and earth.

2. A decision to cast off any attempt at conformity between terrestrial and diurnal rhythm with the celestial five elements and seasonal cycle except to insist on the clockwise motion.

3. A view of energy circulation and anatomy that could not separate in temporal sequence *shao yin tai yang* water and fire since they are structurally connected functions. The new models allow for a continuous three-yin and three-yang entry—exit circulation that are activated in the solar phase by function. The failure of the present system was the inability to remember the cosmographic roots of both systems, and the differentiation between circulation in heaven and anatomical /lunar circulation.

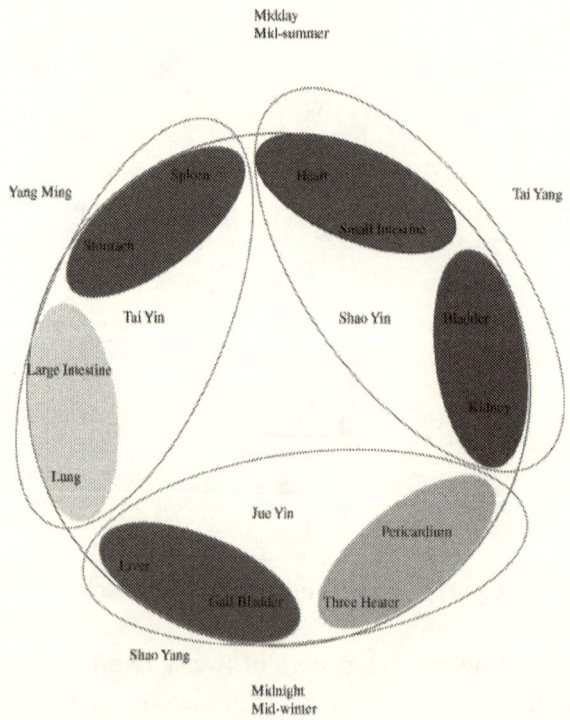

Figure 6. A Three-Dimensional Portrait of the Functions of the Present Chinese Clock.

In every way, the division of six as developed in chapter 6 above describes the same overall cyclic process and relies on the same conception of elements, only realized with greater anatomical precision. The lack of conformity of the traditional mode to cosomographical truth is unacceptable. Figure 3 above, then, shows the distribution of twelve meridians according to the five-element orb sequence. This results in an alternating yin-yang sequence as each orb unfolds as follows, kidney/bladder, liver/gallbladder, pericardium/three heater, heart, small intestine, spleen/stomach, lung, large intestine, and back to kidney. This does not conform to the numerical entry—exit sequence of meridians shown in the standard five-element sequence (Roman numerals) in figure 10.

Figure 7 shows the proposed alternative distribution of meridians according to the same principles but using a three-dimensional special outlook that allows complex blended interactions between elements like wood and fire to develop continuously during any period of rising or declining light. In this mode, the relationships can be simultaneously clockwise and counterclockwise, vertical and horizontal, and they can take into account different adaptive entry-exit point relationships during different solar phases. Most importantly, this model allows for both the conception of continuous cycling of energy within extended blocks of meridians and solar activation of portions of those blocks in phase. This image allows for some revisiting of the fundamental entry-exit sequence that explains how meridians interact with each other.

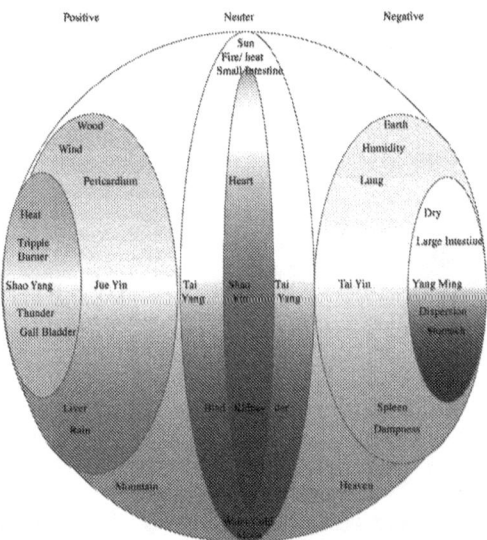

Figure 7. The Three-Dimensional Portrait of the Twelve Meridians Showing Congruent Alignment of the Three-Yin and Three-Yang Blocks with the Clockwise Movement of Five-e Elements

Breaking down the three-dimensional blocks into different working parts results in three different two-dimensional instruments:

1.) a simple three-yin and three-yang sequence that assumes an upright figure and shows the internal sequence of three-yin and three-yang meridian transfers (Fig. 8;)
2). an anatomical sequence that reflects the standing human figure with corresponding element distribution, notably with metal above and earth below the horizon, just as the meridians lay on the body (Fig. 9);
3.) a five-element distribution that places pericardium on the "ministerial fire" rising light side of the mandala (Fig. 10).

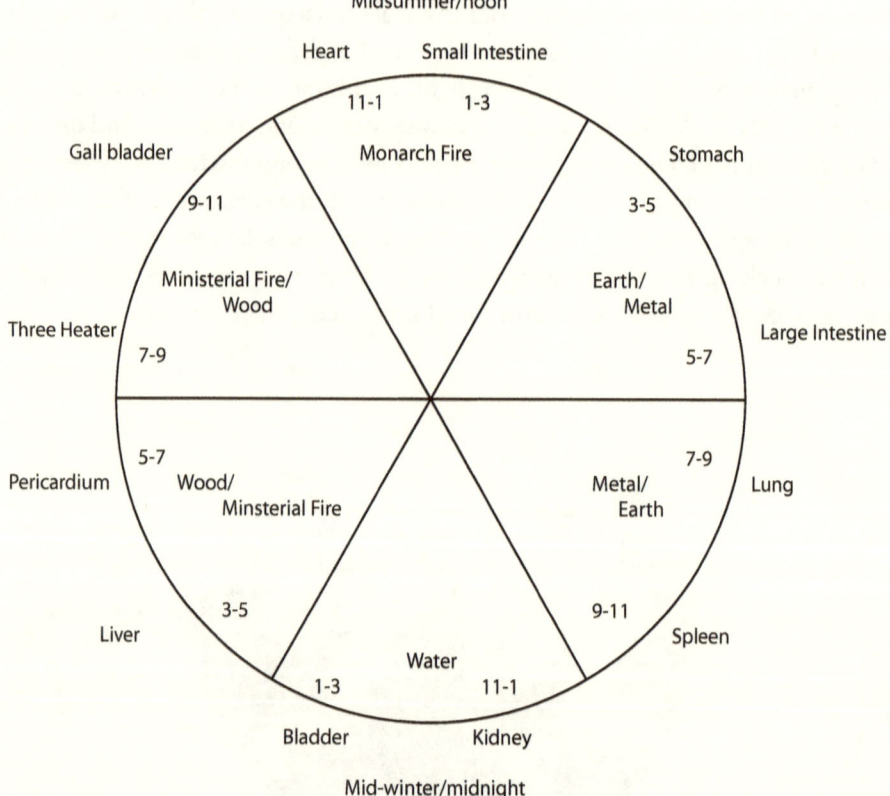

Figure 8. The Three-Yin and Three-Yang Sequence of Twelve Meridians Assigned to Spatial Divisions of the Heavens for either the Annual, Monthly, or Daily Cycle.

American Clock
Clockwise Five Element Phase-Horary Sequence
Counterclockwise Sequence Five Element Entry Exit Points

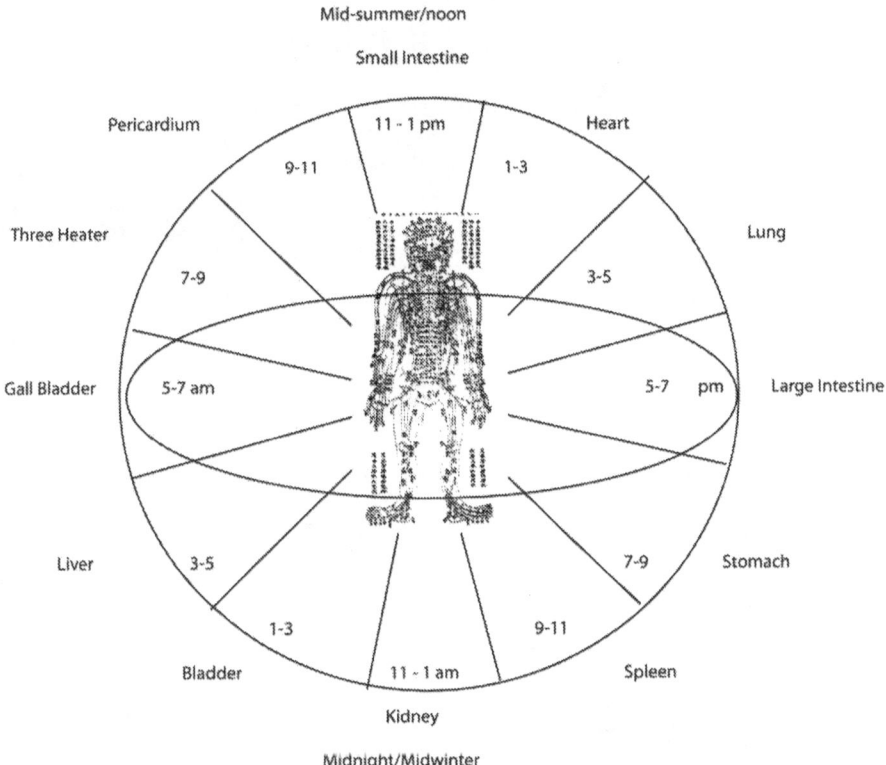

Figure 9. The Anatomical and Spatial Organization of the Cosmographic Phases with Upper Meridians above and Lower Meridians Below.

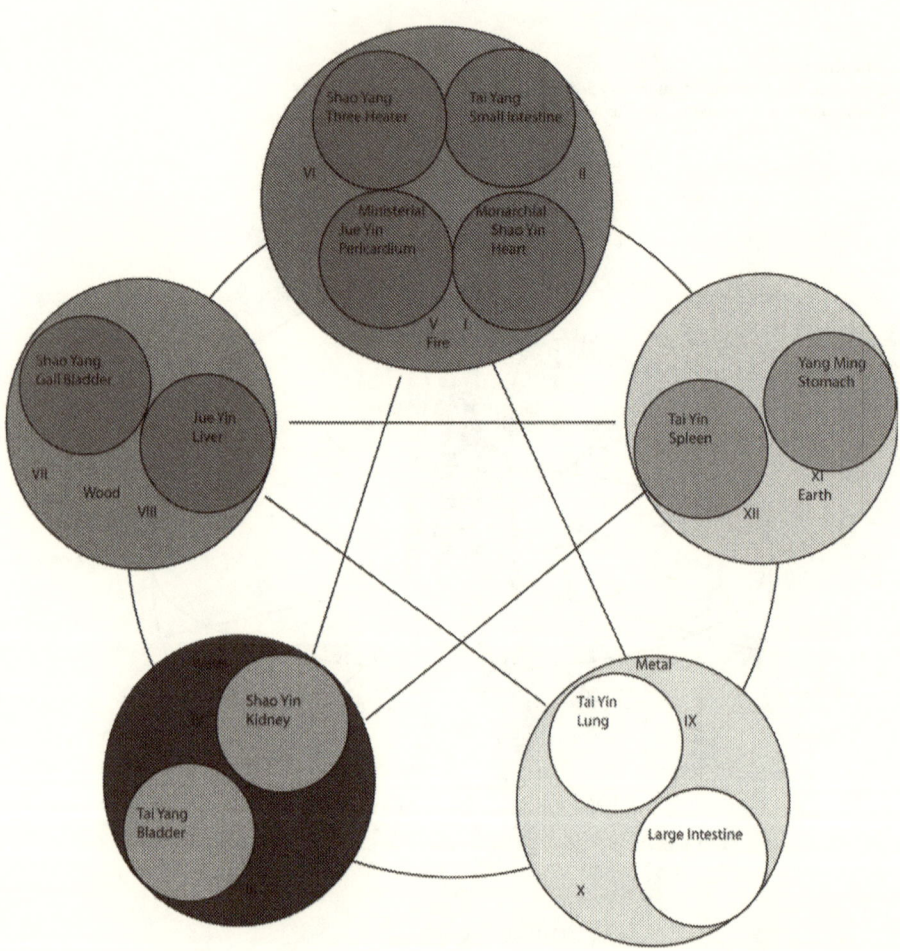

Figure 10. Twelve Meridians in the Five Elements Mandala with Their Standard Sequence in Roman Numerals, Showing the Pericardium/Three Heater Meridians on the Rising Light Side of the Fire Element Orb.

When figure ten is extended into six phases, it resulted in the six or twelve-phase version of the five-element cycle as shown in figure two in one version and figure eleven in the proposed model for the three yins and three yangs. The division of twelve is merely an extension of six, and the proposed model (Fig. 11) solves the riddle of how the stems and branches interact.

Neither the spatial image of the standing man in figures eight and nine, nor the traditional temporal or phase sequence of five-agents in figure three and are fully satisfactory in themselves without remembering the three-dimensional totality that they represent. The valuable thing about them is that they do conform to one or the other ancient standards associated with five-stem or six-branch divisions as they play out physiologically or anatomically. The problem with them is that the models don't connect and communicate cleanly to describe the same natural process. The various sequence of phases and meridians in models for the division by twelve other than figure eleven result in a clock sequence differing from nature or tradition, and the standard Chinese clock, too, is, not, in fact a legitimate model for the number six or twelve divisions because it appeals to the higher authority of the stem distribution of element orbs. This makes it dependent on a clockwise motion while betraying the principles laid down by the cardinal points and the implied counterclockwise motion on earth inherited from antiquity. This compromise with the clockwise solar motion betrays the fundamentally spatial focus of the even number six.

However, the above versions of the cycle are instructive because each of the twelve meridians as projected in models controlled by numbers five or six has clear element affiliations and express important distinct relationships that have meaning in point selection for acupuncture. They must be honored at the same time that the above-to-below distribution of qualities such as hot and cold also are honored. These keep nature in balance over the course of the day or year. The resolution of these models lies ultimately in the three-dimensional representation in figure seven and other three-dimensional models developed below, which give a sense for how these twelve-divisions unfold in their vertical and horizontal complexity.

The division of six already has clarified the distribution of three-yin and three-yang organs in blocks of meridians. These blocks represent the phase and seasonal bias of the whole equilibrium of the cosmos and show the means to create equilibrium through treatment as discussed in chapter six. Figure eleven refines the three yins and three yangs by compressing the three-dimensional content of figure five back into a useful two-dimensional clock. The clockwise element sequence and the counterclockwise six configurations work together and maintain the cosmological principles behind the traditional clock in one image that works congruently in reverse as an entry-exit sequence and forward as a five-element horary sequence.

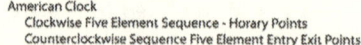

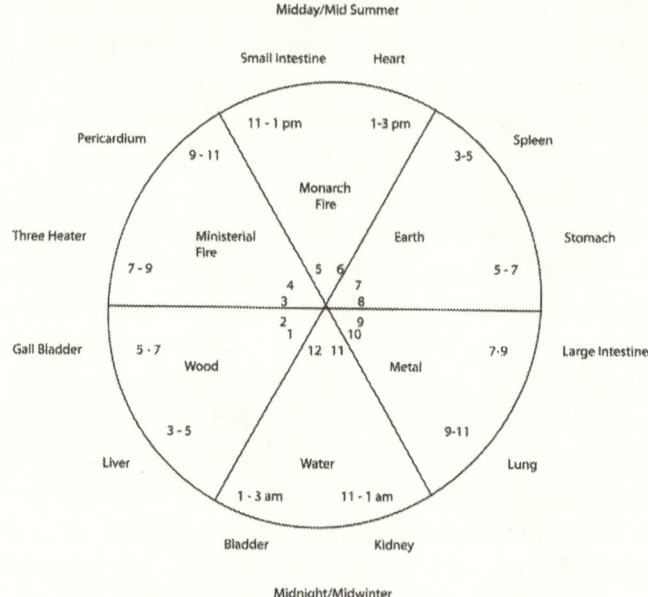

Figure 11. The Clockwise Five Elements and Counterclockwise Three Yins and Three Yangs Consolidated.

This "Chinese clock" is a two-dimensional circle that compresses together the phase-driven element sequence of the stems and the spatial concepts from of the twelve-branch divisions. The model restores the agricultural meanings implicit in both stems and branches to the cycle portrait. The circle implies two distinct motions, that of the clockwise movement of the sun through the zodiac, and the counterclockwise rotation of the earth and apparent motion of the moon. The geometrical patterns of motion found in megalithic observatories are maintained.

Read clockwise, it describes the element-phase sequences appropriate to the daily or annual solar motion. The emerging day begins moving clockwise from the midnight/mid day axis with wood and liver, progresses to ministerial fire, to monarch fire, to earth, then metal, and ends with water. Most of the standard entry—exit relations are preserved in reverse order, and the horary points conform to the standard patterns of chi and light cycles.

Read counterclockwise, the cycle follows with little variation from the three-yin and three-yang meridian entry—exit sequence. The sequence begins with Mmetal lung moving counterclockwise from the midnight/midday axis and moves through the meridian sequence in nearly the standard order. Therefore, the model can be applied with validity to the seasonal, daily, or lunar cycle and accounts both for the clockwise motion of heaven (sun, left to right when facing south) and counterclockwise motion of earth (Moon phases, appear to move counterclockwise to the full moon when facing south) relative to heaven. The proposed model handles both the functional concepts of the ten stems and the directional assumptions of the twelve branches. The solar phase-driven five-element phases are shown in their clockwise relation, and the terrestrial and lunar three-yin and three-yang movements are shown in their counterclockwise sequence. It conforms largely to the traditional clock sequence by preserving the "block sequences" of the three yins and three yangs, but it aligns the sequences with the basic cosmographical and anatomical structures seen in earlier numbered divisions of the same cycles, notably the basic seasonal meanings of the cardinal division by four and its subsequent division of six configuration around the vertical axis of midnight/midday.

The division of twelve does nothing to change the basic set of relationships established by the five elements and the three yins and three yangs, but it does add more detail and dimensions by differentiating anatomically and in month-long-periods associations with greater detail that respect as well the above-to-below underpinning in six.

Figure eleven seems to break down as a model in the transitions between the central heart/kidney block of meridians and the others. The standard meridian entry-and-exit points do not seem to obey the basic necessities of proximity and energetic congruence. Table eleven below describes how the entry—exit sequence can be justified by borrowing from the continuous internal circulation of chi in the central block of meridians, which in turn allows for several access points to and from the heart, small intestine, kidney, and urinary bladder meridians during different periods of solar activation and any transition between blocks.

Figure twelve attempts in another way to display the differences between models based on celestial stems/time and terrestrial branches/space or form. Like figure six, this is a three-dimensional notion that may assist in the understanding the principles behind the Chinese clock. Figure twelve follows figure six by differentiating more overtly the celestial stem system of five-elements from the terrestrial branch configuration of six phases. By placing the five-elements above and the anatomical meridians below, the two can operate in direct counterpoint. The twelve-phase sequence of figure eleven is consistent above and below for the stems and for the branches. In this model, the branches can be conceive either as: (1) circulating clockwise in harmony with the dominant solar stem cycle, or (2) as circulating counterclockwise in an internalized, reactive, and reverse sequence. If the model intends to show the basic cooperation of heaven and earth in causing meteorological and physical change through the seasonal cycle, then both can be conceived as circulating in a clockwise manner (1). If on the other hand the model intends to show the basic cosmological interplay of balanced opposites, the celestial stems beginning at noon or mid summer would circulate clockwise, and the terrestrial branches beginning at midnight and mid-winter would circulate counterclockwise, maintaining balance at all times. The implicit plane of the horizon in the middle is simply the passive location where heaven and earth, each with their specialized phase sequences, interact.

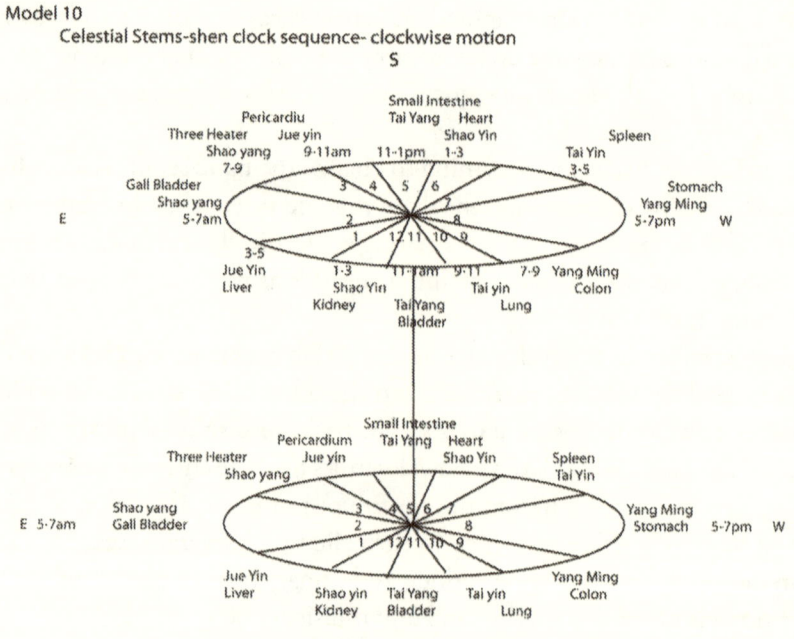

Figure 12. Celestial Stems Above and Terrestrial Branches Below as Two Fixed Contrasting Balanced Motions).

Figure twelve represents the unchanging fixed patterns of cyclic change, and left alone, it contains irresolvable contradictions that lead to the corruptions of the typical Chinese clock. Experientially, every year has differences one from the other, which cannot be explained by this model. Theoretically, the seasons have a regular, progressive, and unidirectional motion that shows little evidence of a reverse motion or reverse effect. And if heaven rules, then changes in meridians and natural elements must conform to the clockwise solar wave of heaven and the sequence of the day and seasons.—Where nature is to conform more closely to the unchanging cyclic order, the clockwise (1) and/ or counterclockwise connections (2) shown in figure eleven—would be the whole story. Tradition simply tries to have it both ways in a two-dimensional representation by reversing the spatial location of the terrestrial blocks of meridians (Fig. 5) and forcing the whole into a clockwise motion. This wrenching of the system also kept water and fire meridians connected in the early afternoon phases, but at the expense of congruity with the natural order (both time phase and anatomical structure).

An alternative stacking of planes is possible that displays the five-phase sequence and entry-exit relations in heaven and a different three-yin and three-yang anatomical sequence below on earth (as in figure eight 8 or structural sequence as in figure nine) allows somewhat for a sense of asymmetry necessary to model change in nature (Fig 13).

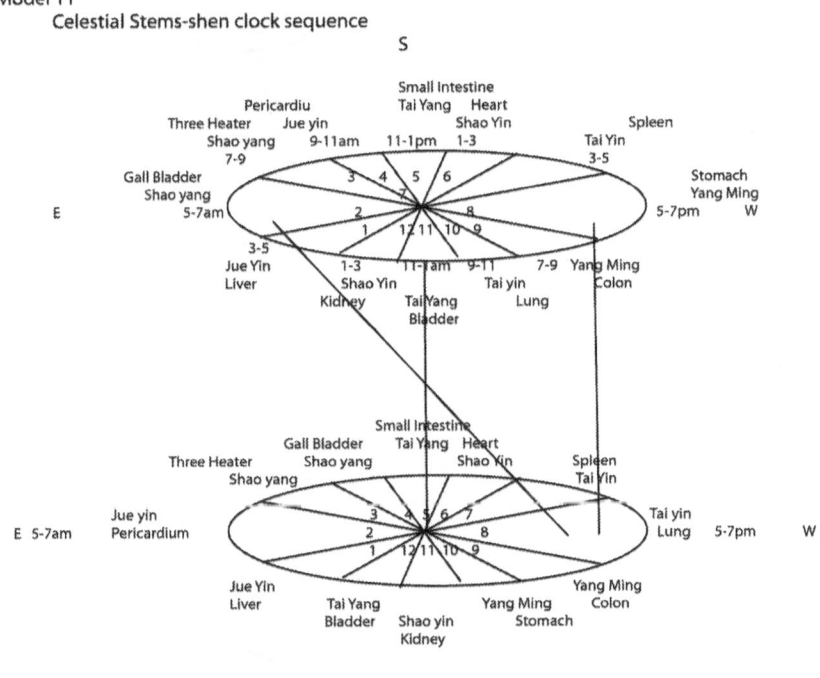

Figure 13. A Version of the Twelve Divisions that Show Anatomical and Three-Yin and Three-Yang orders to the Meridians Below to Account for Asymmetrical Variations in the Annual Cycles.

The stacked planes of figures eleven and twelve and the three-dimensional image of figure five are merely attempts to describe the components of a congruent system of wave-like oscillations in weather and human physiology. Once the multidimensional exterior and interior, vertical and horizontal, spatial and temporal parts are shown as a holistic construct, we have a legitimate framework for understanding the special meanings and uses of the twelve-division system.

Simplified, the new system for twelve mirrors the three yins and three yangs by allowing implicitly for two simultaneous presentations of the declining cycle, one with metal above the waist in conformity with cosmographical space (Fig. 14), the other in conformity with solar and celestial five element phases and time (Fig. 9).

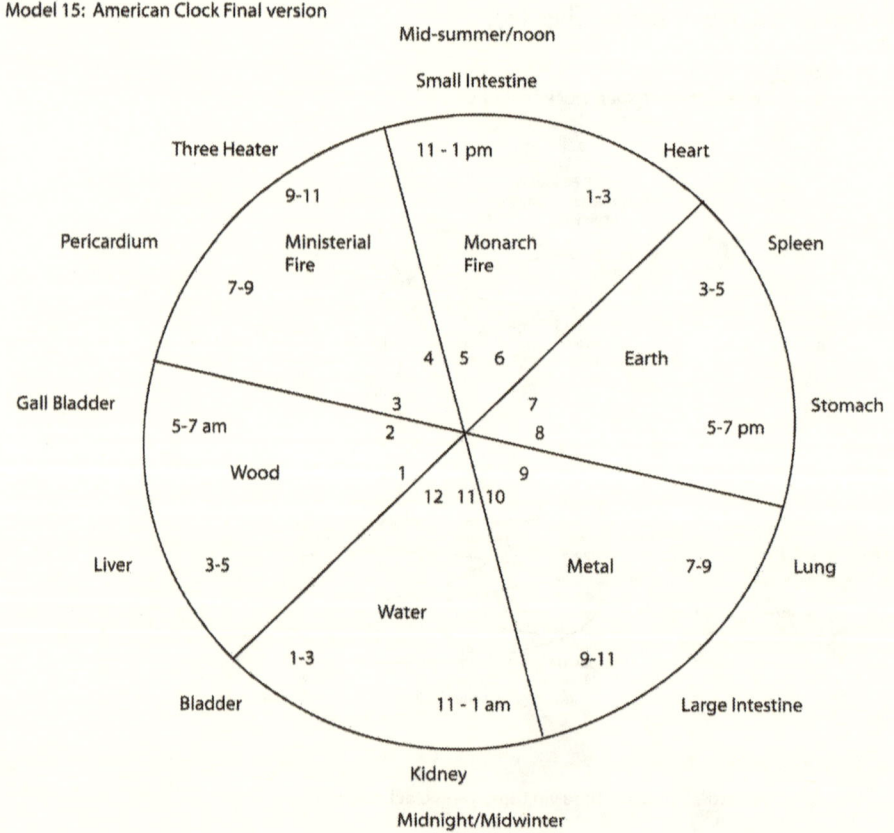

Model 15: American Clock Final version

Figure 14. The Proposed Temporal Model for the Division of Twelve.

These two images, the temporal and the spatial, are interactive. It simply means that in some clinical situations, the solar wave will have an effect on structure that conforms either to the spatial model or the temporal model. The critical thing is to identify the meridian and orb's weak link and to use the above-to-below relationships to support that weak link. In some cases, the lung problem will be aggravated in the afternoon, in others later at night. If in the afternoon, the problem is augmented by increased chi in the lung as an anatomical structure, if late at night it is owing to a chronic deficiency of functional capacity, that could be from various kinds of depletion that involve other orbs as well.

When combined, figures nine and fourteen compress back into two dimensions into figure eleven where their different actions are implied.

Second Version

A natural extension of figure twelve and thirteen is to allow the fixed terrestrial sequence to rotate with the order of heaven in a clockwise motion, but to differentiate the unique stem and branch pairs of the year and their variable conditions as two additional planes, the variable stem rotating clockwise the variable branch rotating counterclockwise. In this way, each variable that was attempted to be included in the traditional system is differentiated and allowed to act congruently to express both the clockwise motions of heaven and the counterclockwise motion of shadows in celestial observatories on earth that seem to explain the balance of nature. Figure fifteen displays the four levels.

The upper and lower planes describe the standard, fixed clockwise circulation of chi in the cosmos and heaven above and internally and in the earth below. The branch rotation on level four 4 from the top is in the time-phase pattern of King Wen to show the basic harmony in the heavenly cause and earthly effects. The upper of the two middle planes (3) that follows shows the variable conditions of a given year whether based on the annual stem branch combination or on experience. The lower of the two middle planes (2) shows the variable conditions acting in counterbalancing motion to heaven driven by special circumstances on earth. This set of twelve divisions is organized structurally and anatomically different from one and two levels and different in sequence as it rotates counterclockwise to show the dynamic counteractive motion of the unique annual lunations. The variable level indicated by an asterik is time phased. The two variable

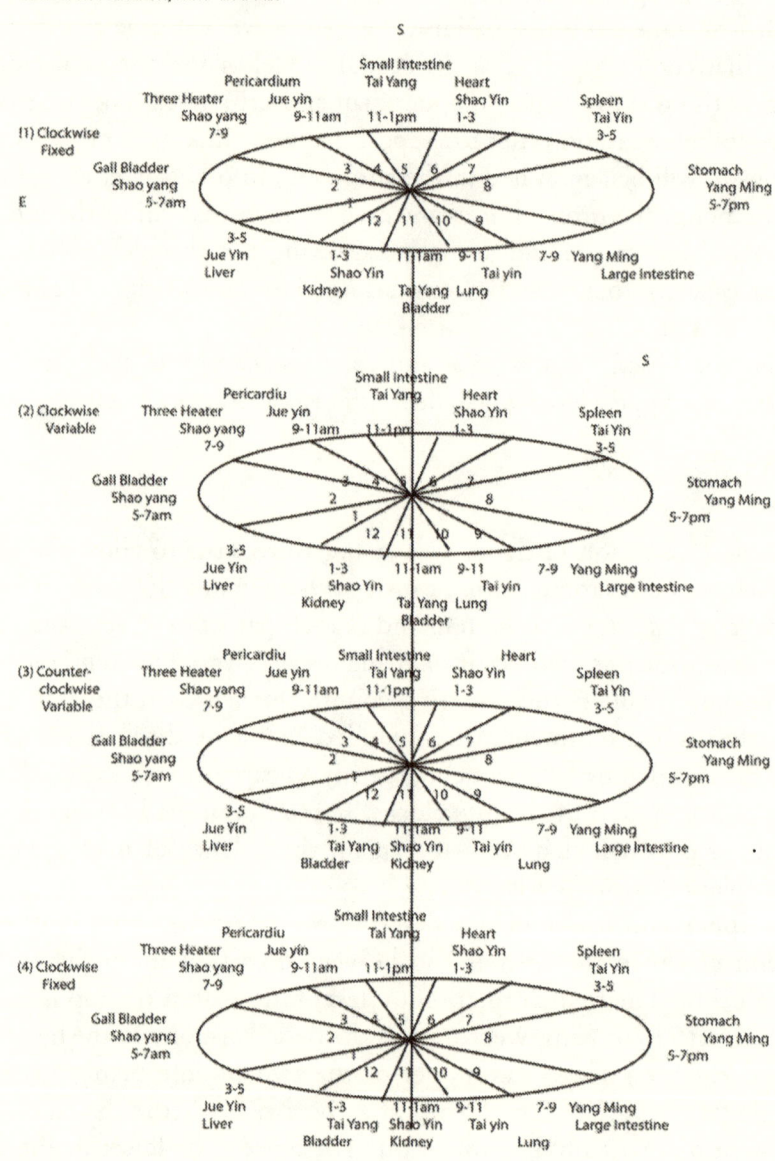

Celestial Stem: Clock sequence- Clockwise

Figure 15. The Four Levels Of Cyclic Interaction Including The Fixed Clockwise Cycle Above, the Fixed Counterclockwise and Internal Circulation Below, and in between the Celestial And Terrestrial Variable Cycles that Follow the Order of Clockwise Circulation in the Upper-Solar Template

cycles are very solar and weather dependent, and reflecting this level, two rotates clockwise while the level three counter-clockwise to compensate. The fixed terrestrial sequence on level one 1 is the same order as figure eight 8 showing the reverse order of entry—exit points working congruently with the forward order of five-elements. The clockwise rotation of the stems and branches together on three of these levels accounts for the element congruence between the five-elements and elements in the six configurations for either twelve months or two-hourly periods. These four levels are replicated with twelve divisions in figure ten 10 to capture the full density of relationships and a system of theoretical times for point activation follows.

II. Medical Philosophy

Figure fifteen offers a good opportunity to study the medical implications of the division by twelve. The first thing to understand is that the full implications of the three yin and three yang are figuratively drawn in a way that includes all the complexity and sophistication of point selection strategies developed in numbers seven, eight, and nine. The twelve divisions divide the body at the waist and include in each of the half meridian, quarter plane divisions the three zones of heaven, earth and human. These in turn may be diagnosed either anatomically or functionally superficially or deeply within the anatomical zones implied by these twelve meridians.

Conforming with other lesser divisions of the cycle there are medical/meteorological connetions made with the world. There are twelve bing xie or disease producing outer evils listed by : Wind, cold, summer heat, dampness, dryness, fire, warmth, phlegm, rheum, worms, toxin, and static blood.[114]

Table one shows some of the complexity and symmetrical harmony in figure nine that develops when the various levels of the three-dimensional cycle play out. Above are the fixed and variable stem driven sequence of phases and below are two terrestrial branch sequences taking into consideration first the structural sequence employed on the counterclockwise variable level one and second, the proposed entry-exit sequence on the fixed clockwise rotating terrestrial sequence. The result is some clear patterns of cyclic congruence and counterpoint from these opposite and complex motions.

Phase I, Jue Yin, Liver is shown in columns one and two to be activated during the liver period between 3 and 5 a.m. These columns show the activation of the synchronized fixed celestial (1) and terrestrial (2) cycles by the solar wave. Arbitrarily, the twelve meridian divisions are allocated both to the heavenly and the terrestrial cycles. The primary focus in the celestial cycle (1) would be the interior aspect of the wood element and the moist and cool qualities it possesses.

Table 1: Table of Motions from Figure Nine.

Phase	Meridian	1 Sun to Fixed to	2 Sun to Fixed Terr	3 Sun to Rotate. Cel Fixed	4 Rot Nadir to	5 Sun to In Phase Terr, Fixed at Phase	6 Rot. Terr. Rot. Terr, Fixed Phase	7 Fixed. Terr. Phase to ... Hvn. Phase in	8 Rot. M.H. to Mid. Cel	9 Rot. Nadr. to Fixed Cel.	10 Phase on to. Fixed. to. Nadir Nadir	11 Rot. Cel. Rot Ter. Fixed in	12 Rot. Ter. Phase to Rot Cel in Phase
P1	Liver	Liv 3-5	Liv 3-5	SI	LI	TH/SI	CO	Liv	CO	CO	Liv	7-9	3-5
P2	Bladder	BL 1-3	BL 1-3	SI	SP	PC	Lu	Bl	GB	Lu	Bl	9-11	1-3a
P3	Kidney	Midn	Midn	SI	Ki	Ki	Ki	Ki	Ki	Ki	Ki	Midnight	
P4	Lung	Lu 9-11	Lu 8-11	SI	TH	Bl	Bl	Lu	GB	Bl	LU	1-3	9-11
P5	Colon	CO 7-9	CO 7-9	SI	Li	Liv	Liv	Co	SI	Liv	CO	3-5	7-9
P6	Stomach	St 5-7	St 5-7	SI	GB	GB	St	St	GB	St	St	Opposites 5-7	5-7
P7	Spleen	Sp 3-5	Sp 3-5	SI	LI	Th	Sp	Ki	Th		Sp	7-9	3-5
P8	Heart/	Ht	Ht	SI	Sp	Pc	Ht	GB	Pc	Ht		9-11	1-3p
P9	SI	SI 11-1	SI 11-1	Ki	SI	SI	SI	SI	SI	SI	SI	Noon	Noon
P10	Pericard.	PC 9-11	Pc 9-11	Pc	SI	Th	Ht	Pc	St	Ht	PC	1-3	9-11a
P11	TH	SI 7-9	Liv 7-9	Liv	Sp	Sp	Bl	Sp	TH		TH	3-5	7-9
P12	Gall.	GB 5-7am	SI 5-7	Ki	GB	St	GB	GB	St		GB	Opposites 5-7pm	5-7

When acting in nature it expresses a hidden but palpable warming direction in the weather and soil, meteorological activation as rain or snow and a general increased dynamism in nature despite the persistence of darkness in the day and winter in the seasonal appearances of things. The yin liver meridian and associated functions begin the process of activating chi and regenerating tissue in preparation for ensuing activities of the day.

Column three shows the rotating celestial cycle Tai Yang small intestine active at every activated phase of the day, which indicates that the celestial rotatation (plane 3 from the bottom in figure 59) matches the sun's motion and that the executive function of the rotating cycle organizes and distributes the solar impulse in a regular set of relationships.

Column four indicates the phase location of the nadir of the rotating terrestrial cycle at the time period activated by the Sun and rotating celestial MH. It is a simple count counterclockwise that matches in reverse the clockwise forward count of the mid-heaven. It becomes the basis for locating column five.

Column five describes the location on the counterclockwise moving terrestrial sequence of plane two that is activated by the sun when it is in Phase 1 at 3-5 a.m. along with the clockwise rotating Tai Yang of the rotating celestial plane three. The sun moves clockwise with the celestial rotation establishing the differential of 9 periods from noon for the solar period of phase 1. However, the terrestrial branches start their sequence at midnight with shao yin kidney. Nine counterclockwise steps from the 11p.m. to 1 a.m. period yields shao yang, three heater as the period (column 4) on the template activated by the nadir when phase 1 is activated by the Sun. The solar activated phase 1 is three phases counterclockwise from the three heater nadir, which on the rotating terrestrial cycle is Yang Ming Colon shown in column 5.

Column six is related to four and five because it is the location on the fixed terrestrial cycle of the rotating terrestrial phase 1, when the fixed phase 1 is activated by the Sun at the standard time. The result is a resonant activation on the fixed cycle at the location of the moving terrestrial phase 1. In the case of Jue yin liver phase 1-3 a.m., the resonant location is at the fixed MH, Small Intestine. It is both interesting and important to see that when the celestial mid heaven strikes phase 1, the terrestrial phase 1 strikes the mid-heaven. This demonstrates a strong mutual resonance and secondary augmented effect of the solar activation.

Column seven shows the time of day and associated fixed phase when Jue Yin rotating in its sequence reaches the mid-heaven of both the fixed celestial and terrestrial cycles. The counterclockwise sequence moves clockwise across the stationary MH point so that the fixed liver reaches the mid heaven five phases after the tai yang small intestine or noon position of the cosmic human. Counting forward five from the noon time yields a time of 7-9 p.m., yang ming colon. Another way to view this activity is to see the clock numbers rotating counterclockwise once for each clockwise advance in the phase rotation. The other way to calculate is to count clockwise from the midheaven on the template to phase 1, liver which yields 9 periods. Then counting nine periods forward from liver reaches yang ming colon once again. The relationships on the template apply both to the celestial and terrestrial fixed planes and the column describes the most time and phase characteristic designated by the sun and rotating mid heaven at the time when a given phase gains prominence in the system. It signals a way to mutually reinforce relationship in this case between Liver and colon during a particular time period.

Column eight also indicates the phase location on the rotating celestial plane of the true MH at any given time period that is designated by the solar activation. When phase 1 of the regular cycle is at the fixed mid heaven at the time 7-9 p.m., colon indicated in column 6 this column also describes the phase and the time when the Tai Yang point of the rotating celestial plane reaches the phase listed. Thus, 7-9 p.m. is when Phase 1 activates the MH and when Tai Yang SI "the midheaven" of the rotating plane (3)and the sun reach the Yang Ming of the arm. This sets in motion a reciprocal strengthening and balancing of the Jue yin of the rising cycle and its counterpart in the descending cycle across the middle spine of the day and body.

Column nine again shows the synchronicity of celestial and terrestrial fixed cycles as the clock face moves in a counterclockwise direction in relation to the fixed cycle. The rotating terrestrial cycle at the mid heaven during Phase 1, Jue yin liver is colon.

Columns ten through twelve focus on the progressive clockwise sequence of terrestrial phases in level one, on the one hand, according to their structural entry/exit sequence and on the other by the counterclockwise rotating cycle. Column 10 establishes the time of day and associated phase when the phase in question, in this case Jue yin liver, reaches the fixed nadir.

Column 11 establishes the phase at the fixed nadir (level 1)by entry/exit sequence at the time of day activated by the phase in this case 3-5 a.m. Jue yin liver.

Column 12 indicates the rotating terrestrial phase at the fixed nadir at the time activated by the phase in this case 3-5 a.m. Jue yin liver. The time when liver arrives at the fixed terrestrial nadir is the time associated with its opposite member in the cycle. This expresses the symmetry of a system organized around both the mid-heaven and nadir, the mid heaven relationship shows a symmetry of parallels; the nadir relationship across the midline explains the oppositions. The terrestrial cycle entry/exit colon at the nadir during phase I, 3-5 a.m. beautifully expresses symmetry with the other ways that colon and Liver phases interact in the progressive cycle. The counterclockwise moving terrestrial phases of the rotating cycle deliver the Jue yin phase to the nadir when the sun activates Jue yin. This again when compared to column 11 mirrors the symmetry between liver and colon on opposite sides of the midline.

The evidence of symmetry and counterbalancing seen in the contemplation of jue yin phase one in table one, becomes even more marked when phases associated with the prime meridian and the east west horizon are select for study the shao yin on the vertical axis of the day and yang ming on the horizontal.

The shao yin kidney phase is activated by the sun's rotation at midnight as shown in columns one and two. Column three tabulates that rotating celestial cycle has kept pace with the Sun's clockwise motion from the midday start point and at midnight this rotating celestial cycle midheaven (plane3) Tai Yang fire, small intestine reaches its furthest extension opposite its natural start point. Column four shows the clockwise motion of the nadir on fixed cycles and at the same time that the celestial MH has moved to the nadir, the nadir has moved to the celestial mid-heaven. Thus, at midnight the sun and celestial motion activates Shao yin and Shao yin rotates to and activates the Mid-Heaven repenting there by the quintessential extreme of yin (column 4), at the same time that the extremity of yin falls under the jurisdiction of the heavenly mandate, yang (column 3, 12). Similarly, the counterclockwise rotation of the terrestrial cycle has rotated the tai yang fire to the fixed nadir below the sun (column 5) and its opposite shao yin phase has risen to the tai yang position delivering the message of earth to the mid heaven on both fixed cycles (column 6, 8, 9). Both clockwise and counterclockwise motions deliver the same emphasis at the transitional mid-point at night between the descending block and the ascending block.

The phase cycles show a direct communication between shao yin and tai yang at midnight, which reinforces the idea that the middle block sequence should be active in its entirety at this time. It also suggests once again that the middle block is an organizing and regulatory door or passageway between two functions. Its umbrella-like sheltering of the process has a timeless component. Its processes create a seamless transition between two tendencies. In this sense the entry exit points that directly link rising to declining blocks have a certain patency either conceived as gall bladder 41 to spleen 1 and stomach 42 to liver 1 or alternatively as from liver 14 to Lung to lung 1 and from spleen 21 to pericardium. Clearly, in any case the midday and midnight relations act as a hinge between the actions of the rising and declining blocks.

The yang ming Stomach phase and meridian on the fixed, spatial portrait of the diurnal cycle falls on the horizon line at sunset between 5 and 7 p.m. Its natural opposition is the rising block's shao yang gall bladder and the two show a parallel symmetry in their relations not unlike that found on the vertical axis. The sun activates the phase (columns 1 & 2) and the rotating celestial MH (column 3) also activates it. The sun has moved four phases from its mid-day position and the nadir has moved four positions to the stomach's opposite member gall bladder (column 4).

The counterclockwise rotation of the rotating terrestrial cycle (level 4 in figure 59) has moved four positions counterclockwise, which places the Sun directly over the rotating terrestrial Shao Yin Kidney (column 5). Kidney activates the earth, brings the whole terrestrial cycle to bear on the Stomach phase at the same time that the rotating celestial cycle brings the mandate of heaven and the sun. Thus, both heaven and earth point to the yang ming giving it a weighted emphasis.

When the rotating terrestrial cycle rotates, the yang ming position when the Sun arrives at the fixed yang ming phase is at tai yang, fire, the mid heaven (column 6). This places it opposite to the Shao yin position, which on the rotating terrestrial cycle has moved four places to activate fixed yang ming. Thus, the center pole remains in balance as yang ming connects both with shao yin and tai yang at the same time that yang ming on the horizon is emphasized by both the celestial and terrestrial advances.

Column seven designates the time of day and associate phase when Yang Ming stomach activates the MH. The clockwise rotation indicates a ten-phase advance and a 5-7 p.m. shao yang gall bladder arrival. So, yang ming is at the fixed and real mid-heaven at dawn when the sun activates its opposite member. This means that the mandate of heaven at dawn, during the rising cycle is controlled by the opposite dominant member of the declining cycle, just as the dominant rising energy shao yang gall bladder occupies the contrasting nadir governing the terrestrial point towards which the declining cycle and yang ming contend. Each places a limitation on the upward or downward tendency of the other.

Column seven again reinforces the reciprocal resonance between the oppositions along the horizon. The clockwise rotation of the fixed sequence of phases brings the Stomach to the midheaven in the time of the gall bladder 5-7 a.m. Column 8 also indicates the phase location on the rotating celestial plane of the true MH at any given time period that is designated by the solar activation, and again that is the Gall bladder. In column 9 the rotating counter clockwise terrestrial phase at the MH during phase 8 stomach is stomach. Thus, the rotating terrestrial stomach, phase8 is on the fixed mid heaven at the time when the rotating mid-heaven reaches the fixed stomach phase.

Columns five and eleven compare reciprocal relations between rotating celestial and terrestrial cycles yielding a symmetrical opposition between celestial and terrestrial cycles at the time when yang ming stomach is in phase.

Column twelve again shows that from various points of view the rotations at ninety degrees from the vertical develop reciprocal with the opposing 90-degree phase. Thus, while the forward moving celestial cycle and the reverse moving terrestrial cycle arrive at the stomach at the same time in the evening to give it and the declining cycle a heavily weighted emphasis, many subtle corollaries in the cycle such as the rotating midheavens and nadirs on levels three and four and the arrival and departures of the rotating phase eight yang ming placements at prominent points of the day show strong connections with the opposite, compensating phase shao yang gall bladder.

These mechanics are striking and they explain features of both cycles and human physiology that otherwise remain obscure.

Shao yin and tai yang act now as a passive hinge between rising and declining and again as a fundamental polarity that maintains order. Shao yin encourages and inspires the movement of decline, Tai yang the movement of ascent. The tai yang in the rotating cycle moves with the sun and the fixed cycle, which symbolically and literally organizes and expresses the solar placement as the mandate of heaven. In the midnight position, heaven and the sun activate the earth, shao yin, but the earth symbolized by shao yin has arrived at the mid-heaven at the same time to express a controlling local emphasis. In the midst of this vertical process the rotating cycles create a weighted emphasis on decline in the evening and a weighted emphasis on raising chi in the morning, by both arriving together at dawn or at sunset at those phases near the horizon. The meeting points of the clockwise and counterclockwise points define the spatial and temporal quadrants of any cycle. Observing the quadrant points on each of the four levels operating together is to see the holographic principle in action at any point in the circle. For each phase contains and refers to all other points, in a regular pattern. The importance of this fundamental link between time and the twelve spatial divisions of the year as they are described by the twelve terrestrial branches is that this sequence is the basis and goal of the meanings for the division of six, itself, which derived conceptually and reciprocally from the twelve branches, whose very existence conveys the whole intent of the numbering system. Above, in the introduction, the root seasonal values that inspired the twelve terrestrial branches were examined in the context of number theory, and it is beyond question that the branches were tied to the regular process of changing conditions of the agricultural growth cycle.

Armed with the new tools from table one, it is possible to anticipate numerous specific uses of the entry-exit points and diurnal clock sequence. The following table offers a reduced set of times favorable for various treatment interventions within specific phases and in keeping with the above discussion it should be used as an aid not as a prescription to discriminate when functions can be accessed for treatment.

With the new tool it is possible to anticipate numerous specific new uses of the entry exit points and diurnal clock sequence. The following table 16 offers a reduced set of essential times for various treatment interventions during specific phases.

Table 2: Essential Times for Treatment of Phases Using the Proposed clock. Horary, EE or Sour Points Can Utilize These Times: I) activated during entire rising and declining phases, ii) augmented in phase only because the function is augmented, and iii) a-temporal in their actions.

#	Phase	sun at phase	sun in phase	Phase at Mid corollary	Phase at Nadir heaven	Rot Celestial phase at MH	Rot Terrestrial phase at Nadir	sun in Rot Ter Phase
5	Tai Yang Small Intestine	11-1pm	11-1pm	11-1pm	11-1am	11-1pm	11-1pm	5-7pm
6	Shao Yin Heart	1-3pm	1-3pm	9-11am	9-11pm	9-11 am	9-11am	11-1am
7	tai yin spleen	3-5pm	3-5pm	7-9am	7-9pm	7-9am	7-9am	1-3am 1-3pm
8	Yang Ming Stomach	5-7pm	5-7pm	5-7am	5-7pm	5-7am	5-7am	
9	Yang Ming Colon	7-9pm	7-9pm	3-5am	3-5pm	3-5am	3-5am	3-5am 3-5pm
10	tai yin Lung	9-11pm	9-11pm	1-3am	1-3pm	1-3am	1-3am	
11	Shao Yin Kidney	11-1am	11-1am	11-1am	11-1pm	11-1am	11-1am	5-7am 11-1pm
12	Tai Yang Bladder	1-3am	1-3am	9-11pm	9-11am	9-11pm	9-11pm	
1	Jue Yin Liver	3-5am	3-5am	7-9pm	7-9am	7-9pm	7-9pm	7-9pm 7-9am
2	Shao Yang gallbladder	5-7am	5-7am	5-7pm	5-7am	5-7pm	5-7pm	
3	Shao Yang Three Heater	7-9am	7-9am	3-5am	3-5am	3-5pm	3-5pm	9-11am 9-11pm
4	Jue yin Pericardium	9-11am	9-11am	1-3pm	1-3am	1-3pm	1-3pm	

When assessing the table data in its entirety, a pattern of complete symmetry and unity emerges. The midday to midnight poles of the cycle act as an exterior to interior, heaven to earth spine that governs the whole at every point in the cycle. Each counterbalancing phase refers in some way to this spine, to its opposite member on the circle and to its bilateral member at the same level across the midline. Tai yang fire of heaven and shao yin water of earth for instance both limit and reinforce each other mutually. The four levels of connection express a basic polarity, heaven to earth, and meteorology to living forms, a time sequence, a communication from heaven to earth and from earth to heaven, a mutual checking of the contrasting impulses to expand and shrink indefinitely, and together they initiate or retard all the other functions in a regular way, each with a slight emphasis determined by their place in the circle.

The Chinese approach to this cycle made it clear that each phase was a precondition for the next phase, just as an implicate realm of potentiality and directionality is implied in any phase of the growth cycle, but every such potential is tied to and regulated by its opposite phase of the day or year (dawn to dusk, spring to autumn, mid-day to midnight, summer to winter, etc.). Heat may be rising, but its degree is determined and conditioned by the cold of the previous epoch and in turn the degree of manifest heat will condition the amount of cold in the subsequent period. This meant to traditional thinkers that present symptoms implied conditions that while invisible either left traces of the past or were anticipating manifestation. The stronger the appearances, the more its opposite is invoked and under stress. If it gets too hot in the summer, storms will bring rain or even hail. Nature is, in short, dynamic and conditions are determined by the direction of change. Change and movement are the basis of the generative model, but it is the directionality of change towards increased light or towards increased darkness that mattered and allowed for meaningful differentiation.

III. Medical Treatments

The basic hypothesis in this section is that any system that purports to describe energy circulation in the body must logically conform to the circulation of energy in the world through the quadrants and divisions of the day or year that align with the topography of the body. This has been developed in previous chapters. When we reconsider tradition and re-impose the seasonal pattern of four cardinal divisions, five-elements or the modified six configurations on the standard division by twelve, we immediately confront the necessity of new horary and entry/exit point relations and mechanics, and more sophisticated ways of treating based on the understanding that blocks of meridians behave interdependently and coherently apart from consideration of the whole cycle. This section proposes an open system of chi circulation that automatically distributes chi in complex ways according to different stimulus either from inside or outside.

The three-dimensional actions of life allow multiple tiers of influence, certainly three, heaven, earth, and horizon, four directions, five transformations, six structural gradations, patterns of lunar development, seven, and interior to exterior relations and these must concordantly work together for balance. To account for these things different models have been developed in previous chapters that help to explain simultaneous adaptations, or whole blocks of adaptation even as the inevitable regular cycle of diurnal and seasonal change continues.

The consequence of open systems of chi circulation that involve multiple meridians at a time is both baffling and intriguing. It offers hints about how to leverage treatment and how to understand the power of the cardinal points of any cycle better. An open system implies an inherent natural intelligence in the system to selectively move energy when and where it is needed. In this model a horary point, properly selected can reorganize the whole system and exit points from a meridian can transfer energy to different entry points or even non-entry points, given different pressures from bodily cycles or external triggers. It is conceivable that some entry and exit points can leverage or activate whole cycles, not merely clear or transfer chi. Some of the redundancies in deep pathways and fluid ions and anomalies in meridian pathways and extra meridian pathways may support this idea. These continuous patent relations between the three blocks of meridians and particularly between the regulatory central block and the others will briefly be examined in several ways by: 1) Reassigning the five element horary point periods, 2) Reassigning the standard acupuncture meridian entry exit points to conform grossly with the celestial stem five-elements sequence of solar activation and the counteracting internal three yin three yang circulation as shown in figure four, 3) Describing the continuous three yin and three yang circulation of chi and twelve division entry exit point patterns within the rising light, declining light, and central blocks of the three yin and three yang sequence, and 4) Focusing on the open system of transmissions between three yin and three yang blocks that allow for system wide equilibrium. This allows a particular seasonal bias to speak for the whole system.

IV. The Traditional Horary Periods
Activated by the Sun and Mid-Heaven

Table 3: According to Celestial Stems and five-elements

Noon		
11 a.m.-1:00 p.m.	Small Intestine Hand Tai Yang	Si 5
1:00 p.m.-3 p.m.	Heart Hand Shao yin	Ht 8
3:00 p.m.-5 p.m.	Spleen Foot Tai Yin	SP 3
Sunset		
5:00 p.m.-7 p.m.	Stomach Foot Yang Ming	St 36
7:00 p.m.-9 p.m.	LI Hand Yang Ming	LI 1
9:00 p.m.-11 p.m.	Lung Hand Tai yin	Lu 8
Midnight		
11:00 p.m. 1:00 a.m.	Kidney foot shao Yin	Ki 8
1:00 a.m.-3:00 a.m.	Bladder Foot Tai yang	UB 66
3:00 a.m.-5:00 a.m.	Liver Foot Jue Yin	Li 1
Dawn		
5:00 a.m.-7:00 a.m.	Gallbladder Foot Shao Yang	GB 41
7:00 a.m.-9:00 a.m.	Three Heater Hand Shao Yang	TH 6
9:00 a.m.-11:00 a.m.	Pericardium Hand Jue Yin	PC 8

Table 4: Periods of Internal Activation of Three Yin and Three Yang Meridians in Solar Sequence.

Noon		
11 a.m.-1:00 p.m.	Small Intestine Hand Tai Yang	Si 5
1:00 p.m.-3 p.m.	Heart Hand Shao Yin	Ht 8
3:00 p.m.-5 p.m.	Spleen Foot Tai Yin	SP 3
Sunset		
5:00 p.m.-7 p.m.	Lung Hand Tai Yin	Lu 8
7:00 p.m.-9 p.m.	Large Intestine Hand Yang Ming	LI 1
9:00 p.m.-11 p.m.	Stomach Foot yangMing	St 36
Midnight		
11:00 p.m.-1 a.m.	Kidney Foot Shao Yin	Ki 8
1:00 a.m.-3:00 a.m.	Bladder Foot Tai yang	UB 66
3:00 a.m.-5:00 a.m.	Liver Foot Jue Yin	
Dawn		
5:00 a.m.-7:00 a.m.	Pericardium Hand Jue Yin	PC 8
7:00 a.m.-9:00 a.m.	Three Heater Hand Shao Yang	TH 6
9:00 a.m.-11:00 a.m.	Gallbladder Foot shao Yang	GB 41

Table 5: Periods of Internal Activation of Three Yin and Three Yang Meridians in Counter Point to the Solar Cycle.

Midnight		
11:00 p.m.-1:00 a.m.	Heart Hand shao yin	Ht 8
1:00 a.m.-3:00 a.m.	Small Intestine Hand Tai yang	Si 5
3:00 a.m.-5:00 a.m.	Spleen Foot Tai yin	SP 3
Dawn		
5:00 a.m.-7:00 a.m.	Lung Hand Tai Yin	Lu 8
7:00 a.m.-9:00 a.m.	LI Hand yangMing	LI 1
9:00 a.m.-11:00 a.m.	Stomach Foot yangMing	St 36
Noon		
11 a.m.-1:00 p.m.	Bladder Foot Tai yang	UB 66
1:00 p.m.-3 p.m.	Kidney foot shao Yin	Ki 8
3:00 p.m.-5 p.m.	Liver Foot Jue Yin	Li 1
Sunset		
5:00 p.m.-7 p.m.	Pericardium Hand Jue Yin	PC 8
7:00 p.m.-9 p.m.	Three Heater Hand Shao Yang	TH 6
9:00 p.m.-11 p.m.	Gallbladder Foot Shao Yang	GB 41

The Division of Twelve: New Horary Point Times and Uses

The following chart shows two possible valid sequences for horary points that would be activated in particular phases according to a) a functional time phase relationship and b) an anatomical space base phase relationship.

Table 6: The Horary Periods Activated by the Sun and Mid Heaven and Their Related Acupuncture Points. According to Phase Activation of Celestial Stems and five-elements

Noon		
11 a.m.-1:00 p.m.	Small Intestine Hand Tai yang	Si 5
1:00 p.m.-3 p.m.	Heart Hand shao Yin	Ht 8
3:00 p.m.-5 p.m.	Spleen Foot Tai Yin	SP 3
Sunset		
5:00 p.m.-7 p.m.	Stomach Foot Yang Ming	St 36
7:00 p.m.-9 p.m.	LI Hand yang Ming	LI 1
9:00 p.m.-11 p.m.	Lung Hand Tai Yin	Lu 8
Midnight		
11:00 p.m. 1:00 a.m.	Kidney foot Shao Yin	Ki 10
1:00 a.m.-3:00 a.m.	Bladder Foot Tai yang	UB 66
3:00 a.m.-5:00 a.m.	Liver Foot Jue Yin	Li 1
Dawn		
5:00 a.m.-7:00 a.m.	Gallbladder Foot Shao Yang	GB 41
7:00 a.m.-9:00 a.m.	Three Heater Hand Shao Yang	TH 6
9:00 a.m.-11:00 a.m.	Pericardium Hand Jue Yin	PC 8

Table 7: The Horary Periods Activated by the Sun and Midheaven According to Solar Phase Activation of the Terestrial Branches with Horary Points. (Clockwise Variable Sequence of Branches.

Noon		
11 a.m.-1:00 p.m.	Small Intestine Hand Tai yang	SI 5
1:00 p.m.-3 p.m.	Heart Hand Shao Yin	Ht 8
3:00 p.m.-5 p.m.	Spleen Foot Tai Yin	SP 3
Sunset		
5:00 p.m.-7 p.m.	Lung Hand Tai Yin	Lu 8
7:00 p.m.-9 p.m.	LI Hand Yang Ming	LI 1
9:00 p.m.-11 p.m.	Stomach Foot Yang Ming	St 36
Midnight		
11:00 p.m.-1:00 a.m.	Kidney Foot Shao Yin	SP 3
1:00 a.m.-3:00 a.m.	Bladder Foot Tai yang	UB 66
3:00 a.m.-5:00 a.m.	Liver Foot Jue Yin	Li 1
Dawn		
5:00 a.m.-7:00 a.m.	Pericardium Hand Jue Yin	PC 8
7:00 a.m.-9:00 a.m.	Three Heater Hand Shao Yang	TH 6
9:00 a.m.-11:00a.m.	Gallbladder Foot Shao Yang	GB 41

Table 8: The Oppositional Horary Periods Activated by the Internal and Countermotion of the Fixed Lunar Terrestrial Cycle With Horary Points.

Noon		
11 a.m.-1:00 p.m.	Small Intestine Hand Tai yang	SI 5
1:00 p.m.-3 p.m.	Heart Hand Shao Yin	Ht 8
3:00 p.m.-5 p.m.	Spleen Foot Tai Yin	SP 3
Sunset		
5:00 p.m.-7 p.m.	Lung Hand Tai Yin	Lu 8
7:00 p.m.-9 p.m.	LI Hand Yang Ming	LI 1
9:00 p.m.-11 p.m.	Stomach Foot Yang Ming	St 36
Midnight		
11:00 p.m. 1:00 a.m.	Kidney foot Shao Yin	Ki 10
1:00 a.m.-3:00 a.m.	Bladder Foot Tai yang	UB 66
3:00 a.m.-5:00 a.m.	Liver Foot Jue Yin	Li 1
Dawn		
5:00 a.m.-7:00 a.m.	Pericardium Hand Jue Yin	PC 8
7:00 a.m.-9:00 a.m.	Three Heater Hand Shao Yang	TH 6
9:00 a.m.-11:00 a.m.	Gallbladder Foot Shao Yang	GB 41

V. Applying Horary Points to Seasonal Treatments

The horary points can be applied to seasonal treatments. Element points used in their diurnal phase when combined with their seasonal phase as well become more congruent and effective. Season vary progressively in location north and south of the equater. The following table has been adapted for the mid-atlantic states of the United states, but dates need to be adjusted for other area.

Table nine describes hypothetical dates of an example sequence of: a) seasonal months and seasonal periods using modified-length months starting a little after the Chinese new year and lasting less than four months for the rising cycle, more than two months for the summer solstice period, less than four months for the declining cycle and more than two months for the winter solstice period. These points can be used in a complementary way with the diurnal open points, and reflect the adapted cycle for the northern hemisphere in the United States at the level of Maryland.

Table 9: Hypothetical Dates for Seasonal Applications including Horary, Element, Source and Luo Points for local area in Western Maryland.

Month Day	Meridian	Zodiac	Horary Pt.
February 5	Jue Yin Liver	Aquarius/Pisces	Li 1
March 15	Shao Yang GB	Pisces/Aries	GB 41
April 20	Shao yangTH	Taurus	TH 6
May 15	Jue Yin Pc Taurus	Pc 8	
June 12	Tai Yang Si	Gemini	SI 1
June 21	Shao Yin Ht	Cancer/Leo	Ht 8
August 15	Tai yin SP	Leo/Virgo	Sp 3
Sept. 30	Yang Ming St	Virgo	St 36
October 20	Yang Ming LI	Libra	LI 1
November 21	tai yin lu	Scorpio	Lu 8
December 1	Shao Yin Ki	Kidney	Ki 10
December 21	Tai yang	Bladder	UB 66

In general analysis shows that the seasonal cycle divisions of six and twelve describe the same general cycle and they rely theoretically on the same progression of the first six branches. The increase and decrease of light is for all practical purposes identical from year to year, and thus a certain sameness or theoretical predictability is to be expected in any depiction. Moreover, over time patterns of action and response in the large system of many years would be predictable based on the laws that govern the more perceptible behaviors within a single cycle, that is a damp year would be likely to be followed by a dry year, a cold (water) year by excessive growth (wood) in the subsequent year, which could justify to some extent the supposed usefulness of the 60 year cycle of stems and branches. However, every year what does change significantly in heaven is the positions of the planets and the unique start point of the first lunation in January or February which puts the growth cycle earlier or later in the unfolding solar cycle and terrestrial conditions vary based on these variable celestial causes and on the consequences of previous cycles on different local conditions i.e. mountainous or low lying terrain, sea side or inland. The Chinese clock offers the possibility of a fully articulated expression of the diurnal and annual cycles. All the gross and subtle relations between meridians derived from the primary eightfold division of the body, and the upper and lower pairs of the six terrestrial configurations become possible in it. The model however is a living dynamic standard not a fixed sequence. It is more effective to understand the dynamic of the system than to accumulate handy point sequences or phases in tables. Their use should be within the context of understanding a person's on-going developmental struggles in their relation to the unfolding cosmic cycles. The reorganized clock provides a consistent and flexile basis for understanding the circulation of energy, annually or daily. I have merely sketched an outline of what could be a very sophisticated method of treatment.

VI. The Division of Twelve Implications for Meridian Entry and Exit Points

The standard entry-exit point sequence in acupuncture is shown in table 1. In general practice these can be used effectively to open the flow of chi to meridians who are abnormally unresponsive using normal five-elements command points or TCM syndrome points. Alternatively, any meridian with a relative excess of chi or qualities to its next in sequence partner as discerned on the pulses and confirmed by symptoms can be drained or equilibrated by using the exit point of the meridian and the entry point of the next. These entry-exit points commonly can in some cases such as combinations liver 14; lung 1 or spleen 21; heart 1 be used to remedy splits from below to above by reconnect yin foot meridians with hand meridians or the reverse in connections above to below in combination such as small Intestine 19 to urinary bladder 1 or large Intestine 21 to stomach 1. The eyes are the source of all the foot yang meridians, so these entry exit points can be instrumental in reorganizing peoples perceptions and helping them to align thoughts and action. Headache problems can sometimes be resolved by opening three heater 22; gallbladder 1. Above all the standard points reflect the conceived pattern of traditional terrestrial branch circulation. Opening these points theoretically will preserve the flow of chi if a patient is symptomatic in a certain phase or function.

Table10: The Traditional Meridian Sequence of Entry-Exit points and Phase Times.

Meridian	Entry Points	Exit Points
Shao Yin Heart	1 Utmost Source	9 Little Rushing In
Tai Yang Small Intestine	1 Little Marsh	19 Listening Palace
Tai Yang Urinary Bladder	1 Eyes Bright	67 Extremity of Yin
Shao Yin Kidney	1 Bubbling Spring	22 Walking on the Veranda
Jue Yin Pericardium	1 Heavenly Pond	8 Palace of Weariness
Shao Yang Three Heater	1 Rushing the Frontier Gate	22 Harmony Bone
Shao Yang Gallbladder	1 Orbit Bone	41 Foot Above Tears
Jue Yin Liver	1 Great Esteem	14 Gate of Hope
Tai Yin Lung	1 Middle Palace	7 Narrow Defile
Yang Ming Large Intestine	4 Joining of the Valleys	21 Welcome Fragrance
Yang Ming Stomach	1 Receive Tears	42 Rushing Yang
Tai Yin Spleen	1 Retired White	21 Great Enveloping
Ren	1 Meeting of Yin	24 Receiving Fluid
Du	1 Long Strength	28 Mouth Crossing

Although all entry-exit points theoretically are important in context, some have wide effects and are more commonly used in the Five Element system:

VII. Heart Kidney Axis Entry—Exit Points

The first two entry exit pairs have wide significance. They are instrumental for restoring balance in any kidney heart imbalance, any fundamental reversal of yin and yang, and any long-term depletion evident in the heart or kidney pulses. The core balance between shaoyin and tai yang is the balance between midday and midnight or between heaven and earth, so this gives these points priority for core balance issues. By insuring the free flow of internal chi in the center, treatment gives leverage and reservoirs of strength to manage day-to-day stress.

In the next section about three yin and three yang patterns of entry exit points, it will be shown that in phase all the heart/kidney axis entry points receive chi from all yang exit points of the foot and yin exit points of the hand, while the exit points deliver chi to the entry points of all yin meridians of the foot and yang meridians of the hand. This wide influence in different phases of the day and year insures that through these entry exit points the equilibrium of the whole body can be supported.

Heart 9 to Si 1: This combination insures that essential purpose is translated into action. si 1 is the first point of tai yang noon, at the highest point of intentionality. If si 1 is not open to internal equilibrium then shen cannot operate throughout the whole body. Blockage here is the first step towards systematic depletion of the heart and systematic despair or departure from purpose. This EE connection can function as or more effectively than the luo points, to link inner to outer. This connection links contemplative wisdom to a path of implementation, and is good for overcoming neurosis and other confused thinking patterns that prevent a productive life.

Bl 67 to Ki 1 This vital link is the key point combination to restore basic disruptions between yang and yin in the whole body. It brings excesses above down, and excesses below up and restores the primordial relationship between heaven above and earth below as expressed in the three yin three yang meridians; tai Yang/shao yin. This balance in the center at the point of contact with the earth the transition at midnight between decline and ascent has tremendous leverage on everything else. This combination establishes a "real" congruent context for other kinds of growth. The equilibrium in the center and below in the unconscious self can have repercussions throughout the rest of the cycle. This imbalance is the imbalance of urban people with not connection to heaven or earth, only the center of activity and social life. Human contact has to be kept in balance with primordial urges and needs (sleep, sex, rest, defecation) on the one hand and quiet contemplation, dream, vision, and spiritual wholeness on the other. This point keeps people from burning out from hyper activity, and restores natural simplicity.

2) *Connections Between the Central Core and Rising and Declining Blocks.*

The next few entry and exit points have to do with links that can be blocked between the core (shao yin tai yang) and the rising cycle (Shao yang, jue yin) and declining cycle (yang ming, tai yin). Using the three yin and three yang structures gives a way to bring balance from outside the phase relations of the five-elements.

Spleen 21 to heart 1: This treatment uses the structural relations of meridians to work around the phase relations of the five-elements (mother son, and control) to accomplish balance. It is very common for heart to be depleted in American patients. This indirectly can lead to neediness a psychological malnourishment and discontent even if people eat good food. The nutrients just aren't distributed, or people get caught in habits having lost touch with their purpose. One way to nourish the heart other than tonifying and drawing support from the mother (wood) which often is under stress in early stages expressed as stagnant liver chi) is to balance relative excess in spleen to the relative deficiency of heart. This connects mother to child, create equilibrium, restore the ascendancy of original nature, breaks destructive habit patterns dependent on a mode of inevitability and decline, and creates a link between postnatal chi and the neutral center.

Ki 22 to Pc1

This vital combination links the core energy to the rising heat and light and self esteem of the rising cycle. This supports the yin of the pericardium creating balance and control of emotional excesses. Indirectly, this is a way of using the five element system to balance disequilibrium between kidney yin and kidney yang. Feeding solid kidney yin energy to Pericardium indirectly strengthens the kidney yang in its functions of maintaining a baseline of percipitative heat in the three jiao. Pericardium and Three Heater meridians are intimately involved in this symbolic relationship as it plays out on the three levels of the torso. An important caveat on this point combination is the recognition that Ki 22 is the traditional exit point precisely because in-phase (the waxing phase generally, and in particular the transition from midnight to early morning time periods) it is close to Pc 1 and maintains a connection between the central core and the imperatives of the waxing light phases. For waning light phases Ki 27 is closer to Lu 1, and the central phase is closer to Ht 1.

3) *This entry exit relation affects the Rising Cycle equilibrium:*

Three Heater 22/23 and GB 1 and or UB 1: Again this corrects an imbalance between the child and its mother by using the structures of the body rather than the phase relations of the five-elements. This is a very essential connection that takes excess rising heat or fire that is causing shao yang headaches and dumping it back down into the foot shao yang partner. This imbalance in the rising cycle arises when people are overwhelmed by external demands and don't know what to do next or are afraid to act because of the multiple or unknown consequences. Reconnecting to gallbladder 1 helps take a problem out of the conceptual level into integration. It is only through the gallbladder with its run down the torso that the "three heaters" can be activated to regulate heat in the torso. The headaches above are a symptom of imbalance in the three burners below. Connecting to UB 1 can reconnect the rising cycle to the neutral center.

Si 19 and UB 1: This vital link in the central block of meridians, is a point where all the heat, light, and consciousness implied by mid day are translated into equilibrium throughout the Sheng cycle. All the Shu points on the back mediate the "will of heaven" but this connection through "Listening Palace" can be broken if people refuse to listen and adapt to the views and competing needs of others. An unusually depleted UB/Kidney pulse will show a blockage here in an otherwise vital portrait. People are expending energy in authoritarian control of outcomes that will never reach fruition without help from outside, either heaven or other people.

4) *This entry exit point highlights an important declining cycle relationship.*

Lung 7 to Li 4: These very important points express the transition between the point of maximum extension of yin into the daytime sky tai yin lung, where it grabs tai yang itself and requires it to yield to the declining half of the light cycle. This power then gets translated into kinetic energy as yang Ming actively governs the trajectory of declining light, increasing cold, the transitional culmination to sunset glory, and night. This is a relationship where internal spiritual virtue must translate into worthy actions. This is a key combination for treatment of upper body and head conditions having to do with blockages in receptivity or the mental communication of these aspirations as idealistic expression. People with post nasal drainage, muzzy headedness, and other head and upper body cloudiness often are those with fixed ideas of how things can be based on self centered shao yang and tai yang cultural rules or ideas of success that are not permeable or adaptive; i.e. capable of letting go, of new inspiration, of allowing for the little or other guy in their emphasis.

5) *Special Rising and Declining Block Connection:*

A unique five-element entry/exit relationship is between liver 14 and lung1 which links the upper entry point to the hand yin meridian for the declining cycle with the upper most point on a yin foot meridian for the rising cycle. This is a strong direct bridge between opposites. Lung and liver is opposites the four cardinal point model where the opposition between metal and wood symbolized the balance between moisture and dryness. They are opposites structurally above to below. They are Ko cycle competitors. liver 14 to lu 1: Apart from the meridian sequence this link bridges below the diaphragm to above the diaphragm and is symbolic of the relationship between air and wind, between the clear air that we breathe and the perception of air moving in the branches and leaves. The liver and gallbladder seem anatomically to be in the shape of a tree standing upright reaching for heaven, where the lungs have the shape of a tree extending downward from above. The two principles interact at the diaphragm. Thus combined with the diaphragm shu and the lung/liver shu on the back this entry exit point combination can be vital for dealing with diaphragm problems, with emotional suppression, with grief weighing down or anger suppressed. It is an important point for American women of the last generation who were inhibited in their emotional expressions.

VIII. The Division of Twelve: The Proposed Entry-Exit Sequence

The proposed entry exit sequence works for the twelve-division sequence of meridians either from figure eight for the structural sequence and nine for the temporal phases and it strikes a balance between five-elements and three yin and three yang principles (Table 2). The sequence is elegant since it also works to explain transfers between meridians on the surface of the wheel and this system works that way. However, the model also assumes that there is an external and internal pattern of circulation and that the internal circulation in each block is continuous, beginning with the yin foot meridian rising to the yin hand partner transferring to the yang hand partner and descending to the foot yang partner, which reconnects with the foot yin partner. This allows for the whole block to operate as a circuit enabling transfers from more than one meridian on the block sending chi to more than one meridian in the block that receives chi. The sequence of phase activation becomes relevant only as a means of discriminating diagnostically between disruptions in flow patterns.

Table 11: Proposed Entry Exit Sequence. [Brackets] Show Implicit or Hidden Transfers that Make the Sequence in the Circle Work. A)Condensed showing Phase Sequence in Figure Nine. B) Combined Sequence For Phase and Structure Showing Internal Connections Between Block Partner Meridians and External Links to Other Blocks of Meridians.

A.

Meridian	Entry Points	Exit Points
Shao Yin Heart	1 Utmost Source	9 Little Rushing In
Tai Yang Small Intestine	1 Little Marsh	19 Listening Palace
[Tai Yang Urinary Bladder	1 Eyes Bright	67 Extremity of Yin] to
[Shao Yin Kidney	1 Bubbling Spring	22 Walking on the Veranda]
And/Or: [Jue Yin Liver	1 Great Esteem	14 Gate of Hope]
Jue Yin Pericardium	1 Heavenly Pond	8 Palace of Weariness
Shao Yang Three Heater	1 Rushing the Frontier Gate	22 Harmony Bone
Shao Yang Gallbladder	1 Orbit Bone	41 Foot Above Tears
Jue Yin Liver	1 Great Esteem	14 Gate of Hope
[Shao Yin Heart	1 Utmost Source	9 Little Rushing In
[Tai Yang Small Intestine	1 Little Marsh	19 Listening Palace]
[Jue Yin Pericardium	1 Heavenly Pond	8 Palace of Weariness]
[Shao Yang Three Heater	1 Rushing the Frontier Gate	22 Harmony Bone]
Tai Yang Urinary Bladder	1 Eyes Bright	67 Extremity of Yin
Shao Yin Kidney	1 Bubbling Spring	27 Storehouse
Tai Yin Lung	1 Middle Palace	7 Narrow Defile
Yang Ming Large Intestine	4 Joining of the Valleys	21 Welcome Fragrance
Yang Ming Stomach	1 Receive Tears	42 Rushing Yang]
Tai Yin Spleen	1 Retired White	20 Encircling Glory
Return to Shao Yin Heart		

B)

Block	Additional Internal Circuit Connections	s e q u e n c e	Meridian	Entry/ Exit Primary Sequence	Point	Additional External Circuit Connections
Central Block						
	From: UB 67,	1	Shao Yin Kidney	Entry	1 Yong Quan Bubbling Spring	From Rising Block: GB 41, From Declining Block: St 45
			Shao Yin Kidney	Exit	22 Bu lang Walking on the Veranda	To Rising Block: Pc 1
		2	Shao Yin Kidney	Exit	27 Shu Fu Storehouse	To Declining Block: Lu 1; secondary [UB 11, GB 1, St 11]
		3	Shao Yin Heart	Entry	1 Ji Quan Utmost Source	From Rising Block: Li 14, From Declining Block Sp 20
		4	Shao Yin Heart	Exit	9 Shao chong	
		5	Tai Yang Small Intestine	Entry	1 Shaoze Lesser Marsh	
		6	Tai Yang Small Intestine	Exit	19 Ting Gong Listening Palace	To: GB 1, St 1
		7	Tai Yang Urinary Bladder	Entry	1 Jing Men Bright Eyes	From Declining Block: LI 21, From Rising Block: TH 23
	To: Ki 1	8	Tai Yang Urinary Bladder	Exit	67 Zhi Yin Utmost Yin, ReachingYin	To Rising Block: Li 1, To Declining Block: Sp 1
Rising Block						
	From: GB 41	1	Jue Yin Liver	Entry	1	From St 42, UB 67
		2	Jue Yin Liver	Exit	14	To: Ht 1, Lu 1,
		3	Jue Yin Pericardium	Entry	1 Tian Chi Heavenly Pond	From: Sp 18, Ki 22
		4	Jue Yin Pericardium	Exit	8 Lao Gong Palace of Wearines or Toil	
		5	Shao Yang Three Heater	Entry	1 GuanChongRushing the Frontier Gate	
		6	Shao Yang Three Heater	Exit	22 Erheliao Harmony Bone	
			Shao Yang Three Heater	Exit	23 Sizhukong Silken Bamboo Hollow	To: UB 1

		7	Shao Yang Gallbladder	Entry	1 Tongziliao Orbit Bone, Pupil Crevice	From LI 21,
	From Li 14		Shao Yang Gallbladder	Entry	22 YuanYe Armpit Abyss	From Ki 27, Sp 20
	To Li 1	8	Shao Yang Gallbladder	Exit	41 Zu lin Qi Rise Above Tears	To Ki 1, Sp 1
Declining Block						
	From: St 42	1	Tai Yin Spleen	Entry	1 Yin Bai Hidden White	From GB 41, UB 67
			Tai Yin Spleen	Exit	18 Tian Xi Heavenly Stream	To: PC 1
	To: Lu 1	2	Tai Yin Spleen	Exit	20ZhouRong Encircling Glory	To: Ht 1.
			Tai Yin Spleen	Exit	21	To: Prime Vertical
		3	Tai Yin Lung	Entry	1 Zhong fu Middle Palace	From: Sp 18, Ki 22
		4	Tai Yin Lung	Exit	7 LieQue Broken Sequence	To:
		5	Yang Ming Large Intestine	Entry	4 He Gu Joining Valley	
		6	Yang Ming Large Intestine	Exit	21 Ying Xiang Welcome Fragrance	To: UB 1, GB 1
		7	Yang Ming Stomach	Entry	1 Chengqi Receive Tears	
	From: Sp 20		Yang Ming Stomach	Entry	12 QuePen Empty Basin	From: Li 14, Ki27
	To: Sp 1		Yang Ming Stomach	Exit	42 Chongyang Rushing Yang	To: Ki 1, Li 1

In figure four the external five-elements solar activation pattern in the rising and declining blocks follows this motion clockwise exactly in reverse, still starting with the foot yin meridian moving to its yang counterpart to the upper body yang counterpart and back to the upper body yin counterpart, which then passes the baton to the next in sequence. This sequence could just as easily have been an alternating sequence: foot yin, foot yang, hand yin and hand yang in as in figure one but the choice was made to emphasize the opposition between he clockwise and counterclockwise motions. Symbolically, all depictions show clear yin and yang balance within each five-element phase or orb. The model breaks down in showing the entry exit relations in the consistent reverse counterclockwise direction of the whole wheel at the critical x times of eleven and three. In the central block heart exiting to Si works as does U B exiting to Kidney or Kidney exiting to UB in sequence. Kidney was placed in the midnight position as the ultimate expression of shao yin that contrasts with Tai Yang SI fire, and because it continues the pattern of placing yin foot meridians at the root of the partner cycle, and other reasons including trajectories, the nature of water, the center, and the infinite loop configuration of chi at the core, but internally the pattern works and it works in the pattern of transfers in the clock sequence.

What works marginally in transfers to or from adjacent "blocks" are the forward transfers in five element phase from pericardium 8 to Si 1, from heart 9 to spleen 1, lung 7 to kidney or in reverse transfers such as: Small Intestine 19 to Pericardium 1, Liver 14 to UB 1. Extensive study of different options all lead to one conclusion; namely, the continuous circulation of chi by block units, which allow for adaptive access.

This means in the above instances that both the forward and reverse motions and sequences can be met. Moving forward in the five-elements solar sequence pericardium eight can exit to small intestine 1, heart 9 can exit to large intestine 1 of the declining cycle directly and to small intestine 1 then descending through UB 67 exit to Spleen 1, or from UB 67 to Ki 1 and from there to exit through kidney 27 to lung 1. The selection of entry exit points for treatment between the central and declining block can be determined by phase or symptoms. That is, if 3 p.m. earth and spleen do UB 67; Sp 1, if 3 p.m. metal and lung symptoms do Ki 27; Lu 1. The next solar sequence example was exit point lung 7 which needs to transfer chi to kidney 1. This can only happen if Lu 7 exits to Si 1 which exits through Si 19 to UB 1, which exit through UB 67 to Ki 1, or, alternatively, Lu 7 exits internally in continuous cycle to Li 1, which exits through Li 21 to UB 1 and thence to UB 67; Ki 1). These access points and conceptual transfers are electrical and therefore virtually simultaneous when phase activation occurs.

The importance of this understanding of the phase activation process is that the five-elements phases are honored, and the entry exit patterns confirm that the shao yin fire heart meridian, generally, and Utmost Source, Heart 1 in particular is the single modulating point through which all shen related, upper body related, and five element related evolutionary functions must pass. Utmost Source organizes internally and small intestine 1 Little Marsh organizes externally the monarch fire and tai yang shao yin equilibrium between qi and fluids at midday and mid-summer. The reference of little marsh and utmost source becomes clearer when it is understood that these points when in phase represent the whole balance of the central block including the water element. Similarly, when the reverse terrestrial cycle is considered it is kidney 1 through which all jing related, structural, and instinctive lunar adaptations must and do pass. From the five element perspective, when the image is superimposed on the cycle, which causes the water element to be removed from the centerline where fire and heaven is dominant, other direct connections between rising and declining blocks of meridians become important. Liver 14 in the standard sequence exits directly to Lu 1. This creates an important equilibrium between above and below the diaphragm, while it connects the rising cycle to the declining cycle. Spleen 18 Heavenly Stream can connect with adjacent Pc 1 Heavenly Pond to create equilibrium between moisture and dryness in the lung and between the seasonal polarities. Lung 7 Narrow Defile in phase can exit to TH 1 Rushing the Frontier Gate. The declining cycle can exit directly to the next in sequence rising cycle, treating the central block as a passive plane from St 42 Rushing Yang to Li 1 Great Esteem. The waxing cycle can surrender directly to the declining cycle treating the central block as a passive plane from Pc 8 to LI 1.

IX. Division of Twelve: Three yin Three Yang Block Transfers

What may not be immediately clear in table 2 about the open-ended way that the energy circulates in phase and between phases is how energy transfers between meridian rising, central, and declining blocks of the terrestrial cycle through entry exit points. This is clarified somewhat by illustrating how chi flows into and out of the central block of tai yang and shao yin meridians. The relation between blocks works similarly to the transfers that are instigated by the solar activated clockwise five-element sequence. For the reverse terrestrial action of the entry-exit point sequence having to do with the branches spleen 20 exits to heart 1 of the central block (Fig. 9). Small intestine exits to UB 1 through UB 67: Li 1 of the waxing block. Liver 14 exits to the Ht 1 of the central block as do its rising partners, GB 41: Ki 1, Pc8: Si 1, and TH 23: UB 1. Kidney exits to Lung 1 of the declining block directly through Ki 27, but indirectly through block partners Ht 9: LI 7, SI 19: ST 1, and UB 67: Sp 1.

All these transfer contacts are important but the critical contribution of this section is the identification from among those entry to exit transfer points that link between the whole blocks of points the most critical transfer points. Obviously the key points are located in-phases on either side of the oppositions of 3 a.m./ 3 p.m.; 11 a.m. /11 p.m. These points differentiate the central from the rising and declining blocks.

Owing to the complexity of higher whole numbers and the reversals in declining cycle models for time or space it is immediately obvious that there must be complex means to make transfers at these critical phases and indeed under environmental pressures at any other time. At the very least in the case of the declining cycle both tai yin lung and spleen must be able to exit to the central block of meridians depending on the model. Table two shows that Lu 7 can exit to SI 1, and that Sp 20 or even 21 can exit to Ht 1 to accomplish this. However, the model in table 2 shows as well that the continual circulation of chi allows St 42 to exit to the central block Ki 1, and LI 21 to exit to UB 1 simultaneously with the others.

The X relations at the transitions between whole blocks of meridians that relate to the rising, central, or declining light periods are very significant because they identify key points that control these whole blocks of functions (Fig. 3). These relations assume that exit points can exit to more than one meridian and entry points can receive from more than one meridian depending on the phase relationships. This concept will appall traditionalists, but it makes sense in a holographic model where there can be dominant patterns of organization and yet capacities for adaptation appropriate to complex human beings.

The master entry exit points for whole blocks are:

Heart 1: which receives from all three of the yin foot meridians; continuously from kidney 27 and in phase from; liver 14 and spleen 20. Given the solar celestial stem and five element perspective Heart 1 and the exit entry sequence Ht 9: Si 1 have the greatest system wide significance for creating movement and receptivity to heaven and all that implies.

Kidney 1: which receives from all of the three yang foot meridians; continuously from UB 67 and in phase from gallbladder 41 and stomach 44. Given the three yin and three yang block sequence approach to the overall diurnal and annual cycles with its emphasis on the water element, and terrestrial branches: kidney 1 and bladder 67 are the most frequently used essential combination for establishing connections with this.

Small Intestine 1: which receives from all three yin hand meridians; continuously from heat 9 and in phase from pericardium 8, and lung 7.

Urinary Bladder 1: which receives from all three yang hand meridians; continuously from small intestine 19; and in phase from three heater 23 and large intestine 20.

Special Entry Exit Relations: Supplementary to these:

Lung1: receives continuously from spleen 20 and in phase from kidney 27 and liver 14. **Pericardium 1:** receives continuously from liver 14 and in phase from kidney22, and spleen18. **Liver 1:** receives continuously from gallbladder 41 and in phase from UB 67 and stomach 42. **Spleen 1:** receives continuously from stomach44 and in phase from UB 67 and gallbladder 44

The root meanings of the three yin and three yang patterns derive from the number three: heaven, horizon earth. All the yang hand meridians end on the head near the eyes, and anatomically the arms generally are positioned over the head with heavenly chi descending to the head. When the arms are at the side however there is a way of viewing the torso as a complete orbit with a heaven as upper torso, horizon as diaphragm to navel, and earth as navel to pubic bone. In this conception there has to be a way for yang and yin meridians of the head to create entry exit closure and the body to do the same. In this conception the internal circuit for the rising cycle would be: liver 1, 14; GB 21, 41 perpetually; for the central cycle: kidney 1, 27; UB 11, 67 perpetually; and for the declining block cycle: spleen 1, 20; St 12 or 15. With this in mind, the exit point of any foot yin meridian can complete a circuit in phase with the entry point of the foot yang meridian, and the reverse, namely the exit point of any yang foot meridian can connect in phase with the entry point of any foot yin meridian. This has already been established. These are useful ways to create balance in the "earth" or "middle" and to open circuits that affect the shoulder area in particular.

X. Division of Twelve Entry Exit Priorities and Uses in Time Phases

Table twelve considers treatment methods that have the possibility of opening, clearing, and emptying entire blocks of meridians for the rising, central or declining cycles. The terms primary, secondary, and tertiary applied to points refer not to the importance of treatment but to the aim of treatment. Primary points are those that stimulate the principal and secondary entry points to a block sequence from one or more of the preceding block's exit points. Secondary refers to the principal entry and exit points of the whole block and tertiary refers to the exit point of the block in combination with the entry point of the next block. Several options are available in each category and the principle behind their use is to activate that pair of points that relate to the block phase that is being activated at the time of treatment.

Table 12: Treating the blocks.

Block	Phase	Movement	Primary	Secondary	Tertiary
Central Block	**Midnight**		Bl67;Li 1 St 42;Ki 1 Lu 7; Si 1	Ki 1;Bl 67 St 42; Ki 1 (Ki 27; Ht 1	Bl 67; Li 1 Ki 27; Pc (LI 20; Bl 67 Ki 1;Pc 1)
Central	**Mid-day**		Li 14; Ht. 1 GB 41; Ki 1 TH 23; Bl 1	Ki 1; Bl 67 (Si 1; Ht. 9) (Ht. 1; Ki 22	Bl 67; Sp 1 SI 22:Bl 1; Bl 67; Sp 1
		Rotation			Ki 1; Bl67
		Up			Ki 1; Ht 9
		Down			Si 1; Bl 67
Rising			UB67; Li 1 Ki 22; Pc 1 St: 42; Li 1- TH 23; Bl. 1 Liv. 14; lu 1	Li 1; GB 41 (Li 14; Pc 1 (Li 14; Pc 1)	GB. 41; Ki 1- GB. 41; Ki 1 Pc 8, si 1
		Rotation			Li 1; GB67
		Up			Li1; Pc8
		Down			TH 1; GB 41
Declining			Bl 67; Sp 1 KI 27; lu 1 St. 42; Li 1	Sp 1; St 4 (Lu 1; Sp 21)	Lu 7: si 1 St. 42; Ki 1
		Rotation			Sp1; St 42
		Up			Sp1; Lu7
		Down			LI 4; St 42

The above chart makes use of cosmographical three yin three yang organization for treatment purposes. The Rotation points treat the primary entry and exit points of the whole block from the beginning to the end. The rotation completes an endlessly repeating internal circuit and it makes the whole circuit available to receive and send chi to the next circuit. The importance of this treatment for gross imbalances of yin and yang cannot be overstated as this treatment should be added to the major clearing treatments in five element acupuncture that preceded causative factor treatment. If the causative factor is part of a block, this treatment makes energy available to the entire block in which it resides and frequently takes care of the mother and son at the same time. This is a highly anatomical conception of the dilemma facing the causative factor and places it in its real physiological context. The general rotation treatment is refined by treating the upward moving chi from below at its critical transition points below and above, and the same with the chi moving downward. This treatment has the possibility of augmenting orthopathic directionality and dealing with problems in extremities-shoulders and hips. The transition treatments have much to do with basic mobility and adaptability at the points where a person makes contact with the world. It explains to some extent the psychology and etiology of movement disorders and where they start.

Treatment t imes for key points in the blocks should be considered. Block transition points can be treated any time, more effectively in solar phase, effectively any time during block phase, or when the nadir or rotating phases strike, or during the opposite phase period as in the example case of jue yin liver:

Liver 1 3-5-a.m.; 3 ñ 11 a.m.; Nadir strikes 5-7 a.m.; 7-9 p.m. opposite 3-5.p.m.

Jue yin may be treated at any time during rising block in accompaniment with active phase and using the preceding exit point Bladder 67 or succeeding entry point in the central block: Ki 1 or Bl 1 or for the declining block Liv 14; lu 1.

The following table provides a list of phase sensitive times to treat the nadir or earth system directly. The opposite times also can access the terrestrial system. To tonify a phase treat the phase 90 degrees or three phases ahead in phase, to disperse treat the phase 90 degrees or three phases behind.

Table13: Key times for treating the nadir earth, and grounding the system.

Number	Meridian Phase	Sun At Phase	Sun in phase Corollary terrestrial	Phase At Midheaven	Phase At Nadir	Rotating Celestial Phase At MH	Rotating Terrestrial Phase At Nadir	Sun In Rotating Terrestrial Phase
5	Tai Yang Small Intestine	11-1pm	11-1pm 5-7pm	11-1pm	11-1am	11-1pm	11-1pm	5-7pm 11-1am
6	Shao Yin Heart	1-3pm	1-3pm	9-11am	9-11pm	9-11am	9-11am	
7	Tai yin spleen	3-5pm	3-5pm	7-9am	7-9pm	7-9am	7-9am	1-3 am 1-3 pm
8	Yang ming stomach	5-7pm	5-7pm	5-7am	5-7pm	5-7am	5-7am	
9	Yangming Large Intestine	7-9pm	7-9pm	3-5am	3-5pm	3-5am	3-5am	3-5 am 3-5 pm

10	Tai yin Lung	9-11pm	9-11pm	1-3am	1-3pm	1-3am	1-3am	
11	Shao yin kidney	11-1am	11-1am	11-1am	11-1pm	11-1am	11-1am	5-7 am. 11-1 p.m.
12	Taiyang Bladder	1-3am	1-3am	9-11pm	9-11am	9-11pm	9-11pm	
1	Jue yin liver	3-5am	3-5am	7-9pm	7-9pm	7-9pm	7-9pm	7-9pm 7-9am
2	Shao yang gallbladder	5-7am	5-7am	5-7pm	5-7am	5-7pm	5-7pm	
3	Shao yang Three heater	7-9am	7-9am	3-5pm	3-5am	3-5pm	3-5pm	9-11pm 9-11pm
4	Jue Yin Pericardium	9-11am	9-11am	1-3pm	1-3pm	1-3pm	1-3pm	

The following Table 14/33 is perhaps the most valuable and useful of all tabulations in this book, because it portrays a highly versatile and comprehensive way to treat all the plausible entry exit point relations that have been discussed above in relation to various mo dels. Each exit point from whole blocks will have an entry point in the next block in sequence and usually the central block as well. This describes a very fluid and adaptive flow of chi in the body, a flow mediated and organized by the eyes (and to a lesser extent the mouth), where all the yang meridians coalesce, the hands and feet where people make their most vivid contacts with the cosmos above and below (intelligence and action), and through the heart/kidney(shao yin and tai yang) axis which maintains the equilibrium of the whole system.

*Table 14: describes the exit points with their various entry points. Entry points to blocks are determined automatically by solar phase and block activation (Shao yang, Jue yin in morning; yang ming, tai yin in afternoon and evening; central at transitions (dawn, midday, dusk, midnight). The first in sequence of possibilities is the primary block relationship. *(Parenthesis) indicates internal to block entry exit points) [Brackets] Indicates alternative block transfers in phase.*

Phase	Yin/yang	Meridian	Point		Entry Exit
1	Jue Yin	Liver	1	Enter From	Bl. 62, St. 42, (GB. 41)
				Exit to	(Pc 1), (GB 22), Lu. 1]
2	Shao Yang	gallbladder	1	Enter From	TH 22, [LI 20],
			41	Exit To	

3	Shao Yang	Three Heater	1	Enter From	Pc 8
			22	Exit To	GB 1
			23	Exit to	Bl 1
			St. 1		
4	Jue yin	Pericardium	1	Enter From	Ki 22, Sp 21, (Li 14)
			8	Exit To	TH 1, si 1, LI 1
5	Tai Yang	Small Intestine	1	Enter From	Ht 9, lu 7, Pc 8
			19	Exit To	Bl 1; Bl 67; Sp 1
					(Bl 1;Bl 67; Ki 1; Ki22; Ht 1)
6	Shao Yin	Heart	1	Enter From	GB41; Ki 1; Ki22; Ht 1
					[Pc 9; si 1-Ki 22]
					(Ki 22)
					[Sp 21]
					[St 42; Ki 1; Ki 22
			9	Exit To	SI 1
7	tai yin	spleen	1	Enter From	Bl 67, St 42, [GB 41]
			21	Exit To	(Lu 1), [Ht. 1] (Pc 1)
8	Yang Ming	Stomach	1	Enter From	LI 20
			42	Exit To	Sp 1 [Ki 1], Li 1
9	Yang Ming	Colon	4	Enter From	Lu 7,
			20	Exit To	St. 1, Bl. 1, [GB 1]
10	tai yin	Lung	1	Enter From	Ki 27, Li 14, (sp 21)
			7	Exit To	LI 4
11	Shao Yin	Kidney	1	Enter From	St 42, (Bl 67), [GB 41]
				Exit To	Pc 1, (Ht 1)
			27	Exit To	Lu 1, GB 22
12	Tai Yang	Bladder	1	Enter From	SI 19, [TH 23], [LI 20]
			67	Exit To	Ki 1, [Li 1], [Sp 1]

Thus, the clock gives a very timely and discriminative way to treat entire systems, and to link functions that already are hardwired into nature and the body. To unite yin leg functions in the body with the remaining terrestrial lower half functions, an alternative to the three yin junction point on the leg (spleen 6) could be the entry and exit points to the circuit of water meridians such UB 1 or Ki 1 (arrival point for all yang leg meridians) and the central block or from the central block as a whole to the declining block treating UB 67; sp 1 in phase or during the afternoon; either of which treatments would emphasize the directionality towards earth and the legs generally. Alternatively the cyclic versus anatomical exit point of the declining block to the entry point of the central block lung 7; small intestine 1 in the circumstance of a relatively full lung pulse on the right wrist distal position and relatively weak left heart pulse on the left distal position could release an upper body or mental blockage preventing energy from descending to earth. The use of the cycles represents a way of thinking about conditions, not a mechanical fix uterine problems or disharmony among the three yin leg meridians.

The recovered system of twelve gives a comprehensive and detailed explanation of how oppositions visible in four, six and eight divisions of the cycle generate motion and the system probably can be applied to the year as well as the day. The consequence would be a capacity for profoundly detailed treatment plans for the more general diagnosis based on lesser systems using five, six or eight as their basis. By applying the model to identify the active and counteractive forces of the month, testing them against general meteorological conditions, then using the cycle for the day, testing the model against the day's meteorological conditions and real bodily balances from pulses and palpation, an exact, cosmographically appropriate selection of point interventions for a given period of day becomes measurably more possible than through the cruder less discriminative models that represent longer duration phases.

These proposed relations offer great insights and possibilities. Most important they meet certain necessary criterion such as theoretical coherence, proximity, connection through known pathways, and functional compatibility. Of great importance has been the recognition of why in the cosmographical model heart1, Utmost Source deserves the name and achieves the distinction of acting as a gatekeeper for the circulation of energy in all the upper body meridians and regulating tidal exchanges of energy between the three blocks of functions. Similarly, bladder 67 and Kidney 1, the extremity of yang has a similarly wide ranging role in regulating yin and yang at the root of the entire system, assuring that yang connects with the yin in all three blocks, and insisting on cohesive functioning of all the leg meridians pairs. These are tremendous insights about just into two points, but insights that are justified by the basic organization of cosmos and body insofar as they dovetail in their responses to the cycles of light.

Certain special meridian anomalies like the role of the upper body kidney points and the strange extensions of meridians topographically past their exit points as in the case of the stomach or three heater meridians, strongly point to the meridian systems as expressions or extensions of great cosmographical circles that locate and describe the movement of chi on the body. Analysis suggests that each arm leg partner with its yin yang counterpart create a complete self-enclosed system whose entry to exit internal circulation differs from its entry to exit circulation to external blocks. Some meridian anomalies get explained as an aside to the examination of this distinction. Kidney 27 can be considered an exit point to heart 1 on an internal cycle or externally to lung 1 of the waning cycle while kidney 22 can be viewed as an exit to lung 1 or spleen 21 of the waning external cycle. Similarly, three heater 23 can be viewed as expressive of a continuous internal circulation to gallbladder 1, which is its shao yang partner, while three heater 21 can be activated during the period of block or phase activation registered by the clock. The widely used conventional transfer in "Five Element Acupuncture" between liver 14 and lung1 is revisited as an important direct link between the waxing block and the waning block of meridians which regulates upward and downward flows of bodily and cosmologic chi in a profound way without affecting the heart/kidney axis. However, this relationship is rejected as part of the standard entry exit sequence because it does not fit the cosmographical sequence and betrays the basic block organization. The transfer of chi from the central block through kidney 22 or 27 to lung1, offers a much more local transfer that cosmographically fits the entry exit relationships between the central block and the descending blockThe proposed revitalized clock sequence harmonizes the five element yin yang sequence with the three yin and three yang sequence of configurations, and by exploring both clockwise and counterclockwise rotations it provides a mechanical explanation for most of the characteristics exhibited by the functional orbs of Chinese medicine. The principles of division by four, five, and six apply directly to the comparable phases of the division by twelve. By emphasizing the overmastering significance of the activating outcome of the morning/ springtime block sequence (jue yin, shao yang), the mediating effect of the middle block (shao yin, tai yang), and the declining effect of the evening block (tai yin, yang ming) of time dependent functions gave an important road map for identifying all important gross tendencies of any time period and therefore of any cosmographical map for an individual. A division of twelve even more than a division of three breaks these gross tendencies into a refined system of relations. An important contribution of this version of the clock is that the specific relations in the division of twelve conform to the gross tendencies evident in cycles with fewer divisions at the same time that the refinements are applied.

The clarification of four levels of action in any cycle allows for an entirely articulate way of explaining relationships, that otherwise seem too complicated or remote. The interaction of the levels gives a mechanical explanation for things like the directional emphasis of yang at dawn and yin in the evening. The discrimination of rotating cycles from template cycles, offers the possibility not only of treating the solar wave through horary points during solar periods, but also by treating points activated by the phase occupying the mid-heaven or nadir when the solar wave is in a position other than noon or midnight. These points will have significant regulatory power on the whole body at those times, and if the meridian associated with the phase has a relative excess or deficiency of chi, such a treatment could be widely beneficial with the right selection of points on that meridian. Indeed, some present treatment benefits may be occurring when practitioners inadvertently treat a point on a meridian that is at that moment fulfilling the mandate of heaven. In five element terms the period when the causative factor occupies the mid-heaven or nadir would be as effective a time to treat the mind/heart of that function in the former and the physical body in the latter as would be time when the horary point is activated by the solar cycle in that function. The four levels give room for exploration for the first time of using the terrestrial cycle as a discriminative basis for medical intervention. Such treatments may be systematically more effective in treating blood and fluid level disorders than points related to the sheng cycle. Nadir activated or terrestrial phases activated by the solar wave may be uniquely suited for treating blood or fluid level disorders as opposed to the more meridian and chi type of disorders. Moreover, clarification of the terrestrial counterclockwise motion gives a theoretical basis for understanding Ko and counteracting energies of the five-phase model, while at the same time offering new points and treatment options for correcting those imbalances when they are identified through pulses and meridian analysis.

Looking ahead a last collection of points to consider in relation to a system that relates profoundly to terrestrial branches is "Ghost points." The thirteen Ghost points of sun si miao are:

Gui gongh, Gizhen, Guitang, Guishi, Guixin, Guitui, Guichuang,Guilei Guixin, Guicu, Guilu, Guifeng.

Thirteen is the next step beyond twelve into the unknown mystery of the outer world with its infinite forms and ultimate death.

Ghost Points: These were points selected to identify on the human body those point of connection with the terrestrial po spirit identity. These points offer a bare outline of material existence reduced of all form except that of the skull which outlines the faint consciousness of identity that remains in the sleep of death. These points tell a story: A) DU 26, Guigong, Ghost Palace is the residual location the now empty seat of consciousness. The ghost has no substance, but all awareness resides in the emptiness. B) DU 16, Guizhen Ghost, Pillow identifies the skull as the repository for spirit or heaven distinct from all below, C) St 6, Guichuang Ghost's bed reinforces both the skull as the outline of identity left in a ghost and refers by position and name to the long sleep of death. These points indicate that the root of life and desire is in the skull, in death, and emphasizes that those desires that drive people though unlimited in dimension are insubstantial and defined only by the structural limits of the skull, which is the only basis of identity. D) DU 23, Guitant, Ghost Hall and ren 24 Guishhi Ghost Market describe the vacant mouth as the unfulfilled desires that keep the ghost attached to the earth. lung1 1 and spleen 1 both are wood points on the same tai yin trajectory and are located at the outer tips of the foot and hand, which are the very limits of the trajectory of yin as the purveyor of the human form. It is tai yin that is responsible for form and structure in living creatures because it is tai yin that is strong enough to bring the invisible and unknowable and formlessness of yin into form, which is itself relatively yin to the formless yang and chi of heaven. E) Lung11, Ghost faith reflects the shadow of the wood hun spirit that remains with a ghost, the seed of possibility of incarnation, but here, in a hand meridian, referring to the hope for consciousness and identity in contrast to spleen 1 or Ghost Fortress which is deep in the earth, in the darkness, below the waist, perfectly symbolizing the hope for form itself and indeed the possibility of form itself, but a possibility rooted in tai yin whose complete cycle with yang ming is all about decline of light and eventual death.

Large Intestine 11, Guitui, Ghost Leg shows by its name the understanding that yang Ming Large Intestine faces down and leads the downward cycle in time even though it is above on the arm structurally and spatially. The earth point refers here in the arm counterpart to spleen 1 the prevalence of the same principle of earth bound form. HP 7 Guixin Ghost Heart, and HP 8 guicu Ghost Cave play on the hollowness of the HP, which mimics the substance, and shen of the heart itself, suggesting that the hand is an empty cave like the pericardium that can never grasp form however much it tries. The hand then is the agent for achieving desires of the heart and the PC governs desire, power, and achievement, all of which end in emptiness.

Urinary Bladder 62: Gui lu, Ghost Path is far down the leg below the waist on the yang and lateral aspect of the leg, which refers to the prime vertical below the horizon. Thus this is a symbol for wandering in darkness, ironically with no counteracting yin substance to give direction. Yin at this level is formless it only has the desire for form, identity, meaning, but it nevertheless wanders in all directions on a tangent with no center. Guifeng Ghost Seal and Guicang Ghost store refer ironically to the memory of authority and reproductive power that the ghost clings to.

These points are useful for treating living ghosts, people who are impoverished financially, in their outlook, or in their needs. People with desires who do not act or who do not follow the path of Hun and Shen, which leads to heaven and consciousness. Ghosts are merely the residual unmet desires and needs and the memory of capacities to achieve them, that are left over after the body goes. The critical thing is to achieve what you legitimately can achieve now according to the will of heaven. These points help to awaken the capacity to connect with our real not our artificial needs, with our humanity, and to learn slowly to embrace true emptiness rather than outer forms with their inevitable futility and desire. Obviously, these points have philosophical implications more than clinical implication, but clearly we get a better understanding of the key locations where the seed of desires and unfulfilled desires reside. Essentially these points treat low self worth or unpursued desires, and help people give a material shape to their lives, to more fully occupy with awareness the empty space occupied by their body. The moral restraint suggested by the sages is necessary to balance and inner development and certainly for psychological health.

The fallibilities and mystery that lie at the root of our fascination with the earth are hinted at by the ghost points, but the cosmographic numbers seven, eight, and nine together show how human structures sustain ever more complex sets of associations resulting from the impulse to incarnateConclusion

Conclusion

The great order and symmetry in nature and life is a structure imposed by divine will upon the entirely irrational void of non existence. Baba Ji Gurinder Singh maintains that we humans are spiritual beings going through a physical experience. While living in the physical labyrinth of life with all its forms, passions, wars, and sorrows, and residing in a fragile body our lives are happier if we understand and respect the symbolic order around us. The basic patterns of life derive from the coherence of the Word and the rhythms of our own spiritual necessities, not from these symbols. Yet, when compared even to music, books, and art, the sounds of nature and the light cycles with their mathematical relations offer a substantial foundation for direct experience, inspiration and even revelation. Geometry is something that anyone can understand and it is profoundly experiential. It is not the whole story, but it provides a basic insight into the ultimate rationality behind the dream like mystery that surrounds us.

By showing how traditional thinkers used whole numbers to understand the body, invent cosmological theory, and organize acupuncture into a system I've established a few key points. 1) Acupuncture is profoundly cosmological and rational, not merely empirical. 2) Acupuncture can be effective because of its geometrical conception of the body not in spite of it. 3) The art of acupuncture is like classical music, it depends on the same proportions and symmetries to be effective. Any meat cutter can throw in huge needles and stimulate nerves but living humans respond well to elegance and balance, to light, to planets, to moods and the moon.

Each number offers a unique vantage point that is complete and comprehensive because the whole world can be seen from it. This means that there really are twelve or more comprehensive systems that individuals can invest their professional acupuncture professions in. A practitioner could focus exclusively on say three or seven and develop to a level of mastery. If someone can be so extreme as to treat everything through the ear or the hand, it is completely conceivable that using the divisions of three with the same focus would yield effective treatments. However, as a generalist, American, I prefer applying the right tool for the particular job. The patients condition and symptoms automatically telegraph which lens to apply and which treatment pattern will work best. However, without understanding something about the meaning of numbers you cannot so effectively switch lenses or apply the right tool to the job. Without numbers you will have to see everything through five elements, through abdominal diagnosis, through pattern diagnosis, or through some other particular system.

Finally, I hope it is clear that acupuncture is a living system with a living connection to the ebb and flow of light in the cosmos. Treatments must use these cosmic cycles to advantage and practitioners need an awareness of light cycles to property advise patients about how to live wisely. That there is a one to one relationship between the body and the world is a revelation in a world of virtual living. That we belong here, that the creation was made and that we are good, the creator did not make a mistake, we are not born in sin, that we have meaningful instincts, conscience, and purpose, that we epitomize something grand and mysterious, so much is at stake in this conception of the human being and in his or her care.

Appendix 1

Table 1. Table Showing the Positions of Points on the Planes or in a Controlling Relation to Planes Owing to their Three-Yin or Three-Yang Trajectories That Control Spatial and Temporal Zones

Cosmographic Plane	Perspective	Particular Aspect:	Phase and Associated Meridians	Point principle	Meridians and Points					
The Prime Vertical **General**										
			Meridians on Plane (Wei intersect)		yin and yang linking, yin chiao yang motility,					
	Treatment	Heaven								
			Meridians on plane		yin and yang linking, (Wei intersect),	yin chiao yang chiao				
	:		Key Points on plane							
				General	Du 20					
			Increasing Light	Particular	Shao Yang/Jue Yin TH 4/ PC 7					
		Horizon	Meridian on Plane		Dai Mo; Yin and Yang Wei; yin and yang qiao					
				Front	Ren 12					
				Back	Du 12					
			Increasing light	Front	Liver 14					
				Back	UB 18, GB 26					
				Front	Spleen, 20, 21					
				Back	UB 20, ST 21,					
		Earth								
			Meridians on plane		yin and yang quiao					

			Key Points on Plane		GB 41, LV 1, 3, [UB 62, KI 6]*
				Front	Liver 9
				Back	GB 30
			Increasing light	Front	Liver 14
				Back	UB 18, GB 24?
			Decreasing light	Front	Spleen 15
				Back	UB, ST 25
The Prime Meridian					
	General				
			Meridians on plane		Ren and du (wei intersect) Chong, tai yang, shao yin
			Key Points on plane		
			Chakras:		Du 26, Mhn 3, 20, 16, 10, 4, 1 — Ren 17, 12, 4 (3,6), 1
			General control points		LU 7, heart 7, KI 6, UB 62, lu 9, 7, LI 4, 7, Li 3, Sp 3, 4,
	Aspect	Heaven	Meridians on Plane		Chong, tai yang, ren and du, shao yin
			Chakras:		Head: du 26, M, Ren 22, 17 Upper jiao; du 10 (11, 1)
			Treatment		Tai yang/shao yin
			Increasing Light	Back	Si 3 (1), (4), UB 14, 16,
				Front	Heart 8, KI 24 (23, 25), KI 27. HT 1
			Decreasing light	Back	Li 4
				Front	LU 7
	Aspect	Horizon	Meridians on Plane		Ren and Du, wei, chong, tai yang/shao yin
			Key Points on Plane		
				Front	Ren 12
				Back	Du 6
			Increasing light	Back	du 6, GB 41, UB 62, UB 48 Yangs key link)
				Front	Ren 8,12, KI 16, Liver — 1, 8 (3), KI 6, Liver 3, KI 19
			Decreasing light	Back	Du 6, ST 40 (36), UB 62
				Front	Ren 12, Spleen 4, KI 19

Aspect					
Earth					
			Meridians on plane		ren and du, chong, tai yang/shao yin
			Key Points on plane	Front	Ren 4 (3,6), 1
				Back	Du 4, 1
			Increasing light	Back	UB 62
				Front	Ren 6 KI 1
			Decreasing light	Back	UB 23 right, UB 62
				Front	KI 16,Ren 4, KI 7
The Great Circle of the Horizon					
Aspect	**Heaven**				
			Meridians on plane		eight extra meridians, all three yins and three yangs leg meridians Yin and yang hand meridians have internal connections but not external connections.
			Key Points on plane	E	Ren 8 (the Navel), South du 4 North, Liver 13 east and west
			Connecting points on plane		Stomach: 25 Spleen 16 Gallbladder, 26, 27, 28 Liver 13, UB 23, KIdney: (16)
		Prime meridian du, cv			
			Control Points:		North above: Si 3, Li 4,
				North below	UB 62, Stomach 42, 36
				South above:	Heart 7, lung 7
				South below	KI 6, Sp 3,4
		Prime Vertical			linking and motility yin and yang linking, yin and yang: yin linking

			Rising Light Yin and Yang linking	East Above when facing south or north	Liver 13 (PC 6, TH 5; GB 41)
			Declining light: Yin and Yang Motility	West Below when facing west or east, connects with	Spleen 16 (Sp 4 ST 42; lu 7)
				when facing north or south	Stomach 25

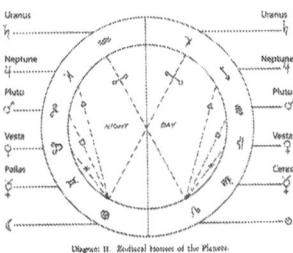

Diagram: 11. Zodiacal Houses of the Planets.

Appendix 2

Structures behind the Three Light Cycles

I. The Three Light Cycles in Nature and Their Structural Foundation

The diurnal cycle, lunar phases, and annual cycles of light and darkness are the root standards for traditional astrological and medical meaning. Calendar stones generally were placed in circles and drawings of the motions of bodies, like the sun and moon, which were always drawn in circles since their major landmarks were repetitive, not merely linear. These representations had references to the major demarcations of heaven above, earth below, dawn and dusk between, half-moon and full moon, equinox and solstice. These landmarks then lead to the discrimination of numbers, the four seasons, twelve lunar months, and approximately 365 days; and these numbers when rounded were the basis of establishing geometrical understandings that allowed for prediction and gave seemingly causative explanations for clusters of events that occur predictably within phases.

These discriminations were arbitrary landmarks in infinity and were perceived as such, but they formed the basis for recognizing three independent and interdependent rational spheres, each having a comprehensive set of standards for locating an individual or event in space and time. These spheres were (1) the local or individual sphere based on the horizon and heavenly vertex directly above any spot, (2) the mundane sphere delineated by equatorial system and the poles with the prime meridian, and (3) the celestial sphere delineated by ecliptic and the prime vertical. These are delineated in figures 10a, 10b, 10c, and 11 (fig. 1).

The Sphere of the Ecliptic
(Mid-Summer afternoon

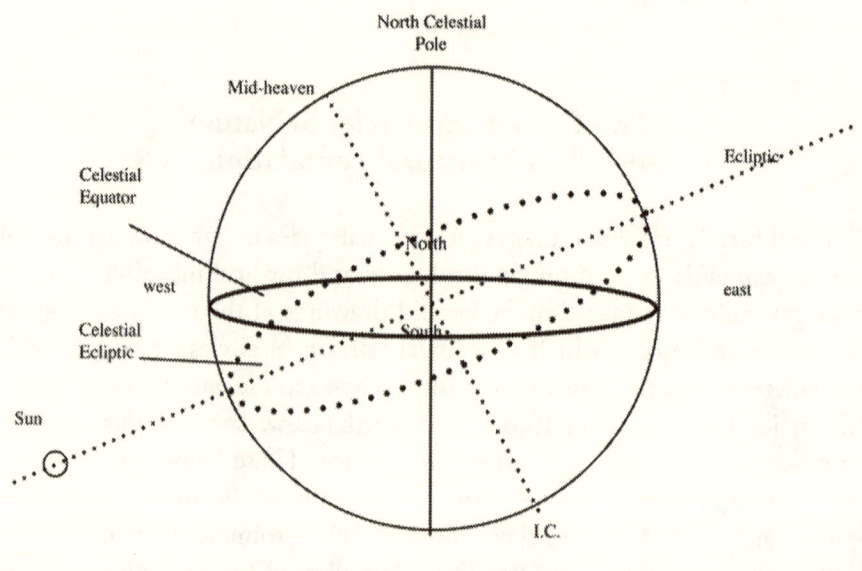

North Celestial Pole

Mid-heaven

Ecliptic

Celestial
Equator

North

west   east

Celestial
Ecliptic

South

Sun

I.C.

——————— Equator

·············· Ecliptic

The Sphere of the Equator
(Mid-Summer afternoon)

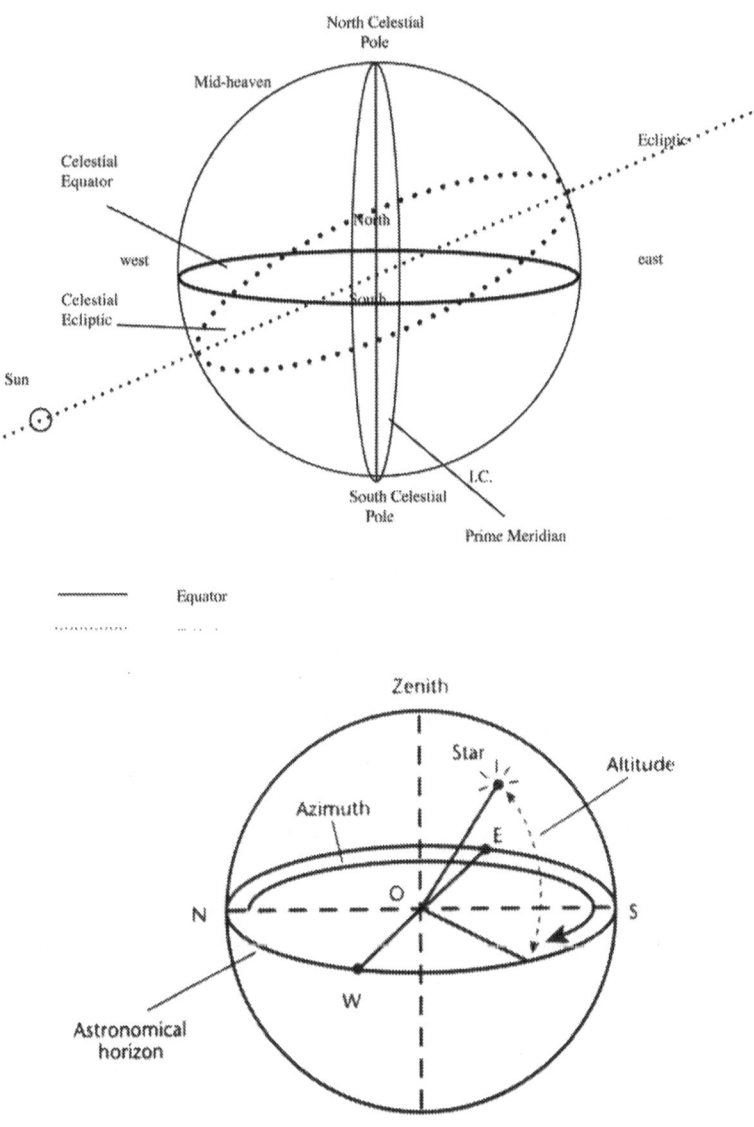

Figure 1. The Three Astronomical Spheres and Three Locational Fra
meworks Implied in the Composition of the Natal Chart: *A*, Celestial Sphere;
B, Mundane Sphere; *C*, Individual or Horizon Sphere

Each sphere provides a clear set of independent demarcations that overlap with one another. In the natal chart, all references to the mundane and individual sphere are visually folded down to two-dimensional positions on the ecliptic or celestial sphere. Nevertheless, all three systems are symbolically present and active, offering for interpretation three different perspectives on life. The three spheres and their perspectives describe visually and structurally the dynamic relations among the three cycles of light.

The celestial sphere defines the conditions that cause the seasonal cycle. The seasons have a special relation to the sun as they are determined by the orbit of the earth around the sun in relation to the equator, which, from the perspective of the ecliptic, is the slant of the earth's rotation. The celestial sphere (fig. 1a) has the sun-earth axis as the primary plane of reference. This solar bias conditions all the experiences and qualitative effects traditionally attributed to constellations visible along it. The earth's orbit around the sun comprises a year, and the temporal focus of this great circle is the annual cycle. The daily cycle can also be roughly identified as one degree of arc in the annual cycle, and the lunation also can be identified approximately as one of the twelve nearly thirty-degree and, therefore, thirty-day phases. The diurnal and solar periods are derivative and convenient but not essential to the system. Owing to the twenty-three-degree angle of the ecliptic to the plane of the equator, we have the variations in the duration of daylight in different parts of the world that account directly or indirectly for most seasonal changes. The prime vertical is a great circle that divides the celestial sphere in half, east and west, and the prime vertical lies east and west of that ninety degrees, differentiating back to front in the cosmic human.

The mundane sphere defines the parameters of the geographical, topographical, biological, and chorographical characteristics of life (fig. 1b). Because of the moon's proximity and earth-centered orbit, the lunations were properly evaluated by their phases and motion through this mundane sphere more than through relation to the ecliptic. Supporting the moon's identity with earth is the tradition in the Mediterranean world, which regarded the moon as the dividing line between heaven and earth. The mundane sphere, with all its activities, was called the sublunary world. All the more distant planets occupied heaven, and their regular cycles (Mars two years, Jupiter twelve years [mirroring the twelve months], Saturn twenty-eight years [mirroring the twenty-eight days of the lunar month] with periods of prominence) were measured along the ecliptic or sun-earth axis. The equator, however, projects its own plane into heaven, and this projection is called the celestial equator. Distances along it are measured in twenty-four divisions of right ascension that mark the periods and degrees of rotation. These are not only divisions in space but also divisions of time, and they are arbitrary and could as easily be twelve two-hour divisions as in the ancient world or twenty-four one-hour divisions. The plane of the celestial equator aligns with the ecliptic twice a year, forming and defining the spring and fall equinoxes in which the day has the same duration as night everywhere on earth.

A line passing through the center of the earth between the north and south pole then extended into space defines the equator's vertical dimension of ninety degrees. The prime meridian is an arbitrarily placed plane at zero degrees that passes through Greenwich, England, which determines longitude east and west. At ninety degrees, space is divided by a plane back to front that creates the prime vertical. These divisions identify basic spatial relations between any positions on the sphere of the earth. The earth completes a complete rotation measured in twenty-four periods of right ascensions in one day—that is, a given point in the earth's rotation takes the period of a day to return to the same point. However, during the diurnal rotation, the point of return projected onto the ecliptic has moved one degree. So for the diurnal starting point projected onto the ecliptic to really return to the original point of reference on the ecliptic, it will take approximately 365 days or one year of such diurnal rotations of an arbitrary point around the poles.

The earth's diurnal rotation, however, completes a cycle that is important to the whole globe, not merely to a point on its surface, and this always has given the equatorial plane a general seasonal implication. This is because when the diurnal rotation occurs in one part of the year or another with different durations of light falling above or below the equator owing to its angle to the ecliptic, seasonal effects become visible.

The moon bridges the gap between the diurnal rotation and seasonal characteristics of the mundane sphere by defining monthly periods of biological change as it rotates around the right ascension of the mundane sphere. The moon, like the earth, moves along the plane of the ecliptic going through the complete cycle each month, but this month-defining motion is dependent on the rate of the earth's orbit for its new monthly start point, which begins nearly thirty degrees in advance of the start point of the previous thirty-day cycle. In this respect, the earth and the moon form a unit whose seasonal temporal periods have the same root but whose relationship changes in a discernible way with each daily cycle. Thus, the lunar phases mediate between diurnal cycle and the annual cycle in a way that uniquely defines the otherwise more indistinct and seamless transitions from day to day and month to month in the annual cycle. The periods of the lunar rotation too match wonderfully with the regular divisions of a circle and with other planetary cycles in ways that deserve awe in any observer in any period.

The moon's mediating effect gives a different center of gravity to the mundane sphere, namely, the monthly biological and seasonal differentiations themselves rather than the diurnal transitions from which they emerge. Even though the diurnal rotation defines the temporal periods of the mundane sphere, it is the twelve lunations that generate the qualitative meanings they depend on. The ancients gave precedence to the twelve annual lunations and applied the division of twelve to the day as well. Moreover, since the earth itself is the focus of this sphere, it is the processes of the whole that must be considered, and that whole is dominated by the earth-moon relationship whose regular cycle is expressed by the lunar periods yet whose ultimate ruler in cosmology, as in myth, remains the sun and the annual cycle.

The individual sphere is defined by the personal and projected celestial horizon and the line formed between a point directly overhead and in the center of the sky known as the zenith, which then passes through the feet to the center of the earth or the nadir below (fig. 1 c). This unique vantage point is that from which everyone observes and defines the daily cycle of light. The personal sphere will align with the mundane sphere when an individual stands on the north pole and at the equator because the zenith will fall on the celestial equator. With the zenith overhead facing south in the Northern Hemisphere, dawn will rise in the left and east, and sunset will descend on your right in the west.

The individual outlook is in most ways private and local. The point where the plane of the horizon and the ecliptic meet, the ascendant, at any given moment expresses the potential for relationship between the individual and the solar system. In this sense, the individual spark or perspective also has collective reference, content, and promise. The individual literally is the point of arbitrary reference on the earth from which the diurnal cycle and every other cycle can be defined and measured. The mediator between all other cycles is the individual vantage point. This perspective is produced ultimately by the diurnal rotation of the earth and the annual rotation of the earth around the sun.

Of particular interest to the mediation of an individual perspective is that these two rotations have different contrasting directions: the earth rotates counterclockwise, giving an apparent clockwise motion to the sun, while the earth orbits the sun in a clockwise direction on the plane of the ecliptic, moving clockwise around the zodiac. Thus, the apparent solar motion through the day corresponds with the real motion around the ecliptic, which result interpretively in the same effects and conclusions even though the mechanics are now understood differently.

The ancients could not see this motion. They merely saw the sun rising and setting in advancing degrees or the midnight zodiac signs advancing in tandem along the ecliptic in a clockwise direction. The structure of the individual sphere, the sphere generated by any location in time and space, however, determines to a great extent the general consequences of phase dynamics for the observer. The duration of light near the north pole in winter is dramatically different from that in its summer. Temperatures on the equator are different from the poles. Summer seasons are shorter and less severe nearer the poles, winter nearer the equator. Thus, for the individual and in the particular case, cyclic effects fall into a different balance, a balance that never can encompass the totality of the whole system.

Locations and individuals will always appear out of balance in some way, though the globe may be considered to be in some kind of balance of qualities. The seasonal range in one location may eliminate snow, so local comprehension of snow will be limited, and the necessary functions of snow in the world economy may be foreign experience and concepts. Even though the individual perspective projected in great circles into heaven may demarcate diurnal and annual cycles, the interpretive vantage point is limited and the conclusions and analogies cannot be projected as referential meanings to the whole system. These obvious conditions nevertheless are important to the differentiation of ascendant meanings from those of the sun, moon, or other bodies in their relations with the three spheres. This outlook from a celestial horizon and zenith is comprised of cosmographical content but otherwise is in every way uniquely private and local.

The Three Spheres

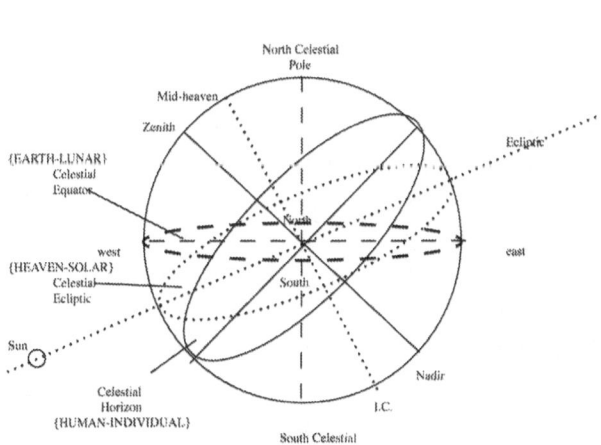

Figure 2. A Combined Rendering of the Three Spheres

The three perspectives embedded in the traditional natal chart and in other cosmographical depictions from ancient stone monuments to traditional medical systems offer opportunities for usefully individualizing and improving on standard chart interpretations and medical understanding. Humans are connected to their world, and that can be described geometrically.

The interpretive meaning of a particular planet in a cosmographical present depends upon which cycle or structure is the lens or object of inquiry. Astrologers assign qualitative standards such as hot and dry to Mars and insist as a result on its aggressive warlike character its tendency to promote fever, its proneness to dehydration, and its emphasis on physical action. An astrologer will then modify his or her interpretation of the degree of bias that Mars symbolically applies in a heavenly map owing to its various geometric relations to a symbolic representation of a great circle, such as the midheaven, or a positional connection to a zodiac sign division of the ecliptic or its positional relation to another planet—all of which exert their bias on that of Mars. However, these approaches are secondary evaluations to those geometric relations of Mars to the sun, moon, and ascendant (the ascendant is a complex representation of the horizon, literally that point where the plane of the horizon intersects the plane of the ecliptic). These connections increase or decrease the original bias significantly. The reason for this is not to be found in the geometric contact itself, to the sun, moon, or ascendant but the recognition behind this contact of the phase mechanics and perspective of the three spheres to which these placements refer.

It is the relation of the functional bias of Mars to the daily, monthly, and annual phase bias of the moment that would reshape any original conception of the effects of the hypothetical qualities associated with Mars. Mars will have an identifiable spatial and phase position in each of the three great spheres—celestial, mundane, and individual—that most radically determines the bias it exerts. The local mediation is as important as any intrinsic characteristics, and the bias of those characteristics can be measured by events in context.

Indeed, because the sun, moon, and ascendant marked on the ecliptic chart are merely points of activation and reference for the three spheres and because the spheres are the greater living extensions of the more static two-dimensional geometry of the of the sun-earth plane, the equatorial plane (with its lunation cycle) and the diurnal horizon, these points of reference, can be said to have meaning primarily as extensions of the three light cycles and their structures not owing to the intrinsic qualities assigned to the planet or its geometric relations in one chart depiction or another. The two-dimensional drawing on the ecliptic hides the presence of the other two cycles and their meanings. The sun for instance has standard qualities that apply to all divisions of the day, month, or year; but it is the phase in which it is placed that determines the bias at a given moment and in turn conditions the secondary bias exerted by Mars or any other body, including the Sun when conceived as merely a heavenly body. The three light cycles are what they are owing to these structures and phase energetics, their geometric relations and the quality of a given phase in any cycle, and essentially their spatial relations and temporal relations.

This recognition that the position in the greater cosmos of the sun, moon, and horizon is depicted as rational points on an ecliptic-based chart now project as parts of a whole-related structure with a phase-determined qualities allows us to expect precise predictable knowledge of the qualitative balance in any moment of a cycle and for precise predictable knowledge of the balance between these cycles at that moment. The standard qualitative knowledge contained in traditional cosmographical models points to this possibility but rarely fulfills it owing to the tendency of the focus to two-dimensional models and prescriptive meanings, which at the very least makes us lose sight of the object.

Yet the implicit three-sphere system of reference can be found in most comprehensive traditional systems (fig. 4). It is to these three spheres that the number three refers when it refers to heaven, earth, and horizon; and because of this understanding, the three spheres and their correlated phases form a legitimate basis for cross comparison because that comparison then has a foundation in the real organization of the solar system. Theoretically then, the position of Mars or any other body—to the sun, to the moon, or to the horizon—when it is viewed as a point on the spheres and in a phase is interpretively more useful than any derivative set of standards. All qualitative standards in systems of traditional medicine and cosmology derive from these structures, not the reverse.

All of Chinese medicine, for instance, down to the level of the functions for individual points, depends on this cosmic model for their empirical effects because the model is a window into the real conditions that generate those effects. The empirical content of any system, therefore, is dependent to some extent on the rational structures implicit in the evident light cycles and ultimately to the natural order or logos to which system and content refer.

Appendix 3

Ten Celestial Stems and Twelve Terrestrial Branch Numbers—A Root Language for Cosmographic Meaning

Early number theory developed pragmatically in relation to the calendar and in particular to the agricultural cycle. Clear evidence for this can be found in the Chinese stems and branches, which are two sets of numeric characters that represent seasonal periods using agricultural images. Why are there two sets of numbers that depict the same agricultural cycle? The stems are associated with heaven, the branches with earth. The core of the distinction lies in this natural image of an inverted tree—an image that is mirrored in the shape of the lungs, the number and shape of the fingers and the toes, and the overall vertical alignment of the body. Heaven and earth form an integrated whole.

Heaven is associated with the five planets, which, in their yang and yin or day and night expressions, yield the number ten. Heaven is the origin of all things below. In the course of this book, it will be shown that the first ten numbers symbolically explain and indeed implement the three-dimensional structure of the body and world. The tree image implies that because the body moves freely on and above the earth, owing to the five digits on the limbs and the five senses that are similarly doubled by the body's bilateral symmetry, the human body is in fact a product of heaven.

From this perspective, the twelve branches refer ultimately to the twelve lunar months and the outer conditions of the world around a person. These conditions too have their effect on the structures of the body, which must adapt to them. Thus, there are twelve divisions on the four fingers, twelve divisions on the body and limbs (three for arms and legs, three on the torso, and three on the face).

The human body is the tree; it is the expression of the roots and branches and is the epitome of both heaven and earth. The two number sequences, one based on five, the other on six, are each doubled to express their yang and yin ramifications and describe the same cyclic process but from different useful vantage points. One deals more with structure, the other with circumstances; and together they effectively unravel the knot of time and space.

Chinese astronomers established a large pattern of sixty days and sixty years to project a manageable set of standards for considering a year or day within a wide yet comprehensible context. The source of the sixty-year cycle is uncertain. Nevertheless, because the year can be divided into six sixty-degree/day periods and because these represented to the traditional mind the perfect union of primary yin (2) and yang (3) principles—namely, the yin principle of two (yin and yang) and the yang principle of three (heaven, earth, and human/horizon)—together these conditions describe the total dynamic of life. On this basis, Chinese astronomers and philosophers could have projected a large pattern of sixty years onto the infinite progression to provide a manageable set of standards for evaluating the traits of a year as they did for the sixty days in each of the three-yin and three-yang patterns.

Other confirmations may have supported this. Agricultural tradition assigned a male positive character to the waxing moon and a female negative character to the waning moon, and this oscillation between positive and negative was projected onto pairs of complete lunations. The first lunation of the New Year was yang, the second lunation negative, and so forth for the twelve lunations. In this context, two lunations each of approximately thirty days, comprised of a positive yang and a subsequent negative yin cycle, established sixty days as the standard for six annual periods. Six such divisions comprise a year.

The fulfilled human life may have been approximately sixty years duration, and this plays out on a grand scale of many years, the typical conception of each day being a complete cycle and, therefore, may be viewed as a year in miniature.

Chinese numbers too suggest that thirty years was considered a generation and that thirty years was that active portion of a life preceded by youth and succeeded by old age. The sixty-year period roughly approximated two complete Saturn cycles (approximately twenty-eight years each, one complete progression of lunations of twenty-eight years), which in turn reflected the lunations and matched the twenty-eight lunar mansions of the Chinese constellations referring to the same. Sixty years round out and projects the duration of three lunar periods of nineteen years each; and this would conform with the heaven, horizon, and earth's cosmology of three that permeates Chinese thought.

The sixty-year cycle works as the sum of all the combinations of ten and twelve symbols or, alternatively, as a combination of the paired ten celestial stems that yield five elements in their combined relation with twelve lunar branches.

Chinese astrology uses the sixty-day and sixty-year patterns as a basis for designating cosmic periods and corresponding temperamental and meteorological patterns in an otherwise infinite chain of possibility. It is entirely rationalistic. Yet it is tempting to anticipate that the world too is more rationalistic than random in its symmetrical patterns of functioning. Each day and year in the cycle of sixty was assigned one of ten characters known as the celestial stems, representing heaven (ten symbols, the structural completion of the human body and for the solar year) and one of the twelve terrestrial branches representing earth (twelve lunar months).

Each day and year, therefore, arrives with a known ontology, a predictable energetic balance, if only the observer understands the symbolic and associative content of the numbers. This system formalizes the words of Chi Po in the *Nei Jing*: "The movements of physical objects (terrestrial branches) and pure energy (celestial root or stem) bear a resemblance to those of roots and branches."

The Chinese New Year too celebrates the arrival of the new impulse from heaven and its unique response from earth with the celebration of the year of the dragon, the tiger, the rabbit, or some other animal representation of the popularized terrestrial branch symbol of the underlying paired stems and branches. The same animal symbols were assigned to designate the lunar months (which are comprised of two fifteen-day segments, their waxing and waning phases) and twelve two-hour periods each day. Since the day epitomizes a year or a lifetime to traditional thinkers, we can see in the twelve two-hour periods the model for the twelve months of the year or, in the six divisions of the cycle, the consolidation of the twelve into yin and yang partners. Clearly, the Chinese applied the same standards to evaluate phases of the seasonal, lunar, and daily cycle.

Within the sixty-day or year cycle, in any given year, the associated stem sets the tone for the annual celestial movement, and the terrestrial branch establishes the tone for the earthly circumstances of the year. Stems are yang; branches are yin. However, within each sequence of ten or twelve respectively, as with any number sequence, the stems and branches alternated between odd and even, masculine and feminine. Thus, the combinations that were supposed to govern a given day or year were considered either harmonious or disharmonious with each other based on gender and phase compatibility. These relations in a given day or year that were thought to produce many of the varied circumstances of life, circumstances which often depart from what should be an ideal regular progression of harmony indicated by the regular circulation of the solar cycle or for the Chinese, are the root patterns in heaven.

These phase relations are ongoing and real, and their effects are palpable in patients. We can regard the larger sixty-year cycle as a reference library to help us identify the real patterns, but it should not be used mechanistically to assume for instance that everyone born in a fire dragon year will have fire element imbalances. Rather, we can recognize the possibility and study the patterns of compensation to achieve a rich individual understanding of a patient's condition. More important is that we recognize from firsthand observation the trend of a year toward heat and dryness and the effects on patients, whether it is a fire year or not. The stems and branches give a very detailed portrait of the annual cycle, and they form a foundation for later more esoteric treatments of number in each of the twelve chapters of this book.

The scholar Joseph Needham viewed the *I Ching* as an important document for demonstrating a movement in Chinese thought from oracle to theory and system.[115] The *I Ching* has an interest in prediction, but the prediction has a strong naturalistic basis. Trigrams extend the basic principle of whole numbers deeply into the details of the equilibrium during any phase of a cycle. In this sense, it is more descriptive than inventive and provides a foundation for the contemplation of relations between phases and experience. The horary or oracular uses were secondary. In figure 1 below for the *ba gua*, I included the basic trigrams to emphasize an early attempt at a rational and empirical basis of the phase relations. Figure 2 shows the trigrams applied to heaven and earth with all their combinations (60) in sequence.

Figure 1. *Ba gua* with Trigrams in a Dynamic Posterior Heaven Configuration

The rational use of number helped free naturalistic medicine from the hands of oracles and shamans. Today, it may do the same by liberating us from the hands of specialists. Evidence for this is seen in the characters for the ten stems and twelve branches with their implied union of number with ancient agricultural meanings. These agricultural meanings are surveyed in detail for the celestial stems in chapter 10 and for the terrestrial branches in chapter 12.

Table 1: Ten Celestial Stems (t'ien Kan) and Twelve Terrestrial Branches ti-chih) Stems and Branches

Numbers	Character	Celestial Stem Name	Stem Character	Terrestraial Branch Name	Branch Character
1. yi	一	jia (3) (chia)	甲	Zi (3,2) (tzu)	子
2. er	二	I	乙	Ch'ou	丑
3. san	三	bing (3) (ping)	丙	yin	寅
4. ssu	四	ting	丁	mao	卯
5. wu (3)	五	Wu(4)	戊	Chen (2) (Ch'en)	戊
6. liu (4)	六	Chi (qi)	己	Si (4) szu	巳
7. qi (1)	七	Geng (1) (Keng)	庚	wu	午
8. ba	八	Xin (1)(hsin)	辛	wei	未
9. jiu (3)	九	Ren (jen)	壬	Shen	申
10. shi	十	Gui (3) (kuei)	癸	You (3) (yu)	酉
11. shi yi	十一			Xu (1) gu(5) hsu戌	
12. shi er	十二		hai	亥	

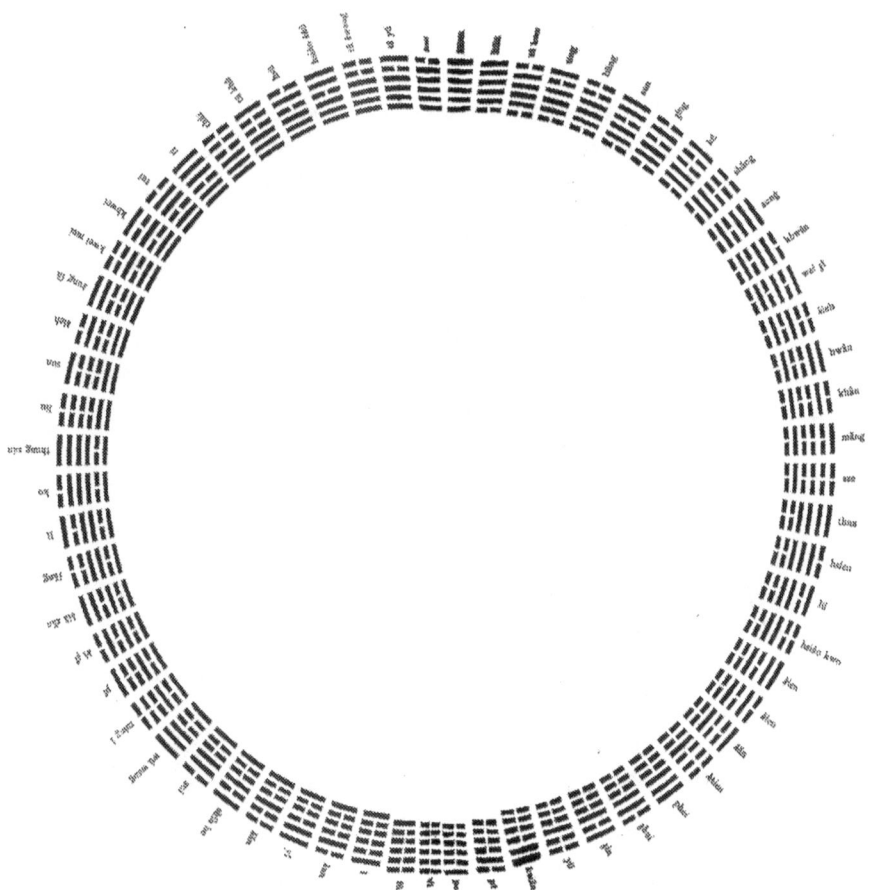

Figure 3. The Sixty-Day/Year Cycle Produced by the Combinations of Ten Celestial Stems with Twelve Terrestrial Branches, Depicted with Matching *I Ching* Hexagrams

III. The Ten Celestial Stems

The characters for celestial stems roughly describe the seasonal growth cycle beginning in winter and returning to winter. The characters are sketchy, but the associations are obvious and unmistakable, and some limited intuitive leaps can be made. On every level, it is important to remember the root cycle and its trends, and the ten stem characters are the way that the ancients viewed the solar year as a metaphor for both the growth cycle and the stages and conditions required for human existence in three dimensions (definitions and terminolgy from Weiger, extrapolations mine):

1. *Jia* (3) [甲] is the the first of the ten stems—hard coverings, scaly plates, a cuirass, the finger nails.[116] *Jia* contains the image for a cultivated field on a staff. The upright line runs through the field, and the field reduces to a square with a cross in it. The upright line can be simply a staff or arrow, but as an upright figure, it can refer to man or as a connection between heaven and earth. It extends below the field, which implies a kind of movement below the surface of the cultivated field. Earth is emphasized. The will of heaven is manifest in the square at the top of the staff, which implies the four directions. *Jia* is the image of a winter landscape that extends everywhere but has life below the surface as heavenly chi extends downward.

    Chia can be viewed as a profound symbol of return and reunion between heaven and earth. The hard, frozen field (extended symbolically as a square, an image of the cosmos, the earth with four directions) above on the staff (implying heaven) symbolically reunites heaven and earth after their death and separation under the tenth symbol *kuei* in the late fall. The earth is closer to heaven in winter because the clouds and sky seem lower; there is less light. Whether it is earth, yin dominating the heavens, or the hardness (yang) of heaven extending below, *chia* symbolizes this mysterious union, which leads to life.

2. *I* [乙] is a cyclic character. To mark (a basic form under the image for a cave, which means to dig and to excavate). Its kind of a seed form or basic shape under the earth, essential space or form. Rad 116, 753[117] *I* is a radical used in many other characters, indicating a basic form of elemental life and movement. The action in the tail of the figure is below the surface. The three levels—heaven, horizon, and earth—are implied in the character, and the movement is from heaven into the earth, where there is a lasting effect. This form is suggestive of the basic movement of chi from heaven into the earth, which then generates life and the ten thousand things.

3. *Bing* (3)[丙] A cyclical character, end of branch, small branch between gates, sheaf of grain, implicit hand holding grain to bind a sheaf of grain but with pillars on either side of symbol, suggesting a container in the earth.[118]
    Ping suggests the act of planting. The emphasis is on human action and on ritual offering. A ritual is repetitive and an act of supplication. Placing a small branch in a container in the earth suggests planting as an annual ritual, an offering of human action in support of life.

4. *Ting* [丁] is a cyclical character, an adult, to mourn, a nail (head and back), line of heaven above stake.[119] This symbol in context indicates a firm leaf appearing above the ground. Heaven's mandate manifest; the horizontal line above at its simplest level implies below, below ground or a stake. The symbol may refer to staking a plant, the roots of a plant extending into the soil and/or a leaf breaking through the soil. It is the penetration of yang into the earth that mysteriously causes birth and emergence. The stake supports the plant and symbolically explains its origin.

5. *Wu (1)* [戊]is a cyclic character, cutting tool below a ledge, emphasis on slope, suggesting a fall.[120] By extension, within a cyclic context, this symbol indicates weeding or pruning, cutting, and falling plants. Chi 36 (3) is one's self. A cyclical character. The ancient character represented the threads of the weft on the weaving loom. On the top two threads is transversal, one thread longitudinal, at the bottom the thread in the shuttle. When *x* was chosen on account of its simplicity to become a cyclical character, the sixth and the ten stems, it was replaced by *y*. It means also chia chieh, a person, one's self, I.

   This ancient symbol for the self shows profoundly that the Chinese concept of self derives from concept of humans as part of nature as an expression of life generated by the five planets and their derivative five elements—wood (spring), fire (summer), metal (autumn), water (winter), and earth (center, late summer, or seasonal transitions)—in their seasonal transformations. Here, Wu indicates the first full manifestation of an individual plant above ground with branches and leaves reaching toward heaven and roots extending into the earth. With emergence and identity, maturation is implied.

6. Chi [己] Oneself. This figure indicates the emergence of the seed above the soil as a defined and self sustaining plant. The character shows the bulk of the plant underground, planted in the earth but moving upwards towards heaven. This is a symbol for movement and change, growth. The two directions and continuity include the concept of yin and yang aspects of the great cycle. The circle is implied, the individual is the focus of this character.

7. Geng (1) [庚] is a cyclic character age (to change, to alter, symbol of wood with cultivated field on top with firm line at level of heaven (upper), straight line to left, upright.[121]

   The emphasis of this character is on the growth and maturation of plants symbolized as reaching for or achieving a height but also of reaching an upper limit, an order in the line above. Tending and ripening is part of the symbolism of the lines on the trunk of the plant, which can be viewed as hands or as proliferating branches.

8. Xin (1) (Hsin) [辛]. There is an implication of ripening in the symbol. Wieger denotes Hsin (1) as bitter, toilsome, the eighth of the ten stems. The character shows a stalk or trunk with heavy burden at the top, indicating a ripe plant, flower, or fruit.[122]

   This is a symbol for work and harvest. The crops are ripe. The focus is on man and the horizon, not on heaven as would be indicated in the ritual planting in the spring where so many other factors condition the fate of the crop. At harvest time, it is industry that is critical.

9. Ren (2) [壬] is a cyclic character (608), great or bad. Image of a man carrying a load with a pole.[123]

   The center pole is man; the longer horizontal line in center of three emphasizes man in the heaven, earth, and man triad—implying work, activity, and harvest. This continues the period when human activity has the greatest bearing on outcomes. The centerline also suggest balance, weights on a pole, and valuation or judgment of matters between heaven and earth. A harvest is weighed and evaluated as good or bad, and the consequences for winter and the outcomes of the years' work and growth cycle are measured and evaluated with a sense of perspective that is implicit in the relations between the three levels.

10. Gui(3) [癸] is a cyclic character. Originally it is the nicely disposed grass on which the ancients poured the libations offered to the manes. The character contains elements of *po* or feet; and because feet refer to deceased ancestors, they indicate separation, divergence, and letting loose. And *fa* is to shoot an arrow.[124]

   When an arrow is shot, it doesn't return. The year's growth cycle is complete, and the idea of separation and transcendence implies the little death of winter for the year is gone and won't return. The feet refer to ancestors and the other world. Life has gone underground into an underworld or above into heaven. In both, a literal and metaphoric sense of heaven and earth are separated physically as well as metaphorically, and their bifurcation causes death or at least the end of the growth cycle.

After ten, cycle after resonant cycle generates harmonics that reduce back to the first nine numbers, while with each cycle the evolution of people and the planet spirals through the same cyclic process but on ever-developing on different level of development and experience, just as each decade of human life strikes a different chord in the human heart. The number eleven and twelve replicate the stages of one and two; eleven, reflecting the divine impulse, represents existence as a separate self, the first year and the first step of life once the child is born. Eleven is a complex transition between the solar template of completion in ten and the lunar template of twelve months. It is a number with a long history in magic because of its symbolism of transition toward incarnation and manifestation. Twelve is the first number to show a complete interlocking directional mechanism for moving around in a world of three dimensions and four directions (12). The rest of the book will focus on the number twelve because of its unique place in astrology and number theory.

II. The Twelve Terrestrial Branches

The terrestrial branch characters from the introduction follow a similar, clear progression to that of the celestial stems. In classical Chinese texts, they also are loosely tied to periods of the growth cycles.

1. Tzu (3) [子] represents a boy, son, sir, seed, a suffix, a cyclic character radical 39.[125]

   This character in the first position implies the whole relation between the seasonal cycles viewed from the perspective of the terrestrial branches. Tzu is a masculine odd number and the start of a cycle, but it implies that the terrestrial branches describe the valued consequences of the union of heaven and earth. The form of the child is the result of the generative and logical cosmic relations outlined in the prebirth meaning of numbers one through ten and of their seasonal counterparts in the ten celestial stems or roots. The celestial stems describe the general consequences of the circulation of chi between heaven and earth; but now the terrestrial cycle will describe the actual outcomes, the particulars, and the son. It is implied that the beloved son starting as a seed is in the earth, invisible below the surface in the winter, and will go through a process of change that includes growth, maturation, productivity, and decline. The terrestrial stems will describe the character of those phases, which are determined by the typical agricultural meanings associated with each of the twelve lunations.

   In some phases, the son and father will supervise this development; and in others, the moon yin will supervise in an alternating rhythm. The cycle starts with a boy, and it is the seed of the previous cycle, the end product of yang, that has penetrated into the earth that eventually will determine the activity in the first phases of the New Year.

2. Ch'ou (3) [丑] the second hour of the day (586) from one to three o'clock in the morning, consists two uprights with the three lines indicating heaven, earth, and man going through. The upper heaven line is short. Wieger's lesson 44 traces the image to ch'ou (3) a hand, bound. To bind, to tie up. It is a cyclical character. Other characters that employ the image contain meanings of handwriting and authority. The horizontal lines in this context indicate fingers.[126]

   The two uprights could simply indicate the number two for the second branch in sequence. They also can indicate limit or containment. The image of the hand combined with the longer horizon and earth lines suggests preparation or preparation for planting, perhaps storage. In ancient urns, the hand would be presenting in a ritual offering; and this meaning corresponds well with the corresponding celestial stems for movement below the surface, soil preparation, prayers for germination, or planting conducted as a ritual offering to heaven and earth.

3. Yin (2) [寅] is a cyclic character meaning to revere. Wieger interprets this as a "man with a stiff bearing, and a cap pays salutation with both hands. Behavior, gait, or ritual politeness."[127]

   Yin is about planting as an annual offering, a ritual offering. The theme of the hands is carried through and applied to the image of a field. The cap could also be a load of seed, an image of a man with a burden walking stiffly. The cap also is a truncated vertical indicating upward movement or growth according to the will of heaven suggesting that outcomes are in heaven's hands. From this perspective, cultivation and planting of this stage is viewed merely as a hopeful prayer or offering for heaven's bounty. A certain strength and wisdom lies in a man who acts correctly in the present in anticipation of a future good. The ethical comparison is to a man taking the action of planting with the assumption of his understanding all the uncertainties of the growth cycle yet placing confidence in order for heaven and natural laws of the growth cycle to achieve an outcome.

   This image appears to have roots in the two downward strokes derived from Ju (4) (Wieger, lesson 15) under a field t'ien (2) a cultivated field with an empty square above the upper line of which has a dot referring to upward motion. This seems a clear indication of the early development of plants with their sprouts and developing root systems, their initial but not in evidence upward motion toward heaven.

4. Mao (3) [卯] is a cyclic character, a term, a mortise, the fourth of the twelve branches, the hour from 5:00 to 7:00 a.m., a term, two leaves of a door

opening, compounded with Mao 3 the Pleiades, also from Ching ( ), a seal or official identifier with two halves that must work together.[128]

This character shows the balance between winter and summer, day and night, at the spring equinox and dawn—which are the meeting of yin and yang or the two interdependent halves of the growth cycle. This phase also suggests the breakdown of the seed that allows for germination and sprouting of the tender plant. Yin yields to yang with the increasing light of spring, and in traditional medicine, it is the *shao yang* gallbladder that acts as the intermediary between inner and outer. The yang-wood principle is responsible for accelerating growth and decisive development. The transition in this phase of the terrestrial cycle is going one way, toward summer and growth as indicated in the dominant left half of the door figure.

5. Ch'en(2) [辰] refers to morning and the hours of 7:00 to 9:00 a.m. Wieger describes this as a part of time, the heavenly bodies. Like the character Wu [戊], a cutting instrument under a ledge. Wieger's lesson 30 connects Ch'en to characters with the meaning of "a woman who bends forward to conceal her shame, menses" and chuan chu "time, epoch, period, menses."[129]

    The association between Ch'en and the heavenly bodies would be very concordant with Wu, the number five, which refers to the circulation of the five planets and five elements between heaven and earth. The image of a cutting instrument is suggestive of the pruning and weeding suggested in Wu, the roughly corresponding fifth celestial stem. Clearly, the plants are in the ground; growth (generative cycle of five elements circulating) is in evidence; and humans are pruning, guiding, weeding, and so forth. The reference to the menses could refer to the menses that were generally associated with the full moon, which corresponds to the midsummer phase of ch'en and szu, or it could refer to early sacrificial ideas of letting blood into the earth to promote growth. Every laborer knows that blood or pain is the price of accomplishment. The image could be as simple as women bending over to tend young growing plants and to chop down weeds. Bending in the fields is like the menstrual cramps, and in early Chinese cultural contexts, this could refer to women's' fieldwork.

6. Szu [巳] is similar to the character han (3): to bud, to put forth buds, to bloom. Wieger calls it a "primitive representing the effort of the blooming of the springing up. Also indistinguishable except in context from chieh (2), meaning the right half part of an official seal."

    Both meanings from han and chien together imply the predominance of the light yang half of the yin yang balance and a full emergence of the identifiable plant above the surface. This is the surging approach to midday and late spring approach to the summer solstice midsummer point of the terrestrial cycle. The three levels—heaven, earth, and human—are represented; but heaven and humans are united in a square, suggesting a field or a combination of labor between heaven and man at this balance point of the growth cycle.

7. Wu (3) [午] Wieger calls it "a cyclic character, noon. This has the image of Kan or a pestle with a right-to-left line indicating action. On a more basic level, it implies a trunk staff or arrow, a vertical line with two lines for yin and yang above."

   The single staff defines a balance point; the left right suggests a pivot or movement between yin and yang at the balance point of noon. The emphasis is action above with a movement downward. This is the yang pestle portion of the grinding action of the annual cycle.

8. Wei (4)[未] Wieger mentioned in lesson 120 that this is "a negation, cyclical character, wood tree symbol below with short line, a tree with its branches superposed."[130]

   This figure has two horizontal lines above like the figure for Wu and linked by the vertical line as well. Here, however, the upper line for heaven and yang is smaller than the line for earth and yin, so the indicator is one of moving toward negation or darkness in the period after noon. The wood imagery suggests branches above and the development of late stage growth of plants in late summer. The negation comes from the entry into the waning or declining phase of any action or cycle.

9. Shen (1) [申] with field above spiritual means double-check man's identity. Wieger calls it: "a cyclic character—to extend, to increase, to report, to state."[131]

   The image for field has an upright line running through the center above and below. This character implies lush growth or abundant growth that extends past the limits of a field, both above and below. Branches and leaves grow toward heaven and roots descend deeper into the earth.

10. You (Yu) (3) [酉] is a cyclic character. In lesson 41g, Wieger states, "From a primitive to represent a vase or amphora for storing fermented liquor."[132] The line above refers to heaven, line at bottom, the dregs, or earth in the amphora—which in its turn can refer to the human body and to the whirling space between and contained by heaven and earth. The vase refers to storage of valuables, fruit, and preserving of essences on the one hand and rotting and ripening—which is the completion of process of digestion and assimilation on the other.

The image describes a yin (year's end, winter) process of enclosure or entombment, the beginning of process leading to death separation of essence from form. It hints at stored foods and fermenting of wine, which are the valued products of the growth cycle. The image of separation of essence from dregs harmonizes with the meanings of the last and tenth celestial stem kuei.

11. XU (hsu) (1)[戌] xu (4) "is the image for a cutting tool in an alcove, suggesting a granary, an enclosure for separating rice from husk or doing other harvest activity. It indicates the end of harvesting and the beginning of storing."[133]

12. Hai(4) [亥] Wieger states in lesson 69k, "With one stroke added to the tail. It is used, in the horary cycle, to designate the time 9:00 to 11:00 p.m. This time, says the glose, is the most propitious for the conception. Hence, numerous different figures that represent two persons, sometimes a man and a woman under heaven that is to say cooperating with the productive action of heaven by begetting children."[134]

Hai (4) shows the link between terrestrial animal and cyclic period by making the symbol explicit. Not all the twelve animals associated with the twelve lunar months have an explicit image in the original early characters, but the boar does. This suggests the possibility that astrological animals were linked to terrestrial branch characters for cyclic phases at a very early stage.

The boar's ancient written character has a downward stroke under the single horizontal line that show activity below, implying that yang is drawn deeply into yin. An extra stroke is added to the tail of the figure, which may imply future birth in the next cycle or a barrier or limit such as physical limitation, death, or disease to this direction of activity. In some forms of the image for hai, there are two lines with the line below for earth being larger than the above line for heaven.

The image implies something rooted in the earth or below the earth. In the standard representation, a dot is placed above a single horizontal line implying heaven, and the single line is the terrestrial line. Thus, the activity under the earth is the activity of winter and death. The boar symbolizes the mystery of sexuality and renewal hidden in the earth. As the last phase of the terrestrial cycle then, the hog epitomizes the terminus, the end of the earthly cycle (in the tail of the hog and its character both of which are curled into a circle or spiral, there always is the implication of renewal and circularity). The boar epitomizes in its nature and shape, the ever-renewing profligacy of the earth itself, and of the generative movement of the terrestrial cycle in general.

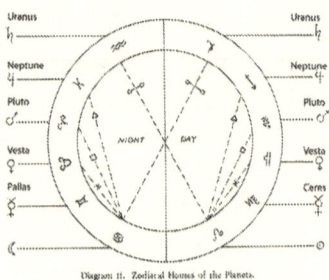

Diagram 11. Zodiacal Houses of the Planets.

Appendix 4

Planetary Weights and Influence

This appendix is for astrologers and acupuncturists who want to give some consideration to the use of planets in analysis of temperament or "causative factor." These relations are approximates based on a limited sample of one hundred patient natal charts. Nevertheless, the conclusions are useful and interesting because they show the possibility of several different kinds of structures or causes for a causative factor diagnosis. A person may be a wood temperamental type because of Saturn in Aries, which inhibits and restricts the wood orb or because of Mars in aspect to the Moon which inflames the passions and overworks the liver, or because of an absence of indicators for wood but strong grandparent emphasis in metal which deprives the wood of chi or grand child earth, which constantly depletes the wood energy. These patterns of element and planetary relationship, then, can be very useful for determining the proper kind of corrective action given the wood causative factor.

Tropical astrology shows two facets of constitutional identity. One facet, closest to personal identity, comprehends all the various natural processes when functioning well—the orthopathy of the self in body, mind, and spirit. Then, from the another perspective, there are those internal and external impediments and struggles that determine character and that explain how a person responds to trauma. Trauma and stress cause deviation from a normal, healthy profile to one that is defined by a reaction that was never corrected fully. The balance between orthopathy and pathology can be spiritual, mental, emotional, physiological, and/or physical aspect. The point of permanent deviation from orthopathic energy circulation is called the causative factor, and it defines the functions that bear the load of this struggle.

It now remains only to prioritize the conditions that determine which element—water, wood, fire, earth, metal, or air—and phase is the one exhibiting the greatest stress and vulnerability both in a celestial chart and in a patients' physiology. This causative factor must ultimately be determined by diagnosing the patient's color, sound, odor, and emotion while considering symptoms, age, habits, and other factors. Nevertheless, sometimes the chart will give indications, particularly when planetary positions show clearly the stress points at the onset of any condition. The prioritization itself will show that patients who present with several different major configurations will deserve a matching sophisticated approaches to treatment. Several example charts and patient histories will further clarify the use of this method of comparison.

To help determine the likely causative factor, I have developed, based on study of natal charts and patient histories, a table of planetary weights (table 26). The sequence in this table is the sequence of factors most likely to create conditions for the causative factor when placing other competing factors aside. The art of determining causative factor using the additional tool of astrology is the art of prioritizing, given complex situations where different competing relationships and conditions are involved.

It always is useful to observe the planetary transits and relations in order to support the intuition of instinctive sensations and trends in the mass mind. When everyone is suddenly possessed with cultural fear about some other group, religious idea, or taboo, a glance at the position of heavenly bodies often will clarify what is behind it. The following are rules of thumb, not quantitative laws. They are based on the study of a limited number of patient records in relation to natal charts. The first chart to consider is the simple relationship of mass to influence.

Table 1. The Order of Planetary Significance by Mass with Sign Correlations

Body	Sign	Element Bias	Health Pos/Neg
Sun/	Leo	Fire	+4
Moon*	Cancer	Metal/water	+2
Jupiter	Sagittarius-Pisces	Wood/fire	+1
Saturn	Capricorn-Aquarius	Earth	-4
Neptune	Pisces-Sagittarius	Wood/metal	-2
Uranus	Aquarius-Capricorn	Water/metal	+1
Mars	Aries-Scorpio	Wood/fire	-3
Venus	Libra-Taurus	Metal/earth	+2
Mercury	Gemini	Metal/water	-1
Pluto	Scorpio-Aries	Metal/water	+1
Chiron	Virgo	Earth	+1
Asteroids	Virgo	Earth	

* The Moon is a special case; because of its proximity, the effect of its mass is out of proportion with all other bodies except the Sun. The Sun and Moon must be considered together as the father and mother of life as we know it.

The following weights have been developed from comparison of medical charts in the energetic system of Chinese medicine and astrology.

Table 2. Points Given to Weighting an Element of Zodiac Signs That Are Occupied by Planets

Element	Planet or combination	Weighted Points
Wood	Moon/Mars	24
Wood	Saturn in Aries	18
Earth	Saturn (fixed signs)	14
Earth	Saturn (any sign)	10
Element	Empty sign	6
Earth	Moon	5
Water		
Fire, Wood	Mars	4
	Ascendant	4
Metal, Water	Uranus	3
Metal	Neptune	3
Fire	Sun	2
Wood	Jupiter	2
Metal/Water	MH	2
	Mercury	1
Metal	Venus	1
Earth	Ceres	1
Metal	Vesta	1
Wood	Pallas	1
Metal/Water	Pluto	1

The sum of the bodies in any western element will be an important predictor of the causative factor. The sun in the Chinese seasonal element will be an important indicator as will be the annual branch element.

Saturn has the greatest bearing on constitutional weakness and strength. The greatest concentration of bodies in a zone of influence may be the causative element or in contrast an empty sign or a weak sign opposite to the fully occupied zone, sign, or phase. In the case of the Sun, Jupiter, or Venus occupying the same element as Saturn, their weights should be subtracted from the value of Saturn because they counterbalance the pressures of Saturn on that element.

Following is a serial explanation for the important designators for the Chinese medicine causative factor in the tropical natal chart. Each of the special case considerations help to clarify the judgment required when applying the rather unambiguous priorities from table 1 in complex real-world conditions. It is impossible to escape the need for clinical judgment and intuitive discrimination either in achieving a medical diagnosis or in synthesizing the factors listed below. However, from the various weights, the practitioner can quickly determine whether a patient's problem is simple or complicated, acute or constitutional, and many of the psychological complexes behind the physical problems will be laid bare.

1. Mars Squares the Moon

The single most definitive determination of a causative factor is the relation between the Moon and Mars. If Mars is 0 degrees (conjunct), 90 degrees (square), or 180 degrees (opposite)—that is, in the same quadruplicity as the moon—the causative factor will always be wood. It indicates serious emotional imbalance and stress on the liver. This single aspect relation among the numerous ones available overrides the influence of Saturn. Saturn placement in the same chart may well indicate another area of dysfunction that needs consistent support.

2. Saturn in Zodiac Signs

Other things being unremarkable, the placement of Saturn in a given zodiac sign will indicate the causative factor to be the Chinese element associated with that sign (table 22). Saturn symbolically indicates a disciplining restriction or inhibition of function that frequently results in complex compensations physiologically and psychologically. Opposite phases need to be considered if Saturn's negative effects are buffered, for stress invariably affects the whole circuit of polarized qualities indicated. Saturn also symbolized change, enforced and constant change, and this puts pressure on a person's constitution (orthopathy, i.e., sun and moon). Frequently, it indicates the system most likely to weaken or break down in the face of trauma. It is the position of greatest strength or endurance and greatest possible weakness, i.e., causative factor for pathology. Saturn has a bias toward the earth element owing to its yin structive focus, so any aspects of 0, 90, or 180 degrees will bias the other planet toward the earth element. Following is a list of the causative factor for different placements with different degrees of certainty.

Table 3. Saturn in the Following Signs Causes CF

Western Sign	Element	Causative Factor Meridian	Ch El	Probability
Aries	Fire	Gallbladder/liver	Wood	very high
Taurus	Earth	Earth /spleen		high
		Pericardium	Minister Fire	moderate
Gemini	Air/metal	Small Intestine	Fire	high
			Metal*	slight
Cancer	water	Stomach	Earth	moderate
		Water/bladder	water*	Moderate
			Fire	Moderate
Leo	Fire	Heart	Fire	High
Virgo	Metal	Large intestine	Metal	Moderate
		Earth/Stomach/spleen	Earth	Moderate
			Fire*	Slight
Libra	Metal	Large intestine	Metal	Moderate
		Liver/KIdney	Wood	Moderate
			KI yang	Moderate
Scorpio	Water	KIdney Yin	Water	High
			Metal	Moderate
Sagittarius	Fire		Wood	Moderate
			Fire*	Moderate
Capricorn	earth		Earth	Good
			Water	Good
Aquarius	Air	Lungs	Metal*	High
			Water	Good
Pisces	Water		Wood	Moderate
			Water	Moderate

* Chinese elements affiliated with Saturn placement in table 2 very frequently have element affiliation of the zodiac sign opposite in position.

1. Saturn

Saturn generally will indicate the causative factor but not if either the Sun, Jupiter, and/or Venus occupy signs in the same western element or triplicity.

Were Saturn to be in Leo but Jupiter and the Sun also in Leo or in Sagittarius or Aries, a secondary point of stress could become the predominant causative factor. High among the candidates in this case would be the element and phase associations of the opposite sign or an empty element (one not having any planetary placement). Chinese elements affiliated with Saturn placement in table 2 very frequently have element affiliation of the zodiac sign opposite in position. This very much conforms with the notions of yin and yang and responds with the checks and balances evident in the seasonal cycle.

There is no wood equivalent in elements of Western astrology, but Aries has proved to be a direct counterpart. Saturn is most likely to be the definitive causative factor (wood) in a chart when in Aries. Not does this placement almost invariably produce a wood causative factor, but it is an indicator with one of the highest degrees of precision and certainty. Saturn in Aries is a constitutional bias that overrides any solar bias.

Saturn, when in fixed signs, almost always is a causative factor for the element associated with the signs. The wood element is least likely to be the definitive causative factor in zodiac signs that show moderate tendencies toward wood—notably Pisces, Sagittarius, and the sign opposite to Aries, Libra. Moreover, if in Sagittarius or Pisces, other factors in the chart will determine which element of the two possible ones, fire or wood, will be the causative factor. Generally, it will be Mars and Saturn together that determine the tilt toward wood versus water or fire.

Saturn square the moon is a strong indicator of an earth imbalance or pathology. It will definitely display as a problem in the digestive process and is likely to be the causative factor unless Saturn is in Aries, Taurus, Leo, or Scorpio.

Saturn opposite or square Mars is a strong indicator of internal stress in whatever signs they are in. It indicates a fundamental conflict between yin and yang productive of pathology either emotional or physical depending on a person's solar temperament. The ninety-degree relationship gives a valuation and directionality similar to the equilibrium of the equinoxes, where the days are of equal length but where there is an uncompromising bias forcing movement toward greater light or greater darkness. The relationship is more dynamic than a simple opposition because of the contrasting aims of whole phases and quadrants that are brought into tension. The ninety-degree aspect has a character similar to those generated by the ko cycle conflict between the elements because of the different quadrant conflicts implied. Saturn in a square aspect generally will bias toward a wood or earth/metal orb imbalance because it cannot affect directly by the aspect alone the *shao yin-tai yang* equilibrium. The exception would be with Saturn or Mars placed directly in the noon or midnight position or the midheaven, nadir position in phase, which would allow them to dictate stress to the whole system.

398 | William Wadsworth

2. Mars

Mars, as the ruler of Aries, tends to cause pathologies related to the wood or fire element, depending on the placement. The more aggravated or pronounced the Mars energy, the more likely is a wood imbalance. In whatever sign it is placed, that element is prone to periodic short-term imbalance and flare-up but is unlikely alone to cause chronic disease. The one exception is the position of Mars ninety degrees from the Moon in any chart, which yields invariably a constitutional bias toward pathology of wood (see below under Moon).

3. Empty Elements

If there are no planets in a tropical chart element and no strong indicators for wood imbalance, the empty element may be regarded as a chronic lack of resource. As a clinical matter, transfers should not be made through meridians associated with empty elements as they will neutralize the energy and not pass it on effectively.

If two elements lack planets, generally this concentrates the planetary activity in some element, which then becomes the causative factor.

However, the whole pattern should be viewed synergistically to see if the elements involved are complementary, i.e., fire and air (yang) or earth and water (yin), which will accentuate the lack of yang or yin. This creates conditions as determining as the idea of causative factor conception and would beg treatment before the causative factor could be considered.

4. Uranus and Neptune

Uranus or Neptune—prominent and in a 90 or 180 relation to the Moon, Mars, or Saturn—are strong indicators of possible metal element stress, which means eventual lung and colon pathologies. These planets effect the nervous system as well. So with a supported Saturn and Uranus and/or Neptune situated in a cluster of planets or strongly situated in Aquarius or Libra with conflicts to Mars, this may indicate a causative factor.

5. Gemini, a Special Case

With Saturn in Gemini or fire signs when it is combined in a chart with Mars, Sun, Mercury, Jupiter, or Uranus, the masculine planets, which will activate the sign and element, the causative factor can be narrowed from a fire element to the sorter official and meridian, the small intestine.

6. Moon

The moon is the most important regulator of human physiology, and invariably, it strives and generally succeeds in supporting the orthopathy. However, since its qualities are cool and moist, the Moon is more susceptible than the Sun to life-degrading chi, particularly cold and damp pathologies that are responsible for most brief illnesses and that, under conditions of lowered immune function, can cause lingering, chronic conditions. Certain element and aspect relations to the Moon can determine the causative factor if other primary factors are neutral.

Table 3. Weighted Lunar Contributions to the Count Toward Causative Factor

Moon	In aspect to 0, 90, 180	Element	Weights
moon	Mars	Wood	24
moon	Saturn	Earth	8
moon	Uranus	Metal	6
moon	Neptune	Metal/Water	6
moon	Scorpio	Water	3
moon	in Sagittarius	Wood	3
moon	Conjunct Sun	Any Element	2
moon	in Capricorn	Earth	2
moon	in Aries	Wood	1

I've listed these positions in the order of their pathologic influence. These lunar positions each place significant stress on the associated element. It is difficult to quantify just how much, but in a complex case that involves the above elements, these relations shift the balance.

7. Multiple Planets in an Element

If more than five planets occupy one element, this creates an excessive emphasis in the meridians associated with that element. Generally, if the Sun, Venus, Mercury, and Jupiter are involved, depending on rulerships, this will support the orthopathy; but if accompanied by an opposite empty element, the otherwise positive pattern will hog energy and take what little is available in an empty element. If the relationship is sequential with the planetary majority falling in earth, for example, and the empty element being the preceding element in the seasonal sequence, fire, this too can be draining. The causative factor usually will be the empty element in this kind of pairing.

If, on the other hand, the multiple planets include Saturn and/or Mars and negative placement by sign or geometric aspect to the Moon, then the cluster of planets will shift the causative factor from the empty sign to the element that is emphasized by the majority of planets, particularly, if the sum contains Saturn.

8. Mars

Mars will only become the causative factor if Saturn is supported by the Sun and Jupiter; and Mars is strongly emphasized in a position of vulnerability, such as being angular in a day chart, being the only planet in an element (or triplicity), square or opposite Saturn, or the planet that is ruler to the chart (the zodiac sign ruled by Mars is on the ascendant, the intersection of the horizon with the ecliptic).

9. The Sun

The Sun supports life, but if in a strong combination conjunction, square, or opposite to Mars or Pluto in a warm sign, this can cause wood and fire imbalances that affect the constitution. Furthermore, the Sun activates the zone that it occupies, which, over time, wears that function down. Late in life, the element in which the Sun is placed no longer can give energy to compensate for Saturn placements or other deficits, which means that either the element associated with the Sun or other elements that depend on solar support to maintain functioning will start competing with the causative factor in the generation of medical symptoms and signs. The Sun runs through the rather limited resources in the mutable signs most quickly. This last may, because of the greater imbalance between light and darkness in these signs/phases, create more than normal stress on the constitution without consideration of planetary factors.

10. The Ascendant

The element of the ascendant (near the horizon) is significant, and it accounts for some of the error in my early studies where planet positions were looked up in an ephemeris rather than from chart calculations. Mars is strengthened in the ascendant and less of a malefic, but Saturn becomes more harmful. Cardinal signs support orthopathy; fixed signs tend to cause the most problems by impeding the free flow of energy. Rulership patterns also are particularly important. The planetary rulerships with their symmetrical relations to the solstice, midday, and midnight axis may be seen in figure 1.

Figure 1. Rulership Arrangement from Firmicus Maternus Adapted to Invisible Planets, Planetoids, and Asteroids of Contemporary Astrology (each has associated characteristics for day and night placements in a natal chart)

The Western horoscope provides several very important insights into Chinese diagnosis and human physiology. Foremost among them is the reintegration of planetary affinities and relations into the dynamics of phase energies. They reintroduce the importance of planetary elevation, angularity (proximity to the horizon or midheaven), and celestial geometry into diagnosis. The weight of emphasis, both positive and negative, in an element (using the adapted posterior heaven mode) does much to clarify stress points in the affiliated meridian system. Also of great importance, perhaps of greatest importance, is the recognition that patients can be ranked constitutionally in three categories based on the complexity evident in their charts as follows:

1. A patient whose causative condition is determined by a straightforward moon connected to Mars, prominent Saturn with no support, or an empty element with other factors in balance is usually easy to diagnose and has a clear path of treatment, namely to calm or support the weak link. These causative factors can be diagnosed with a certainty of over 90 percent.

2. A second layer of diagnosis is required in several instances where competing factors in the natal chart indicate two strong competitors for the role of causative factor. People sometimes have more than one defining trauma or condition. Both competing areas will need to be treated, either by sedating or stimulating, but in every case, one causative factor will take priority. The following examples are the most frequent situations in which elements compete and where analysis among several nearly equal priorities is needed:

    a. Where the weakening Saturn placement is compensated for by a supportive placement of the Sun or Jupiter and sometimes Venus in the same element. This means that a problem lurks in the element and that compensations usually neutralize the harm but that energy is tied up in the element in a way that can allow a weakness elsewhere or lead to long-term constitutional weakening. If angularity, prominence, sign, or aspect emphasizes Saturn, it can override other supports, generate a bias at the level of causative factor, but this is not usually the case. Secondary element weaknesses need to be examined to find the causative factor in most cases.

b. In the case of the wood causative factor determined by more than five planets being in the signs Pisces and/or Sagittarius that include either Saturn or, if Saturn is elsewhere and supported, Mars. The Western horoscope shows very clearly that the wood element and liver will operate positively unless massively interfered with. There are three types of wood causative factors. First are those with moon square Mars, which is a most destructive position for it affects the liver, which, if disrupted by chemical imbalances and emotional pressures, can lead to dysfunction and psychosis. The next is Saturn in Aries who is the born warrior, executive, or strategist with a chronic set of emotional repressions and disciplines. Apart from these two specific types of impaired wood, the natural buoyancy of springtime and liver energy wood is difficult to impair, except by an overwhelming number of stresses and obligations indicated by numbers of connection in the chart.

c. When no planet or only one planet—particularly Mars, Venus, or Mercury—occupies a sign whose associated element is one of the four western elements, that empty or weak place can be the causative factor; but it will not be the causative factor if a clear, problematic Saturn or other strong indication is present. The empty element, if opposite by phase to a strong Saturn or multiple planet occupation, can cause serious problems to compound those of the dominant opposite member. Emptiness is a weak causative factor based on an absence or deficiency rather than a positive harm. So if Saturn is somewhat compensated by the Sun, Jupiter, or Venus and you have an empty element, the two choices can compete for the dominant role. Similarly, if you have five bodies in a single element, particularly in one sign, and emptiness in another element, the two will work synergistically—the one hogging energy, the other empty to create an imbalance in one or the other element. The moon-to-Mars relationship alone will take priority over an otherwise clear causative from Saturn placement. However, the inhibiting and problematic actions of Saturn do not go away even in this case. Thus, in all the examples above, it is possible as a diagnostic response to assign two causative factors to the patient as the single causative factor even in a case where clear clinical predominance does not adequately describe the configuration or lend itself to remedial treatment because two elements are competing for the honors as harm producers. Both elements and correlations will have to be treated. By looking at the pattern of compensation and opposition in the Chinese terrestrial model, it is possible to see ways to correct the imbalance.

3. Naturally, the third category is those patients for whom it is extremely difficult to determine the causative factor owing both to the chart and the medical indicators. There are those people, particularly elderly people, who have suffered a decline in function in several organ systems for whom it is difficult to be a diagnostic purist. You may need to treat several different, competing conditions all equally grave and operating in relation to one other, a situation faced by every doctor who has patients with multiple conditions. Yet some people have exceedingly complicated charts that forewarn of these multiple problems. You can have, for instance, a partially supported Saturn, an empty element, and several clustered planets in one element, Mars and Saturn prominent, and squares of either not to the moon but damagingly to planets related to different organ systems. Such a complex will erode the orthopathy across the board in several organ/meridian orbs and that continuously. The priorities I have given generally clarify with careful study what takes precedent, but some cases require great clinical experience and expertise to interpret just as they do to diagnose. I have known several patients who were treated legitimately on three or more causative factors by different practitioners who had justly noticed significant imbalances in each set of elements. Such a history may not indicate a failure in diagnosis by the practitioner so much as the emergence of different layers of internalized stress in different circumstances or phases of life. A complicated chart has the advantage over having no chart of providing a detailed map of functional stresses before they become symptomatic so that the practitioner needs not grope to find a single causative factor that never will be obvious, yet a valid long-term strategy becomes possible in light of the interactive balance of correlative factors.

Individual zodiac signs preserve meanings nearly identical to those in some Chinese officials. This suggests that there is an underlying template of conditions, perhaps embedded in the commonalities of human physiology and organ function with their mental connections. This suggests that the celestial cycles have much in common owing to their tendency toward abstraction. The above comparison uses the fully developed Chinese medical system as a stable point from which to view the Western horoscope. The Western horoscope lacks the equivalent terrestrial model, so by intruding that of the Chinese into the void and by retaining the diurnal structure and essential seasonal meanings of the tropical system, it achieves a greater clinical usefulness and immediacy.

One other important component for weighing the effect of planets in a chart remains to be examined. Modern technology has provided precise knowledge of the arrangement and scale of the solar system. Planetary mass and size is now known, where before it was unknown. Mass is not the only consideration, but it is important. Speculation about the attributes or locations of undiscovered planets or the significance of asteroids can introduce ideas that complete the astrological portrait of the psyche, but they should not entirely supersede mass. Proximity compensate for mass to some extent, and small bodies may carry important symbolic and geometric freight; but mass and physical influence bear a direct relation, and most likely, it is mass that has the most bearing on human physiology and consequently on human psychology. In consciousness, neither time nor space may have a great bearing. Nevertheless, mass is independently worthy of consideration as a gross measure of likely influence. Mass may be a material equivalent for metaphysical influence if the esoteric law, as above and so below, is applied (table 44).

The difficulty arising in a direct attribution of influence by mass lies not so much with the qualifying factors of speed and distance, but with the difficulty of defining the character of that influence. The hierarchy of influence by mass forces a reconsideration of traditional attributions. If the Sun/Moon and Jupiter are the most influential by mass and proximity, what is the character of that influence? The qualitative consequences of Saturn that transits frequently are easier to observe than those of the Sun, Moon, and Jupiter, but it is salutary to consider what kind of life, and we would have without either particularly in the details of their "invisible because pervasive and necessary" life-giving functions. Orthopathy is difficult to detail, but the evidence from the observation of cycles thus far is that Sun, Moon, and Jupiter placements in phase generally support life and orthopathy and ward off pathology in contrast to Saturn, which is a primary indicator for pathology. Life and health need direct study to clarify the functional dynamics that support it. However, in the mean time, due emphasis should be given to those bodies with greatest mass and influence, and their functional characteristics should be studied.

Table 4. The Order of Planetary Significance by Mass with Sign Correlations

Sun/Moon*	Weight for Orthopathy	Weight for Pathology
Leo/Cancer	9	1
Jupiter Sagittarius	7	3
Saturn Capricorn	5	5
Neptune Pisces	3	5
Uranus Aquarius	5	2
Mars Aries	2	4
Venus Libra	2	0
Mercury Gemini	1	0
Pluto Scorpio	0	2
Chiron/Virgo	1	0
Asteroids/Virgo	1	0

* The Moon is a special case; because of its proximity, the effect of its mass is out of proportion with all other bodies except the Sun. The Sun and Moon must be considered together as the father and mother of life as we know it.

How planets might mechanically activate the phase energetic is simply unknown, but they appear to do so sufficiently to create a bias in the qualitative effects of phases. The ancients extrapolated from the light cycles and suggested that the light of the planets was the causative agency. Early modern scientists discredited this notion and the idea that the mass and gravity were sufficient. Kepler's idea that humans have an innate positional sense for geometric relations, including a subconscious awareness of planetary positions, was largely forgotten by history. That humans respond to the light available through lunations and the fact that light influences brain activity and behavior is now well documented by research but not to the degree that the light of the planets and stars could be credibly introduced. However, the phase descriptions for diurnal and seasonal cycles to which humans are very obviously linked by a wide range of physiological and functional responses do revitalize the possibility that the major light cycles and cycles of human physiology work synergistically. It may be that the degree of sympathy between the world system and the body has a high degree of sensitivity. How high is unknown, but hints from studies of biological change in similar populations separated by great distances suggest that sensitivity is high. If humans are as tightly wired in as the meridian system with its cosmographical (spatial and temporal) links suggest, then the world system is responsive to other nearby solar system bodies. The other bodies do minutely affect orbital paths, and even the solar cycles too seem tied to oppositions of planets. Jupiter's motions and oppositions to Saturn have relations to solar flare activity. It seems plausible to postulate that some qualitative effects on the light cycle result from related solar system activity. The meteorological cycles that respond to the increase and decrease of light may as well be influenced dynamically if minutely by solar flares and other electromagnetic phenomena that, at some level, are quantifiable or that has a bearing on periods, particularly those phases of any cycle that are most vulnerable to minute shifts. A few days of cloud covers in the late autumn in northern latitudes caused by a shift in the jet stream and linked to pressures in the upper atmosphere can make a great difference to environmental conditions below.

Planetary influence remains a far-fetched idea from the scientific perspective, but it is less far-fetched than it was before Einstein or before quantum mechanics or before the studies by Ott, Cleve Backster, and Malcolm Gladwell; and daily, it seems more scientifically plausible as instrumentation improves.[135] Nobody knew until satellite measurements made it possible to know that the Appalachian Mountains move six inches every day in response to the Moon's gravitational pull. That pull also happens minutely in every other object including humans, perhaps, at the very least, affecting in small degrees things like fluid pressure and osmolarity. Scientific speculation aside the cogency and predictability of causative factor determination using the combined understanding from two separate traditional systems that study human responses to the cycles of light suggest that these cycles and the impact of planetary cycles on them are an object worthy of more detailed study when the right instrumentation becomes available. In the meantime, the theory is useful in the clinic as an aid to functional diagnosis and treatment.

Endnotes

1   Levi Strauss, *Structural Anthropolgy*, translated by Jacobeth, Claire and Schoepf, Brooke G. (New York: Basic Books, 1963), pp. 93, 94.

2   Cornford, *Plato's Cosmology: The Timaeus of Plato*, (Cambridge: Hackett, 1979), p. 39.

3   Mi, Huang-fu, The Systematic Classic of Acupuncture and Moxibustion. Boulder: Blue Poppy Press, 1993), p. 129.

4   S. K. Heniger, *The Cosmographical Glasse: Renaissance Diagrams of the Universe* (San Marino: The Huntington Liberary, 1977), p 6.

5   Clegg, Frederick W. "This algorithm has exponential complexity. We say that this (and similar problems) are nonpolynomial-complete, or NP-complete for short. There are algorithms that do a "pretty good" job in solving the traveling salesman problem, BUT ... there is no known algorithm that totally solves this problem, yielding a mathematically provable minimum solution, other than just running through all these cases." web.archive.org/web/20041104033210/http://www.engr.sjsu.edu/mcroc/CmpE126/chap01.htm. This article is concerned with modeling data efficiently and finds that mathematical models for determining real conditions such as the number of steps in the Eiffel tower, or the likely positions of objects being hunted can be prohibitively complex mathematically and in computer algorithms.

6   Julian Jaynes, *The Origin and History of Consciousness*, (Princeton: The Bollingen Series); Jaynes, Julian. The origin of consciousness in *the breakdown of the bicameral mind* (Boston : Houghton Mifflin, 1990); Levi Srauss, Structural Anthropology, Organized village

7   A vast literature reduced to a few examples: Dhammapada Sambhava, Padma, Thurman, Robert, Ed., *The Tibetan Book of the Dead: Chapter Eight: The Natural Liberation Through Naked Vision*, (New York: Bantam Books, 1994); , Julian Johnson, *The Path of the Masters*, (Beas:Sawan Service Leage, 1939); Andrew Cohen, Living Enlightenment: *a Call for Evolution Beyond Ego*, (Lenox: Moksha Press, 2002).

8   Don Beck and Christopher Cowan, *Spiral Dynamics: Mastering Values, leadership and Change*.

9   Pierce, Charles Sanders, Hoopes, James editor., *Peirce on Signs: Writings on Semiotics by Charles Sanders Pierce* ;_____ *Philosophical Writings of Peirce.* _____. *The Essential Peirce: Selected Philosophical Writings*, 1893-1913; L.M. & E.T. Nozawa: Unpublished Summary of Piercian Semiology.

10   Backster, Cleve, Primary Perception:Biocommunication with Plants, Living Foods, and Human Cells, (Anza, Ca. White Rose Millennium Press, 2003); Csikszentimihalie, Mihaly, *Creativity: Flow and the Psychology of Discovery and Invention*, ; George J. Bugh, *Spin Wave Technology* (Fort Worth: Vasant Corporation, 2002). Teilhard de Chardin, *Phenomenon of man*, (New York: Harper 1981); Peter Tompkins and Christopher Bird, *The Secret Life of Plants* (New York Harper Collins, Perennial, 1989); Don E. Beck and Christopher Cowan, *Spiral Dynamics: Mastering Values, Leadership and Change. Ken Wilbur, A Theory of Everything: An Integral Vision for Business, Politics, Science, and Spirituality.* (New York: Crown, 2003); Joseph Chilton Pearce, *Magical Child*; _____. *The Biology of Transcendence, a Blueprint of the Human Spirit.* Thomas Lewis, *Lives of a Cell: Notes of a Biology Watcher.*

11   Hammer, Leon, *Chinese Pulse Diagnosis: A Contemporary Approach* (Seattle: Eastland Press, 2001).

12   Burton, Robert, *Anatomy of Melancholy*, p. 23; modeled on historical writings about Saturn and melancholy, see: Raymond Klibansky, Erwin Panofsky, Fritz Saxl, *Saturn and melancholy: Studies in the History of Natural Philosophy, Religion, and Art*, (New York: Basic Books, 1964;

13   Vicari, E. Patricia, *A View From Minerva's Tower* (Toronto: Univeristy of Toronto Press, 1989), pp. 3, 4. Other important texts on cosmography include: Robert Westman, "The Melanchthon Circle, Rheticus, and the Wittenbuerg Interpretation of the Copernican Theory," *Isis* 66 (1975): 165-93.; John North, "Celestial Influence the Major Premiss of Astrology, " *Stars, Minds, and Fate: Essays in Ancient and Modern Cosmology* (London: Hambledon Press, 1989) , pp. 243-301; Lynn Thorndike, *The Shere of Sacrobosco and Its Commentators* (Chicago: University of Chicago Press, 1949; Margarita Bowen, *Empiricism and Geographical Thought from Francis Bacon to Alexander Humbot* (Cambridge: Cambridge University Press, 1981); E. G. R. Taylor, *Tudor Geogrpaphy*, 1485-1583 (New York: Octagon Books, 1968); Claudii Ptholemei Alexandrini, *Philosophi Cosmographia* (Rome, 1278), introduction by R.A. Skelton (Amsterdam: Theatrum Orbis Terrarum, 1966); Francis Johnson, *Astronomical Thought in Renaissance England* (New York: Octagon Books, 1968: George Perrigo Conger, *Theories of Macrocosm and Microcosm in the History of Philosophy* (New York: Columbia University Press, 1922; E. J. Aiton, "Celestial Spheres and Circles, *Hist. Sci.*, XIX (1981): 75-114.

14   Blundeville, Thomas, *The Theoriques of the Seven Planets, shewing all their diverse motions, and all other accidnts, called passions . . . whereunto is added . . . a briefe extract of Maginus his Theoriques . . . the making . . . of . . . instruments for sea men.* (London, 1602).

15 David Pingree, Ed, Trans, The Yavanajataka of Sphujidhvaja, (Cambridge: Harvard University Press, 1978), pp. 16, 17.

16 *Sources of Chinese Thought*, edited by W. Theodore Voll De Bary and Irene Block (New York: Columbia University Press, 1949).

17 Edmund Spenser's Poetry:Authoritative Texts Criticism, Selected and edied by Mclean, Hugh and Prescott, Anne, New York: Norton, 1993), p. 461.

18 Critchlow, Keith and Bull, Rod, *Time Stands Still: New Light on Megalithic Science*, (Floris Books, 1989). P 10.

19 Hesiod, The Homeric Hymns and Homerica, edited and translated by, Hugh G. Evelyn-White, and Glenn W. Most (Loeb Classical Library #57, 1981), p. 31

20 ibid., p. 33

21 Hamilton, Edith and Cairns Huntington, editors, *The Collected dialogues of Plato*, *Timaeus* trans by Benjamin Jowett,(New York: Bollingen Series LXXI, Pantheon Books, 1963), p. 1174.

22 Maternus, Firmicus, *Matheseos Libri VIII*, in *Ancient Astrology Theory and Practice*, Translated by jean Rhys Bram: Park Ridge: Noyes Press, 1975., p. 102.

23 *A Complete translation of The Yellow Emperor's Clasic of Internal medicine and the Difficult Classic*, Book 1, Chapter 4, Trans Lu, Henry C. Vancouver: Academy of Oriental Heritage , 1978)

24 Su wen Nei Jing p. 24

25 Lu Gwei-Djen and Joseph Needham, , p. 41

26 Manfred Pokert, Theoretical Foundations of Chinese Medicine, (Boston: MIT Press, 1978, p. 105.

27 Wieger, L., Chinese Charactersd: Their origin, etymology, history, classification and signification, New York: Paragon 1965), p. 26.

28 Plato, *Timaeus and Critias*, trans. By Desmond Lee, (New York: Penguin, 1965), p 43.

29 Plato, Timaeus and Critias, p. 48.

30

31 Needham and Lu, Celestial Lancets

32 Unschuld, *Nei Jing*, p. 84. Translates wu xing as fiveelements referring to the historical reference to elements as officials. This helps explain the link to "celestial stems." Respectfully, I will use this notion occasionally but in general retain "phase" and "element" because of western familiarty and the useful in the physical ramifications in context.

33 Wieger, Ibid., p. 28.

34 Unschuld, p. 88.

35 Lee, 48, Image in Figure 1 b. p. 77; Cornford, p. 68.

36 Ibid., figure 1b), p. 77; Cornford

37 Cornford. p. 238.

38 The Nei P'ien of Ko hung. Translated by James Wave, (Cambridge: MIT, 1960), p. 120.

39   Woods, John E., *Sun, Moon and Standing Stones*, (Oxford: at the University Press, 1978), p 76.

40   Mantak Chia and Maneewan Chia, *Fusion of the Five Elements*, (Huntington, NY.: Healing Tao Books, 1989), p. 5.

41   Needham

42   Wieger, Ibid., p. 29.

43   Needham

44   Wieger, Ibid., p. 30.

45   Desmond Lee, p. 97.

46   Wieger, Ibid. pp. 29, 118.

47   Ashmand, J. M., *Ptolemy's tetrabiblos*, (North Hollywood, Symbols and Signs, 1976), pp. 21, 22.

48   Needham, Celestial Lancets, Ibid. p. ;

49   Lu trans, Yellow Emperors Classic of Internal Medicine.

50   su wen or Porkert

51   *A Complete translation of The Yellow Emperor's Clasic of Internal medicine and the Difficult Classic*, Book 1, Chapter 4, Trans Lu, Henry C. Vancouver: Academy of Oriental Heritage , 1978), p 24.

52   Unschuld, *Neijing*, p. 92.

53   *A Complete translation of The Yellow Emperor's Clasic of Internal medicine and the Difficult Classic*, Book 1, Chapter 4, Trans Lu, Henry C. Vancouver: Academy of Oriental Heritage , 1978), p 14.

54   Porkert, p. 51.

55   Porkert, ibid., Fig. 12, p. 69.

56   Mantak and Maneewan Chia, *Fusion of the Five-elements, pp. 33-76.*

57   Wieger, Ibid., p. 107

58   Mantak and Maneewan Chia, *Fusion*, pp. 20, 40, 101.

59   Mantak and Maneewan Chia, *Fusion*, p. 145.

60   Lonny Jarrett, Nourishing Destiny, The Inner Tradition of Chinese Medicine p. 147, 153.

61   Unschuld, *Su Wen*, p. 134.

62   Porkert, ibid., Fig. 11, p. 67.

63   Lu Gwei-Djen, and Needham, J. *Celestial Lancets: A History and Rationale of Acupuncture and Moxa.* (Cambridge: Cambridge University Press, 1980), p. 47. Six pointed start each side representing a yin and yang devision of the circle, shows the relation between the three yin and three yang and the twelve branches.

64   Wieger, Ibid.,

65   Lu, Su Wen, p.

66   Sources of Chinese Tradition, px

67   Unschuld, p. 140.

68 Modeled on Porkert

69 Yoshio Omura, Acupuncture Medicine. 133.

70 Needham, *Science and Civilization.*, p. 351;Porkert, ibid., pp. 64, 65

71 Needham, Ibid., p. 354.

72 Needham, Ibid., p. 351.

73 Needham,

74 Porkert, ibid., p 67.

75 Porkert, pp. 65, 66.

76 Porkert, p. 65, 66.

77 Porkert, figure 14, p. 74.

78 William Lowe, pp.

79 Nei Jing: above

80 *Chinese Medical Chinese: Grammar & Vocabulary*, Nigel Wiseman Ye FengTerms, (Brookline, Mass. : Paradigm Publications, 2002), p. 85

81 lilly, Ptolemy

82 Lu and Needham, Science and Civilization Vol. 2, Pseudo Sciences, p. 350.

83 Wikipedia

84 Lu and Neeham, p. 350.

85 Thom. A., and Thom, A.S., "A megalithic lunar observatory in Orkney', *Journal for the History of Astronomy*, 4, 111-23 (1973); Wood, John E., Sun, Moonand Standing Stones. (Oxford: at the University Press, 1978), pp. 71-80.

86 Wood, John E., Sun, Moon, and Standing Stones, (Oxford: at the University Press, 1978), pp. 66, 67.

87 Woods, p. 72.

88 Woods, p. 76; Hoyle, F. *Stonehenge: An Eclipse Predictor.*

89 Wieger, p 26.

90 Porkert, *ibid*, p. 72.

91 William Butler Yeats, **Vision**, ; Phases of the Moon. Pp. 96-99.

92 Dane Rudhyar, The Lunation Cycle (1967), p. 15

93 Rudhyar, p. 17

94 Rudhyar, p. 24

95 Wieger, Ibid. p 57.

96 Birch, Stephen and Matsumoto, Kiko, Extraordinary Vessels

97 Plato, *Timaeus and Critias*, p. 49.

98 Legge, James trans., I Ching Book of Changes, edited with Introduction ch'u Chai with Winberg Chai, New Hyde Park: University Books, 1964), p. xiv.

99 Lu Gwei-Djen and Joseph Needham, *Celestial Lancets: A History and Rationale of Acupuncture and Moxa*, (Cambride: Cambridge University Press, 1980), p. 40.

100 Lu Gwei-Djen and Joseph Needham, *Science and Civilization*, p. 358.

[101] Firmicus Maternus, *Ancient Astrology Theory and Practice*, trans. JeanR. Bram, (Park Ridge, NJ), p. 45.

[102] Wieger, Ibid., p. 34.

[103]

[104] Unschuld, Paul, *HuangDi Nei Jing: Nature Knowledge, and x in An Ancient Chinese Medicine*,(Berkeley: University of California Press, 2003), p. 480.

[105] Unschuld, Ibid., p. 480.

[106] Wieger, Ibid., p. 68.

[107] (Lu, Su-wen, IV, p. 24)

[108] Lu Gwei-Djen and Joseph Needham, *Science and Civilization.*

[109] Theodora Lao, *The Handbook of Chinese Horoscopes*, (New York: Harper Collins, 1995), Intro. xxvi.

[110] Needham

[111] Manfred Porkert

[112] Theodora Lao, astrol

[113] Modified from Porkert, ibid., p. 72; Needham, Celestial Lancets, p.

[114] Chinese Medical Chinese: Grammar & Vocabulary, Nigel Wiseman Ye FengTerms, p 86.

[115] Celestial Lancets pp.

[116] /Wieger p. 577.

[117] Wieger , p. 422

[118] Wieger p 124

[119] Wieger

[120] Wieger

[121] Wieger

[122] Wieger

[123] Wieger p. 608

[124] Wieger, p . 614

[125] Wieger

[126] Wieger

[127] Wieger

[128] Wieger

[129] Wieger, p. 87.

[130] Wieger

[131] Wieger

[132] Wieger, p 663.

[133] Wieger

[134] Wieger, p.175

[135]

Bibliography

Agrippa, C. *The Occult Philosophy.*

Blundeville, Thomas. *The Theoriques of the Seven Planets.*London, 1602.

Burton, Robert. *The Anatomy of Melancholy.* Edited by Thomas C. Faulkner, Nicolas K. KIessling, and Rhonda L. Blair. Introduction by J. B. Bamborough. Oxford: The Clarendon Press, 1989.

Carpenter, Nathanael. *Geography Delineated Forth in Two Bookes, Containing the Spherical and Topicall Partes Thereof.* Oxford: John Litchfield and William Turner . . . for Henry Cripps, 1625.

Cunningham, Charles. *The Cosmographical Glasse: Renaissance Diagrams of the Universe.* Edited by S. K. Heniger. San Marino: The Huntington Library, 1977.

Herbermahn, Charles, ed. *The Cosmographia Introductio of Martin Waldseemuller, in Facsimile followed by Four Voyages of Amerigo Vespucci with their translation into English and Waldsemuller's two World Maps of 1507.* Introduction by Joseph Fisher and Franz Von Wiesen. New York: The United States Catholic Historical Society, 1907.

Heylyn, Peter. *Microcosmus or a Little Description of the Great World.* Oxford, 1621.

Hues, Robert. *A learned treatise of globes, both coelestiall and terrestriall: with their several uses.* London, 1659.

Kepler, Johannes. *Mysterium Cosmographicum: The Secret of the Universe.* Translated by A. M. Duncan. Introduction by E. J. Aiton. New York: Arabis Books, 1981.

Maternus, Firmicus. *Ancient Astrology Theory and Practice: Matheseos Libri VIII.* Translated by J. R. Bram. Park Ridge, NJ: Noyes Press, 1975.

Ptholemei, Claudii. *Alexandrini Philosphi Cosmographia.* Introduction by R. A. Skelton. Amsterdam: Theatrum Orbis Terrarum, 1966.

_____. *Tetrabiblos or Quadripartite: Being Four Books of the Influence of the Stars.* Translated by J. M. Ashmand. North Hollywood: Symbols and Signs, 1976.

Secondary Sources

Aiton, E. J. "Celestial Spheres and Circles," *History of Science* 19(1981): 75-114.

Allen, D. C. *The Star-Crossed Renaissance: The Quarrel about Astrology and Its Influence in England.* New York: Octagon Books, 1973.

Andersson, S., and T. Lundeberg. "Acupuncture—from empiricism to science: functional background to acupuncture effects in pain and disease." *Med Hypotheses* 45, no.3 (Sept 1995): 271-81.

Babb, Lawrence. *The Elizabethan Malady: A Study of Melancholia in English Literature from 1580 to 1642.* East Lansing: Michigan State College Press, 1951.

Backster, Cleve. *Primary Perception: Biocommunication with Plants, Living Foods, and Human Cells.* Anza, CA: White Rose Millennium Press, 2003.

Bamborough, J. B. "Robert Burton's Astrological Notebook," *Review of English Studies* NS 32 (1981): 267-85.

Bensky, D., and A. Gamble. *Chinese Herbal Medicine: Materia Medica.* Seattle: Eastland Press, 1986.

Birch, S. *Naming the Unnameable : A Historical Study of Radial Pulse Six Pulse Position Diagnosis.* 1992.

Bowen, Margarita. *Empiricism and Geographical Thought from Francis Bacon to Alexander Humbolt.* Cambridge: Cambridge University Press, 1981.

Briggs, Katherine M., ed. *The Last of the Astrologers: Mr. William Lilly's History of his Life and Times from the Year 1602 to 1681.* Introduction by William Lilly. Ilkey: The Scolar Press Republished Flklor Society, 1974.

Browne, Robert M. "Robert Burton and the New Cosmology," *Modern Language Quarterly* 13 (1952): 1311-48.

Capp, B. *English Almanacs 1500-1800: Astrology and the Popular Press.* New York: Cornell University Press, 1979.

Cassirer, Ernst. *The Individual and the Cosmos in Renaissance Philosophy.* Translated by Mario Domandi. New York: Harper and Row, 1964.

Chen, E. M. *The Tao Te Ching.* New York: Paragon House, 1989.

Chia, Mantak, and Fusion Maneewan. *The Five Elements.* Huntington, NY: Healing Tao Books, 1989.

Cleary, T. *The Inner Teachings of Taoism.* Boston: Shambala, 1986.

Cohen, A. *Embracing Heaven and Earth.* Lenox, MA: Moksha Press, 2001.

Colie, Rosalie. *Paradoxia Epidemica: The Renaissance Tradition of Paradox.* Princeton: Princeton University Press, 1966.

Conger, George Perrigo. *Theories of Macrocosm and Microcosm in the History of Philosophy.* New York: Columbia University Press, 1922.

Deadman, P., Mazin Al-Khafaji, and Kevin Baker. *A Manual of Acupuncture.* Eastland Press, 1998.

Debus, Allen G. *The Chemical Philosophy Paracelsian Science and Medicine in the Sixteenth and Seventeenth Centuries.* New York: Science History Publications, 1977.

Doob, Penelope. *The Idea of the Labyrinth from Classical Antiquity through the Middle Ages.* Ithaca: Cornell University Press, 1990.

Eckman, P. *In the Footsteps of the Yellow Emperor: Tracing the History of Traditional Acupuncture.* San Francisco: Cypress, 1996.

Ellis, Wiseman N., and K. Boss. *Grasping the Wind.* Brookline, MA: Paradigm, 1989.

Feingold, Mordechai. *The Mathematician's Apprentice.* Cambridge: Cambridge University Press, 1981.

Field, J. V. *Kepler's Rejection of Numerology, Occult, and Scientific Mentalities in the Renaissance.* Edited by Brian W. Vickers. Cambridge: The Cambridge University Press, 1984.

_____. *Geometrical Cosmology.* London: Athelone Press, 1988.

Fox, Ruth. *The Tangled Chain: The Structure of Disorder in the Anatomy of Melancholy.* Berkeley: University of California Press, 1976.

Frank, Robert G. *Harvey and the Oxford Physiologists: A Study of Scientific Ideas.* Berkeley: University of California Press, 1980.

_____. *Science, Medicine, and the Universities of Early Modern England: Background and Sources, History of Science.* 11 1973. pp. 194-216.

French, Peter J. *John Dee: The World of an Elizabethan Magus.* London: Routledge & Kegan Paul, 1972.

Gardiner, Judith K. "Elizabethan Psychology and Burton's Anatomy of Melancholy," *JHI* 37 (1977): 377.

Garin, E. *Astrology in the Renaissance: The Zodiac of Life.* Boston: Routeledge and Kegan Paul, 1983.

Ginzburg, Carlos. *The Cheese and the Worms: the Conviction of a Sixteenth-Century Miller.* Translated by John and Ain Tedeschi. New York: Penguin, 1982.

Girardot, N. J. *Myth and Meaning in Early Taoism.* Berkeley: University of California Press.

Hamilton, Edith, and Huntington Cairns, eds. "Timaeus" from *The Collected Dialogues of Plato*. Translated by Benjamin Jowett. New York, Pantheon Books, 1961.

Hammer, L. *Dragon Rises, Red Bird Flies*. Barrytown, NY: Station Hill Press, 1990.

_____. *Chinese Pulse Diagnosis: A Contemporary Approach*. Seattle: Eastland Press, 2001.

Henderson, J. B. *The Development and Decline of Chinese Cosmology*. New York: Columbia University Press, 1984.

Hicks, S. *Catalogue of Acupuncture Point Translations*. Columbia, MD: Traditional Acupuncture Institute, 1985.

Hicks, A., J. Hicks, and P. Mole. *Five Element Constitutional Acupuncture*.

Homann, R., trans. *Pai Wen Pien or the Hundred Questions: A Dialogue Between Two Taoists on the Macrocosmic and Microcosmic System of Correspondence*. Leiden: E. J. Brill, 1976.

Jarrett, L. S. *Nourishing Destiny: The Inner Tradition of Chinese Medicine*. Stockbridge, MA: Spirit Path Press, 1998.

Johnson, Julian. *The Path of the Masters*. Beas, India: Radha Soami Society, 1985.

KIdwell, Carol. *Pontano: Poet and Prime Minister*. London: Duckworth, 1991.

Klibansky, Raymond, Erwin Panofsky, and Fritz Saxl. *Saturn and Melancholy: Studies in the History of Natural Philosophy, Religion, and Art*. New York: Basic Books, 1964.

Kocher, Paul H. *Science and Religion in Elizabethan England*. San Marino: University of California Press, 1953.

Larre, C., and E. Rochat de la Vallee. *Rooted in Spirit: The Heart of Chinese Medicine*. Barrytown, NY: Station Hill Press, 1993.

Larre, C., J. Schatz, and E. Rochat de la Vallee. *Survey of Traditional Chinese Medicine*. Columbia, MD: Traditional Acupuncture Institute, 1986.

Lau, Theodora. *The Handbook of Chinese Horoscopes*. New York: Colophon Books, 1979.

Legge, James, trans. *I Ching: Book of Changes*. Introduction by Ch'u Chai. New Hyde Park: University Books.

Little, Lucy, trans. *Astrosynthesis: The Rational System of Horoscope Interpretation According to Morin de Villefranche*. New York: Zoltan Mason Emerald Books, 1974.

Lonsdale, W., ed. *Star Rhythms: Readings in a Living Astrology*. Richmond, CA: North Atlantic Books, 1979.

Low, Royston. *The Secondary Vessels of Acupuncture*. New York: Thorsons, 1985.

_____. *The Celestial Stems: Acupuncture Theory and Practice in Relation to the Influence of Cosmic Forces upon the Body*. New York: Thorston, 1985.

Lu, Gwei-Djen, and J. Needham. *Celestial Lancets: A History and Rationale of Acupuncture and Moxa*. Cambridge: Cambridge University Press, 1980.

Manaka, Yoshio. With Kazuko Itaya, and Birch Stephen. Brookline: Paradigm, 1995.

Manilius, M. *The Five Books of M. Manilius Containing a System of the Ancient Astronomy and Astrology*. London: Jacob Tonson at the Judges Head, 1697; republished in 1953 by National Astrological Library, Washington DC.

Matsumoto, K., and S. Birch. *Five Elements and Ten Stems: Nan Jing Theory, Diagnostics, and Practice*. Brookline: Paradigm Publications, 1983.

_____. *Hara Diagnosis: Reflections on the Sea*. Brookline: Paradigm, 1998.

Maternus, Firmicus. "Matheseos Libri VIII" in *Ancient Astrology Theory and Practice*. Translated by Jean Rhys Bram. Park Ridge: Noyes Press, 1975.

May, Margaret. *Galen on the Usefulness of the Parts*. Ithaca: Cornell University Press, 1968.

Moss, J. D. *Novelties in the Heavens: Rhetoric and Science in the Copernican Controversy*. Chicago: University of Chicago Press, 1993.

Nauert, Charles G., Jr. *Agrippa and the Crisis of Renaissance Thought*. Urbana: University of Illinois Press, 1965.

Needham, Joseph. *Science and Civilization in China*. Cambridge: Cambridge University Press, 1954.

Nicolson, Marjorie H. *Science and Imagination*. Ithaca: Cornell University Press, 1971.

Noonan, G. C. *Classical Scientific Astrology*.

North, J. D. "Astrology and the Fortunes of Churches," Centaurus 24 (1980): 186.

_____. *Horoscopes and History*. London: The Warburg Institute, 1986.

_____. *Stars, Minds, and Fate: Essays in Ancient and Modern Cosmology*. London: W. V. Hambledon Press, 1989.

O'Connor, J., and D. Bensky, trans. *Acupuncture: A Comprehensive Text*. Shanghai College of Traditional Medicine. Seattle: Eastland Press, 1987.

Omura, Y. *Acupuncture Medicine: Its Historical and Clinical Background*. Tokyo: Japan Publications, 1982.

Ong, Walter J. *Interface of the Word: Studies in the Evolution of Consciousness and Culture*. Ithaca: Cornell University Press, 1977.

Pagel, Walter. *Paracelsus: An Introduction to Philosophical Medicine in the Era of the Renaissance*, 2nd ed. Basel, 1982.

Parker, Derek. *Familiar to All, William Lilly, and Astrology in the Seventeenth Century*. London: Jonathan Cape, 1975.

Parr, Johnstone. *Tamburlaine's Malady and Other Essays on Astrology in Elizabethan Drama*. University of Alabama Press, 1953.

Pingree, D., trans. *The Yavanajataka of Sphujidhvaja* (Harvard Oriental Series). Cambridge: Harvard University Press, 1978.

Porkert, M. *The Theoretical Foundations of Chinese Medicine*. Cambridge, MA: MIT Press, 1982.

_____. "The Essentials of Chinese Diagnostics." *Acta Medicinae Sinensis*. Zurich, Switzerland: Chinese Medicine Publications.

Renaker, David. "Robert Burton and Ramist Method." *Renaissance Quarterly* 24 (1971): 210-220.

Rudhyar, Dane: *Astrological Signs: The Pulse of Life*. Boulder: Shamballa, 1978.

Schmitt, Charles B. *Renaissance Aristotelianism: Studies in Renaissance Philosophy and Science*. London: Variorum Reprints, 1981.

Shumaker, Wayne. *Occult Sciences in the Renaissance*. Berkeley: University of California Press, 1972.

Sun Bear, Wabun. *The Medicine Wheel: Earth Astrology*. Englewood Cliffs, NJ: Prentice Hall, 1980.

Stock, Brian. *Myth and Science in the Twelfth Century: A Study of Bernard Silvester*. Princeton: Princeton University Press, 1972.

Sun Tzu. *The Art of War*. Hong Kong: Grand Cultural Service Company, 1973.

Taylor, E. G. R. *Tudor Geography 1485-1583*. New York: Octagon Press, 1968.

_____. *The Mathematical Practitioners of Tudor and Steuart England*. 1485-1714.

Temkin, Owsei. *Galenism: Rise and Decline of a Medical Philosophy*. Ithaca: Cornell University Press, 1973.

Thorndike, Lynn. *The History of Magic and Experimental Sciences*. (6 vols.) New York: Columbia University Press, 1923-1931.

_____. *The Sphere of Sacrobosco and Its Commentators*. Chicago: University of Chicago Press, 1949.

Unschuld, P. U. *Medicine in China: A History of Ideas*. Berkeley: University of California Press, 1985.

Vicari, E. Patricia. "Saturn Culminating, Mars Ascending: The Fortunate/ Unfortunate Horoscope of Robert Burton" In *Familiar Colloquy: Essays Presented to Edward Barker*. Edited by Patricia Bruckmann. Ontario: Oberon Press, 1978.

_____. *"Anatomy of Melancholy,"* in *The View from Minerva's Tower: Learning and Imagination*. Toronto: University of Toronto Press, 1989.

Volgune, Alexander. *Lunar Astrology: An Attempt at a Reconstruction of the Ancient Astrological System*. Translated by John Broglio. New York: ASI Publishers, 1972.

Wallace, William A. *Causality and Scientific Explanation*. Ann Arbor: University of Michigan Press, 1972.

_____. *From a Realist Point of View: Essays on the Philosophy of Science*. New York: University Press of America, 1983.

Westman, Robert S. "The Astronomers's Role in the Sixteenth Century," *Science History Publications* 18 (1980): 105-47.

_____. "The Melanchthon Circle, Rheticus and the Wittenberg Interpretation of the Copernican Theory," *Isis* 66 (1975): 165-93.

_____. "Nature, Art, and Psyche" In *Occult and Scientific Mentalities in the Renaissance*. Edited by Brian Vickers. Cambridge: Cambridge University Press, 1984.

White, Lynne, Jr., "Medical Astrologers and Late Medieval Technology." *Viator* 6 (1975): 295-305.

Wieger, L. *Chinese Characters: Their Origin, Etymology, History, Classification, and Signification*. New York: Paragon Book Reprint, 1965.

Wilber, K. *A Theory of Everything: An Integral Vision for Business, Politics, Science, and Spirituality*. Boston: Shambala, 2000.

Wittkower, Rudolph. *Heiroglyphics in the Early Renaissance*. Albany: The State University of New York Press, 1968.

Wood, John E. *Sun, Moon, and Standing Stones*. Oxford: Oxford University Press, 1978.

Worsley, J. R. *The Meridians of Ch'I Energy: Point Reference Guide*. Columbia, Md.: Traditional Acupuncture Institute, 1979; Tisbury: Element Books, 1982.

Yates, Frances A. *Giordano Bruno and the Hermetic Tradition*. Chicago: University of Chicago Press, 1964.

_____. "The Hermetic Tradition in Renaissance Science" in *Art Science, and History in the Renaissance*. Edited by Charles S. Singleton. Baltimore: Johns Hopkins Univerisity Press, 1968.

Yeats, William Butler. *A Vision*. New York: Collier, 1967.

www.ingramcontent.com/pod-product-compliance
Lightning Source LLC
Chambersburg PA
CBHW031814170526
45157CB00001B/56